59. Alghamdi, M. Familial Mediterranean fever, review of the literature. *Clin. Rheumatol.* **2017**, *36*, 1707–1713. [CrossRef] [PubMed]
60. Gülez, N.; Makay, B.; Sözeri, B. Long-Term Effectiveness and safety of canakinumab in pediatric familial Mediterranean fever patients. *Mod. Rheumatol.* **2018**, 1–13. [CrossRef] [PubMed]
61. Laskari, K.; Boura, P.; Dalekos, G.N.; Garyfallos, A.; Karokis, D.; Pikazis, D.; Settas, L.; Skarantavos, G.; Tsitsami, E.; Sfikakis, P.P. Long-term beneficial effect of canakinumab in colchicine-resistant familial Mediterranean fever. *J. Rheumatol.* **2017**, *44*, 102–109. [CrossRef]
62. Başaran, Ö.; Uncu, N.; Çelikel, B.A.; Taktak, A.; Gür, G.; Cakar, N. Interleukin-1 targeting treatment in familial Mediterranean fever: An experience of pediatric patients. *Mod. Rheumatol.* **2015**, *25*, 621–624. [CrossRef]
63. Cetin, P.; Sari, I.; Sozeri, B.; Cam, O.; Birlik, M.; Akkoc, N.; Onen, F.; Akar, S. Efficacy of Interleukin-1 Targeting Treatments in Patients with Familial Mediterranean Fever. *Inflammation* **2015**, *38*, 27–31. [CrossRef] [PubMed]
64. Özçakar, Z.B.; Özdel, S.; Yılmaz, S.; Kurt-Şükür, E.D.; Ekim, M.; Yalçınkaya, F. Anti-IL-1 treatment in familial Mediterranean fever and related amyloidosis. *Clin. Rheumatol.* **2016**, *35*, 441–446. [CrossRef] [PubMed]
65. Houten, S.M.; Kuis, W.; Duran, M.; de Koning, T.J.; van Royen-Kerkhof, A.; Romeijn, G.J.; Frenkel, J.; Dorland, L.; de Barse, M.M.J.; Huijbers, W.A.R.; et al. Mutations in *MVK*, encoding mevalonate kinase, cause hyperimmunoglobulinaemia D and periodic fever syndrome. *Nat. Genet.* **1999**, *22*, 175–177. [CrossRef] [PubMed]
66. Ter Haar, N.M.; Jeyaratnam, J.; Lachmann, H.J.; Simon, A.; Brogan, P.A.; Doglio, M.; Cattalini, M.; Anton, J.; Modesto, C.; Quartier, P.; et al. The phenotype and genotype of mevalonate kinase deficiency: A series of 114 cases from the Eurofever registry. *Arthritis Rheumatol.* **2016**, *68*, 2795–2805. [CrossRef] [PubMed]
67. Bodar, E.J.; Kuijk, L.M.; Drenth, J.P.H.; van der Meer, J.W.M.; Simon, A.; Frenkel, J. On-demand anakinra treatment is effective in mevalonate kinase deficiency. *Ann. Rheum. Dis.* **2011**, *70*, 2155–2158. [CrossRef] [PubMed]
68. van der Hilst, J.C.H.; Bodar, E.J.; Barron, K.S.; Frenkel, J.; Drenth, J.P.H.; van der Meer, J.W.M.; Simon, A. International HIDS Study Group Long-term follow-up, clinical features, and quality of life in a series of 103 patients with hyperimmunoglobulinemia D syndrome. *Medicine* **2008**, *87*, 301–310. [CrossRef]
69. Galeotti, C.; Meinzer, U.; Quartier, P.; Rossi-Semerano, L.; Bader-Meunier, B.; Pillet, P.; Koné-Paut, I. Efficacy of interleukin-1-targeting drugs in mevalonate kinase deficiency. *Rheumatology* **2012**, *51*, 1855–1859. [CrossRef]
70. Tanaka, T.; Yoshioka, K.; Nishikomori, R.; Sakai, H.; Abe, J.; Yamashita, Y.; Hiramoto, R.; Morimoto, A.; Ishii, E.; Arakawa, H.; et al. National survey of Japanese patients with mevalonate kinase deficiency reveals distinctive genetic and clinical characteristics. *Mod. Rheumatol.* **2018**, *29*, 181–187. [CrossRef]
71. Ozen, S.; Kuemmerle-Deschner, J.B.; Cimaz, R.; Livneh, A.; Quartier, P.; Kone-Paut, I.; Zeft, A.; Spalding, S.; Gul, A.; Hentgen, V.; et al. International retrospective chart review of treatment patterns in severe familial Mediterranean fever, tumor necrosis factor receptor-associated periodic syndrome, and mevalonate kinase deficiency/hyperimmunoglobulinemia D syndrome. *Arthritis Care Res.* **2017**, *69*, 578–586. [CrossRef]
72. Ter Haar, N.; Lachmann, H.; Özen, S.; Woo, P.; Uziel, Y.; Modesto, C.; Koné-Paut, I.; Cantarini, L.; Insalaco, A.; Neven, B.; et al. Treatment of autoinflammatory diseases: Results from the Eurofever Registry and a literature review. *Ann. Rheum. Dis.* **2013**, *72*, 678–685. [CrossRef]
73. Vitale, A.; Insalaco, A.; Sfriso, P.; Lopalco, G.; Emmi, G.; Cattalini, M.; Manna, R.; Cimaz, R.; Priori, R.; Talarico, R.; et al. A snapshot on the on-label and off-label use of the interleukin-1 inhibitors in Italy among rheumatologists and pediatric rheumatologists: A nationwide multi-center retrospective observational study. *Front. Pharmacol.* **2016**, *7*, 380. [CrossRef]
74. Rossi-Semerano, L.; Fautrel, B.; Wendling, D.; Hachulla, E.; Galeotti, C.; Semerano, L.; Touitou, I.; Koné-Paut, I. Tolerance and efficacy of off-label anti-interleukin-1 treatments in France: A nationwide survey. *Orphanet J. Rare Dis.* **2015**, *10*, 19. [CrossRef]
75. Topaloglu, R.; Batu, E.D.; Orhan, D.; Ozen, S.; Besbas, N. Anti-interleukin 1 treatment in secondary amyloidosis associated with autoinflammatory diseases. *Pediatr. Nephrol.* **2016**, *31*, 633–640. [CrossRef]
76. Gattorno, M.; Piccini, A.; Lasigliè, D.; Tassi, S.; Brisca, G.; Carta, S.; Delfino, L.; Ferlito, F.; Pelagatti, M.A.; Caroli, F.; et al. The pattern of response to anti-interleukin-1 treatment distinguishes two subsets of patients with systemic-onset juvenile idiopathic arthritis. *Arthritis Rheum.* **2008**, *58*, 1505–1515. [CrossRef]

77. Brachat, A.H.; Grom, A.A.; Wulffraat, N.; Brunner, H.I.; Quartier, P.; Brik, R.; McCann, L.; Ozdogan, H.; Rutkowska-Sak, L.; Schneider, R.; et al. Early changes in gene expression and inflammatory proteins in systemic juvenile idiopathic arthritis patients on canakinumab therapy. *Arthritis Res. Ther.* **2017**, *19*, 13. [CrossRef]
78. Feist, E.; Quartier, P.; Fautrel, B.; Schneider, R.; Sfriso, P.; Efthimiou, P.; Cantarini, L.; Lheritier, K.; Leon, K.; Karyekar, C.S.; et al. Efficacy and safety of canakinumab in patients with Still's disease: Exposure-response analysis of pooled systemic juvenile idiopathic arthritis data by age groups. *Clin. Exp. Rheumatol.* **2018**, *36*, 668–675.
79. Grom, A.A.; Ilowite, N.T.; Pascual, V.; Brunner, H.I.; Martini, A.; Lovell, D.; Ruperto, N.; Abrams, K.; Leon, K.; Lheritier, K.; et al. Rate and clinical presentation of macrophage activation syndrome in patients with systemic juvenile idiopathic arthritis treated with canakinumab. *Arthritis Rheumatol.* **2016**, *68*, 218–228. [CrossRef]
80. Ilowite, N.; Porras, O.; Reiff, A.; Rudge, S.; Punaro, M.; Martin, A.; Allen, R.; Harville, T.; Sun, Y.-N.; Bevirt, T.; et al. Anakinra in the treatment of polyarticular-course juvenile rheumatoid arthritis: Safety and preliminary efficacy results of a randomized multicenter study. *Clin. Rheumatol.* **2009**, *28*, 129–137. [CrossRef]
81. Kimura, Y.; Grevich, S.; Beukelman, T.; Morgan, E.; Nigrovic, P.A.; Mieszkalski, K.; Graham, T.B.; Ibarra, M.; Ilowite, N.; Klein-Gitelman, M.; et al. Pilot study comparing the Childhood Arthritis & Rheumatology Research Alliance (CARRA) systemic Juvenile Idiopathic Arthritis consensus treatment plans. *Pediatr. Rheumatol. Online J.* **2017**, *15*, 23.
82. Quartier, P.; Allantaz, F.; Cimaz, R.; Pillet, P.; Messiaen, C.; Bardin, C.; Bossuyt, X.; Boutten, A.; Bienvenu, J.; Duquesne, A.; et al. A multicentre, randomised, double-blind, placebo-controlled trial with the interleukin-1 receptor antagonist anakinra in patients with systemic-onset juvenile idiopathic arthritis (ANAJIS trial). *Ann. Rheum. Dis.* **2011**, *70*, 747–754. [CrossRef]
83. Ruperto, N.; Brunner, H.I.; Quartier, P.; Constantin, T.; Wulffraat, N.; Horneff, G.; Brik, R.; McCann, L.; Kasapcopur, O.; Rutkowska-Sak, L.; et al. Two randomized trials of canakinumab in systemic juvenile idiopathic arthritis. *N. Engl. J. Med.* **2012**, *367*, 2396–2406. [CrossRef] [PubMed]
84. Ruperto, N.; Quartier, P.; Wulffraat, N.; Woo, P.; Ravelli, A.; Mouy, R.; Bader-Meunier, B.; Vastert, S.J.; Noseda, E.; D'Ambrosio, D.; et al. A phase II, multicenter, open-label study evaluating dosing and preliminary safety and efficacy of canakinumab in systemic juvenile idiopathic arthritis with active systemic features. *Arthritis Rheum.* **2012**, *64*, 557–567. [CrossRef]
85. Ruperto, N.; Brunner, H.I.; Quartier, P.; Constantin, T.; Wulffraat, N.M.; Horneff, G.; Kasapcopur, O.; Schneider, R.; Anton, J.; Barash, J.; et al. Canakinumab in patients with systemic juvenile idiopathic arthritis and active systemic features: Results from the 5-year long-term extension of the phase III pivotal trials. *Ann. Rheum. Dis.* **2018**, *77*, 1710–1719. [CrossRef] [PubMed]
86. Hedrich, C.M.; Bruck, N.; Fiebig, B.; Gahr, M. Anakinra: A safe and effective first-line treatment in systemic onset juvenile idiopathic arthritis (SoJIA). *Rheumatol. Int.* **2012**, *32*, 3525–3530. [CrossRef]
87. Horneff, G.; Schulz, A.C.; Klotsche, J.; Hospach, A.; Minden, K.; Foeldvari, I.; Trauzeddel, R.; Ganser, G.; Weller-Heinemann, F.; Haas, J.P. Experience with etanercept, tocilizumab and interleukin-1 inhibitors in systemic onset juvenile idiopathic arthritis patients from the BIKER registry. *Arthritis Res. Ther.* **2017**, *19*, 256. [CrossRef] [PubMed]
88. Kearsley-Fleet, L.; Beresford, M.W.; Davies, R.; De Cock, D.; Baildam, E.; Foster, H.E.; Southwood, T.R.; Thomson, W.; Hyrich, K.L. Short-term outcomes in patients with systemic juvenile idiopathic arthritis treated with either tocilizumab or anakinra. *Rheumatology* **2019**, *58*, 94–102. [CrossRef] [PubMed]
89. Lequerré, T.; Quartier, P.; Rosellini, D.; Alaoui, F.; De Bandt, M.; Mejjad, O.; Koné-Paut, I.; Michel, M.; Dernis, E.; Khellaf, M.; et al. Interleukin-1 receptor antagonist (anakinra) treatment in patients with systemic-onset juvenile idiopathic arthritis or adult onset Still disease: Preliminary experience in France. *Ann. Rheum. Dis.* **2008**, *67*, 302–308. [CrossRef] [PubMed]
90. Nigrovic, P.A.; Mannion, M.; Prince, F.H.M.; Zeft, A.; Rabinovich, C.E.; van Rossum, M.A.J.; Cortis, E.; Pardeo, M.; Miettunen, P.M.; Janow, G.; et al. Anakinra as first-line disease-modifying therapy in systemic juvenile idiopathic arthritis: Report of forty-six patients from an international multicenter series. *Arthritis Rheum.* **2011**, *63*, 545–555. [CrossRef] [PubMed]
91. Pardeo, M.; Pires Marafon, D.; Insalaco, A.; Bracaglia, C.; Nicolai, R.; Messia, V.; De Benedetti, F. Anakinra in systemic juvenile idiopathic arthritis: A single-center experience. *J. Rheumatol.* **2015**, *42*, 1523–1527. [CrossRef] [PubMed]

92. Romano, M.; Pontikaki, I.; Gattinara, M.; Ardoino, I.; Donati, C.; Boracchi, P.; Meroni, P.L.; Gerloni, V. Drug survival and reasons for discontinuation of the first course of biological therapy in 301 juvenile idiopathic arthritis patients. *Reumatismo* **2014**, *65*, 278–285. [CrossRef]
93. Vastert, S.J.; de Jager, W.; Noordman, B.J.; Holzinger, D.; Kuis, W.; Prakken, B.J.; Wulffraat, N.M. Effectiveness of first-line treatment with recombinant interleukin-1 receptor antagonist in steroid-naive patients with new-onset systemic juvenile idiopathic arthritis: Results of a prospective cohort study. *Arthritis Rheumatol.* **2014**, *66*, 1034–1043. [CrossRef]
94. Marrani, E.; Burns, J.C.; Cimaz, R. How should we classify Kawasaki disease? *Front. Immunol.* **2018**, *9*, 2974. [CrossRef]
95. Principi, N.; Rigante, D.; Esposito, S. The role of infection in Kawasaki syndrome. *J. Infect.* **2013**, *67*, 1–10. [CrossRef]
96. De Rosa, G.; Pardeo, M.; Rigante, D. Current recommendations for the pharmacologic therapy in Kawasaki syndrome and management of its cardiovascular complications. *Eur. Rev. Med. Pharmacol. Sci.* **2007**, *11*, 301–308.
97. Rigante, D.; Valentini, P.; Rizzo, D.; Leo, A.; De Rosa, G.; Onesimo, R.; De Nisco, A.; Angelone, D.F.; Compagnone, A.; Delogu, A.B. Responsiveness to intravenous immunoglobulins and occurrence of coronary artery abnormalities in a single-center cohort of Italian patients with Kawasaki syndrome. *Rheumatol. Int.* **2010**, *30*, 841–846. [CrossRef]
98. Rigante, D.; Andreozzi, L.; Fastiggi, M.; Bracci, B.; Natale, M.F.; Esposito, S. Critical overview of the risk scoring systems to predict non-responsiveness to intravenous immunoglobulin in Kawasaki syndrome. *Int. J. Mol. Sci.* **2016**, *17*, 278. [CrossRef] [PubMed]
99. Agarwal, S.; Agrawal, D.K. Kawasaki disease: Etiopathogenesis and novel treatment strategies. *Expert Rev. Clin. Immunol.* **2017**, *13*, 247–258. [CrossRef] [PubMed]
100. Blonz, G.; Lacroix, S.; Benbrik, N.; Warin-Fresse, K.; Masseau, A.; Trewick, D.; Hamidou, M.; Stephan, J.-L.; Néel, A. Severe late-onset Kawasaki disease successfully treated with anakinra. *J. Clin. Rheumatol.* **2018**, 1. [CrossRef]
101. Cohen, S.; Tacke, C.E.; Straver, B.; Meijer, N.; Kuipers, I.M.; Kuijpers, T.W. A child with severe relapsing Kawasaki disease rescued by IL-1 receptor blockade and extracorporeal membrane oxygenation. *Ann. Rheum. Dis.* **2012**, *71*, 2059–2061. [CrossRef] [PubMed]
102. Guillaume, M.-P.; Reumaux, H.; Dubos, F. Usefulness and safety of anakinra in refractory Kawasaki disease complicated by coronary artery aneurysm. *Cardiol. Young* **2018**, *28*, 739–742. [CrossRef]
103. Sánchez-Manubens, J.; Gelman, A.; Franch, N.; Teodoro, S.; Palacios, J.R.; Rudi, N.; Rivera, J.; Antón, J. A child with resistant Kawasaki disease successfully treated with anakinra: A case report. *BMC Pediatr.* **2017**, *17*, 102. [CrossRef] [PubMed]
104. Shafferman, A.; Birmingham, J.D.; Cron, R.Q. High dose Anakinra for treatment of severe neonatal Kawasaki disease: A case report. *Pediatr. Rheumatol. Online J.* **2014**, *12*, 26. [CrossRef]
105. Koné-Paut, I.; Cimaz, R.; Herberg, J.; Bates, O.; Carbasse, A.; Saulnier, J.P.; Maggio, M.C.; Anton, J.; Piram, M. The use of interleukin 1 receptor antagonist (anakinra) in Kawasaki disease: A retrospective cases series. *Autoimmun. Rev.* **2018**, *17*, 768–774. [CrossRef] [PubMed]
106. Kaiser, C.; Knight, A.; Nordström, D.; Pettersson, T.; Fransson, J.; Florin-Robertsson, E.; Pilström, B. Injection-site reactions upon Kineret (anakinra) administration: Experiences and explanations. *Rheumatol. Int.* **2012**, *32*, 295–299. [CrossRef] [PubMed]
107. Sota, J.; Vitale, A.; Insalaco, A.; Sfriso, P.; Lopalco, G.; Emmi, G.; Cattalini, M.; Manna, R.; Cimaz, R.; Priori, R.; et al. Safety profile of the interleukin-1 inhibitors anakinra and canakinumab in real-life clinical practice: A nationwide multicenter retrospective observational study. *Clin. Rheumatol.* **2018**, *37*, 2233–2240. [CrossRef]

MDPI
St. Alban-Anlage 66
4052 Basel
Switzerland
Tel. +41 61 683 77 34
Fax +41 61 302 89 18
www.mdpi.com

International Journal of Molecular Sciences Editorial Office
E-mail: ijms@mdpi.com
www.mdpi.com/journal/ijms

Peroxisome Proliferator-Activated Receptors (PPARs): A Themed Issue in Honor of Prof. Walter Wahli

Peroxisome Proliferator-Activated Receptors (PPARs): A Themed Issue in Honor of Prof. Walter Wahli

Guest Editors

Hervé Guillou

Manuel Vázquez-Carrera

MDPI

Basel • Beijing • Wuhan • Barcelona • Belgrade • Novi Sad • Cluj • Manchester

Guest Editors
Hervé Guillou
INRAE Toulouse
Toxalim Unit
Toulouse
France

Manuel Vázquez-Carrera
Department of Pharmacology,
Toxicology and Therapeutic
Chemistry
University of Barcelona
Barcelona
Spain

Editorial Office
MDPI AG
Grosspeteranlage 5
4052 Basel, Switzerland

This is a reprint of the Special Issue, published open access by the journal *Biomolecules* (ISSN 2218-273X), freely accessible at: https://www.mdpi.com/journal/biomolecules/special_issues/780I4JKB7V.

For citation purposes, cite each article independently as indicated on the article page online and as indicated below:

Lastname, A.A.; Lastname, B.B. Article Title. *Journal Name* **Year**, *Volume Number*, Page Range.

ISBN 978-3-7258-5529-2 (Hbk)
ISBN 978-3-7258-5530-8 (PDF)
https://doi.org/10.3390/books978-3-7258-5530-8

Contents

About the Editors

Hervé Guillou

Hervé Guillou obtained his PhD in 2004 in Rennes. There, he studied a critical gene in fatty acid metabolism (FADS2) through cellular biology and biochemical approaches. Then, for three years he has been a post-doctoral scientist in Dr. Len Stephens and Dr. Phill Hawkins' group (The Babraham Institute, Cambridge, UK) where he has worked on PI3-Kinase signaling. He joined INRAE (Toulouse) in 2008. His work aims at developing approaches that could allow a better understanding of how changes in hepatic lipid metabolism (MASLD) may result from or influence endocrine disruption. His group focuses on mechanisms involving changes in gene expression and the roles of nuclear receptors such as PPARs. Since 2024, he has been the director of the Toulouse research center in food toxicology (Toxalim, INRAE and Toulouse University).

Manuel Vázquez-Carrera

Manuel Vázquez-Carrera is a full professor at the Faculty of Pharmacy and Food Sciences of the University of Barcelona, where he leads the research group "Pharmacological targets in inflammation and metabolic diseases." After completing his PhD in Pharmacology and Therapeutic Chemistry at the University of Barcelona in 1994, he carried out a postdoctoral stay at the University of Lausanne under the supervision of Prof. Walter Wahli, where he studied peroxisome proliferator-activated receptors (PPARs) and their role in inflammation. Since becoming an associate professor in 1997, he has directed his research toward understanding how inflammation contributes to insulin resistance, type 2 diabetes mellitus, metabolic dysfunction-associated steatotic liver disease (MASLD), and diabetic cardiomyopathy. His group has made contributions to uncovering the mechanisms by which PPARs, particularly PPARβ/δ, regulate metabolic and inflammatory pathways and their potential as therapeutic targets.

biomolecules

MDPI

Editorial

Peroxisome Proliferator-Activated Receptors (PPARs): A Themed Issue in Honor of Prof. Walter Wahli

Hervé Guillou [1,*] and Manuel Vázquez-Carrera [2,3,4,5,*]

1 INRAE Toulouse, Toxalim Unit, 180 Chemin de Tournefeuille, BP93173, 31027 Toulouse, France
2 Department of Pharmacology, Toxicology and Therapeutic Chemistry, Faculty of Pharmacy and Food Sciences, 08028 Barcelona, Spain
3 Institute of Biomedicine of the University of Barcelona (IBUB), University of Barcelona, 08028 Barcelona, Spain
4 Spanish Biomedical Research Center in Diabetes and Associated Metabolic Diseases (CIBERDEM)-Instituto de Salud Carlos III, 28029 Madrid, Spain
5 Pediatric Research Institute-Hospital Sant Joan de Déu, 08950 Esplugues de Llobregat, Spain
* Correspondence: herve.guillou@inrae.fr (H.G.); mvazquezcarrera@ub.edu (M.V.-C.)

It is with great pleasure that we introduce this Special Issue dedicated to Peroxisome Proliferator-Activated Receptors (PPARs), honoring the pioneering contributions of **Professor Walter Wahli**. Over the course of more than four decades, Prof. Wahli has significantly contributed to our understanding of nuclear receptor biology, metabolic regulation, and the intersection of lipid signaling and inflammation. His work spans from foundational mechanistic insights to concepts that underpin today's translational efforts in metabolic, inflammatory, and cardiovascular diseases. Beyond his discoveries, Prof. Wahli's enthusiasm, generosity, and commitment to collaborative science have inspired generations of researchers worldwide.

1. From Vitellogenin to Nuclear Receptor Mechanisms

Prof. Wahli's scientific journey began with elegant studies in Xenopus laevis that established the liver vitellogenin genes that encode yolk proteins as hormone-responsive target models. As a Ph.D. researcher in Rudolf Weber's laboratory (University of Berne, Switzerland), he was the first to isolate the vitellogenin mRNA [1] using sucrose gradients, and, in collaboration with a postdoc in the laboratory, Gerhart Ryffel, he titrated vitellogenin mRNA accumulation in the liver of estrogen-treated male *Xenopus laevis*, using C_0t curves [2]. These two papers were the core of his PhD thesis at the University of Berne, which was completed in only two years (1975–1977) and for which he received a "*Summa cum laude*". After his Ph.D., he pursued his quest for the molecular mechanism of estrogen action during his postdoctoral stage in Igor Dawid's laboratory (Department of Embryology of the Carnegie Institution, Baltimore, MD, USA) and then at the National Institutes of Health, Bethesda, where he cloned vitellogenin mRNA [3] and built the first Xenopus genomic library. From this library, he isolated the vitellogenin genes and found, a big surprise at the time, that these genes were organized into not less than 33 exons and 32 introns [4–6]. This finding was highlighted in the front-page picture of the Cell issue in which his 1980 paper [5] was published.

Upon returning to Switzerland as a Full Professor and Director of the Institute of Animal Biology at the University of Lausanne, Walter Wahli's group defined the canonical estrogen response element (ERE) from a comparative study of the promoter of the estrogen-responsive vitellogenin and apo-VLDLI genes, characterized its sequence requirements and cooperative receptor binding, and demonstrated hormone-dependent transcription using a technically demanding homologous in vitro system [7–14]. Collectively, these landmark contributions showed that the estrogen receptor resides in the nucleus in an inactive state and, upon ligand binding, drives transcription by engaging EREs in target promoters, principles that became pillars of nuclear receptor biology.

2. Founding and Expanding the PPAR Field

In the late 1980s, Prof. Wahli and one of his postdoctoral fellows, Greg Krey, in collaboration with Christine Dreyer (Tuebingen, Germany), applied the estrogen receptor DNA-binding domain as a probe to uncover novel receptors, leading to the discovery and characterization of the PPAR subfamily (PPARα, PPARβ/δ, and PPARγ) [15], for which a first member was discovered independently by Stephen Green [16]. This work, together with seminal ligand studies, established that fatty acids and eicosanoids directly modulate gene expression through PPARs, marking a paradigm shift that linked nutrient lipids to transcriptional control [17,18]. This significant finding opened a new field of investigation of the highest biomedical significance. A postdoc in Walter Wahli's research group, Pallavi Devchand, was the first to demonstrate that the fibrate hypolipidemic drugs, already on the market for some time, can directly bind to PPARα, which unveiled the mode of action of these drugs in lowering circulating lipids and uncovered the involvement of PPARα in the control of inflammation [19]. These findings prompted them to investigate the metabolic response mammals have evolved to survive long periods of energy deprivation. Indeed, a postdoctoral fellow in Wahli's group, Sander Kersten, demonstrated that PPARα plays a pivotal role in regulating energy stores during fasting. By modulating gene expression, PPARα stimulates hepatic fatty acid oxidation, providing energy-rich substrates that can be metabolized by peripheral organs [20].

3. Physiological Breadth: From Tissue Repair to Cancer and Cardiometa-bolism

Prof. Wahli's research continuously broadened the physiological canvas of PPARs. In the skin, a novel discovery by his group, initiated by postdoctoral fellow Liliane Michalik, revealed impaired wound healing in PPARα- and PPARβ/δ-deficient mice [21]. Building on this, and in collaboration with Andrew Tan, they identified a homeostatic control of keratinocyte proliferation and differentiation mediated through a PPAR-driven response in dermal fibroblasts via interleukin 1 (IL-1) signaling [22]. These findings have also opened an avenue for investigating the possible implications of PPARβ/δ in controlling tumor development [23].

Expanding from skin to metabolic tissues, genetic tools developed in collaboration with Pierre Chambon and Daniel Metzger, within the framework of a Human Frontier Science Program project directed by Walter Wahli, allowed them to demonstrate that PPARγ was required in differentiated, mature white and brown adipocytes for their survival [24]. In parallel, in the vasculature, collaboration with Philippe Boucher uncovered that PPARγ protects against calcification by antagonizing Wnt Family Member 5A (WNT5A) signaling, a finding with direct clinical relevance for atherosclerosis and valvular disease [25]. Extending into skeletal muscle physiology, Profs. Wahli and Chambon's teams demonstrated that PPARβ/δ plays a crucial role in maintaining oxidative muscle fibers, with its deficiency leading to fiber-type switching, obesity, and diabetes [26]. Likewise, in the heart, collaboration with Manuel Vázquez-Carrera revealed that PPARβ/δ activation counteracts phenylephrine-induced cardiomyocyte hypertrophy in neonatal rat ventricular cardiomyocytes [27]. Finally, in the endocrine pancreas, Wahli's group demonstrated that epithelial PPARβ/δ constrains β-cell mass and exocytosis, unveiling a regulatory checkpoint in insulin secretion [28]. In addition, sex-dependent repression by sumoylated PPARα in the female liver offered a mechanistic framework for estrogen-linked cholestatic disease [29]. Together, these studies demonstrate how PPARs orchestrate cellular and systemic physiology across diverse tissues and organs.

4. Developmental and Nutritional Axes

A hallmark of Prof. Wahli's contributions is the integration of developmental, nutritional, and epigenetic dimensions. He uncovered a glucocorticoid receptor–PPARα axis in fetal liver that equips neonates for milk lipid catabolism, including epigenetic gating of fibroblast growth factor 21 (FGF21) responsiveness [30]. In collaboration with Hervé Guillou and Catherine Postic, they revealed a specific PPARα–carbohydrate-responsive element-binding protein (ChREBP) cross-talk required for glucose-mediated FGF21 induction and endocrine control of sugar intake [31] and established hepatocyte PPARα as a protector against metabolic dysfunction-associated steatotic liver disease (MASLD) and hypercholesterolemia during aging [32]. In the intestine, collaboration with Philippe Sansonetti demonstrated that a high-fat diet disrupts PPARγ-dependent epithelial-microbial homeostasis [33]. In contrast, caloric restriction polarizes mucosal programs with PPARα and interferon-stimulated gene factor 3 (ISGF3) as potential candidate regulators [34]. Extending these insights to systems physiology, collaborative work with Sven Pettersson demonstrated that the gut microbiota can influence skeletal muscle mass and function [35].

5. PPARs as Nutrient-Sensing Transcriptional Hubs Coordinating Inter-Organ Communication and Stress Cytokines

In recent years, Prof. Wahli's work, mainly collaborations with Hervé Guillou and Manuel Vázquez-Carrera, has focused on PPAR control of systemic metabolism and the regulation of stress cytokines. These collaborations have built on the understanding that intact PPARα activity in hepatocytes is required for the crosstalk between adipose tissues and the liver during fat mobilization [36] and why metabolic control by PPARα in hepatocytes plays a crucial role in host defense against infection [37]. In addition, these collaborations have contributed to elucidating that growth differentiation factor 15 (GDF15) mediates the metabolic effects of PPARβ/δ by activating AMPK [38] and that PPARβ/δ prevents inflammation and fibrosis during diabetic cardiomyopathy [39]. All these studies confirm that PPARs serve as nutrient-sensing transcriptional hubs that coordinate inter-organ communication to maintain metabolic and inflammatory homeostasis. Prof. Wahli's mentorship and scientific imagination remain central drivers of these achievements.

6. About This Special Issue

The articles assembled in this Special Issue mirror the diversity and translational promise of the PPAR field. Contributions span basic mechanisms of receptor–ligand interaction and endogenous activation; structural and epigenetic regulation; tissue- and species-specific roles (e.g., lung, heart, skin, and ray-finned fish); natural ligand and modulator pharmacology; and broad therapeutic or translational applications in areas such as cardiovascular function, metabolic disorders, skin pathology, and clinical perspectives on PPAR biology. In curating these works, our aim is twofold: to celebrate Prof. Wahli's foundational role and to chart the questions that will define the next chapter of the PPAR research journey (precision targeting of PPAR networks, context-aware modulation, and integrative systems approaches bridging molecules to medicine).

7. Concluding Remarks

Prof. Walter Wahli's scientific legacy is distinguished by originality, rigor, and a gift for connecting dots across biology (from amphibian vitellogenesis to mammalian lactation, and from nuclear receptor mechanics to organismal homeostasis). Equally, it is a story of collaboration. Many of the advances highlighted here were made possible by enduring partnerships (including Béatrice Desvergne, Liliane Michalik, Pierre Chambon, Nguan Soon Tan, Hervé Guillou, and Manuel Vázquez-Carrera) that exemplify the best of our community. On behalf of all contributors, we express our deep gratitude to Prof. Wahli for his leadership, inspiration, and friendship. May this Special Issue honor his achievements and energize future discoveries.

Conflicts of Interest: The authors declare no conflict of interest.

References

1. Wahli, W.; Wyler, T.; Weber, R.; Ryffel, G.U. Size, complexity and abundance of a specific poly(A)-containing RNA of liver from male *Xenopus* induced to vitellogenin synthesis by estrogen. *Eur. J. Biochem.* **1976**, *66*, 457–465. [CrossRef] [PubMed]
2. Ryffel, G.U.; Wahli, W.; Weber, R. Quantitation of vitellogenin messenger RNA in the liver of male *Xenopus* toads during primary and secondary stimulation by estrogen. *Cell* **1977**, *11*, 213–221. [CrossRef] [PubMed]
3. Wahli, W.; Dawid, I.B.; Wyler, T.; Jaggi, R.B.; Weber, R.; Ryffel, G.U. Vitellogenin in *Xenopus laevis* is encoded in a small family of genes. *Cell* **1979**, *16*, 535–549. [CrossRef] [PubMed]
4. Wahli, W.; Dawid, I.B. Isolation of two closely related vitellogenin genes, including their flanking regions, from a *Xenopus laevis* gene library. *Proc. Natl. Acad. Sci. USA* **1980**, *77*, 1437–1441. [CrossRef]
5. Wahli, W.; Dawid, I.B.; Wyler, T.; Weber, R.; Ryffel, G.U. Comparative analysis of the structural organization of two closely related vitellogenin genes in *Xenopus laevis*. *Cell* **1980**, *20*, 107–117. [CrossRef]
6. Wahli, W.; Dawid, I.B.; Ryffel, G.U.; Weber, R. Vitellogenesis and the vitellogenin gene family. *Science* **1981**, *212*, 298–304. [CrossRef]
7. Walker, P.; Germond, J.-E.; Brown-Luedi, M.; Givel, F.; Wahli, W. Sequence homologies in the region preceding the transcription initiation site of the liver estrogen-responsive vitellogenin and apo-VLDLII genes. *Nucleic Acids Res.* **1984**, *12*, 8611–8626. [CrossRef]
8. Martinez, E.; Givel, F.; Wahli, W. The estrogen responsive element as an inducible enhancer: DNA sequence requirements and its conversion to a glucocorticoid responsive element. *EMBO J.* **1987**, *6*, 3719–3727. [CrossRef]
9. Martinez, E.; Givel, F.; Wahli, W. A common ancestor DNA motif for invertebrate and vertebrate hormone response elements. *EMBO J.* **1991**, *10*, 263–268. [CrossRef]
10. Martinez, E.; Wahli, W. Cooperative binding of estrogen receptor to imperfect estrogen-responsive DNA elements correlates with their synergistic hormone-dependent enhancer activity. *EMBO J.* **1989**, *8*, 3781–3791. [CrossRef]
11. Theulaz, I.; Hipskind, R.; ten Heggeler-Bordier, S.; Green, S.; Kumar, V.; Chambon, P.; Wahli, W. Expression of human estrogen receptor mutants in *Xenopus* oocytes: Correlation between transcriptional activity and ability to form protein–DNA complexes. *EMBO J.* **1988**, *7*, 1653–1660. [CrossRef] [PubMed]
12. Green, S.; Kumar, V.; Theulaz, I.; Wahli, W.; Chambon, P. The N-terminal DNA-binding 'zinc finger' of the oestrogen and glucocorticoid receptors determines target gene specificity. *EMBO J.* **1988**, *7*, 3037–3044. [CrossRef] [PubMed]
13. Corthésy, B.; Hipskind, R.; Theulaz, I.; Wahli, W. Estrogen-dependent in vitro transcription from the *Xenopus* vitellogenin promoter in homologous liver nuclear extracts. *Science* **1988**, *239*, 1137–1139. [CrossRef] [PubMed]
14. Schild, C.; Claret, F.X.; Wahli, W.; Wolffe, A.P. A nucleosome-dependent static loop potentiates estrogen-regulated transcription from the *Xenopus* vitellogenin B1 promoter in vitro. *EMBO J.* **1993**, *12*, 423–433. [CrossRef]
15. Dreyer, C.; Krey, G.; Keller, H.; Givel, F.; Helftenbein, G.; Wahli, W. Control of the peroxisomal β-oxidation pathway by a novel family of nuclear hormone receptors. *Cell* **1992**, *68*, 879–887. [CrossRef]
16. Issemann, I.; Green, S. Activation of a Member of the Steroid Hormone Receptor Superfamily by Peroxisome Proliferators. *Nature* **1990**, *347*, 645–650. [CrossRef]
17. Krey, G.; Braissant, O.; L'Horset, F.; Kalkhoven, E.; Perroud, M.; Parker, M.G.; Wahli, W. Fatty acids, eicosanoids and hypolipidaemic agents identified as ligands of peroxisome proliferator-activated receptors by coactivator-dependent receptor ligand assay. *Mol. Endocrinol.* **1997**, *11*, 779–791. [CrossRef]
18. Kliewer, S.A.; Sundseth, S.S.; Jones, S.A.; Brown, P.J.; Wisely, G.B.; Kobel, C.S.; Devchand, P.; Wahli, W.; Willson, T.M.; Lenhard, J.M.; et al. Fatty acids and eicosanoids regulate gene expression through direct interactions with PPARα and PPARγ. *Proc. Natl. Acad. Sci. USA* **1997**, *94*, 4318–4323. [CrossRef]
19. Devchand, P.R.; Keller, H.; Peters, J.M.; Vazquez, M.; Gonzalez, F.J.; Wahli, W. The PPARα–leukotriene B4 pathway to inflammation control. *Nature* **1996**, *384*, 39–43. [CrossRef]
20. Kersten, S.; Seydoux, J.; Peters, J.M.; Gonzalez, F.J.; Desvergne, B.; Wahli, W. Peroxisome proliferator-activated receptor α mediates the adaptive response to fasting. *J. Clin. Investig.* **1999**, *103*, 1489–1498. [CrossRef]

21. Michalik, L.; Desvergne, B.; Basu-Modak, S.; Escher, P.; Peters, J.M.; Kaya, G.; Gonzalez, F.J.; Zakany, J.; Metzger, D.; Chambon, P.; et al. Impaired skin wound healing in peroxisome proliferator-activated receptor (PPAR)α and PPARβ mutant mice. *J. Cell Biol.* **2001**, *154*, 799–814. [CrossRef]
22. Chong, H.C.; Tan, M.J.; Philippe, V.; Tan, S.H.; Tan, C.K.; Ku, C.W.; Goh, Y.Y.; Wahli, W.; Michalik, L.; Tan, N.S. Regulation of epithelial–mesenchymal IL-1 signaling by PPARβ/δ is essential for skin homeostasis and wound healing. *J. Cell Biol.* **2009**, *184*, 817–831. [CrossRef] [PubMed]
23. Michalik, L.; Desvergne, B.; Wahli, W. Peroxisome proliferator-activated receptors and cancers: Complex stories. *Nat. Rev. Cancer* **2004**, *4*, 61–70. [CrossRef] [PubMed]
24. Imai, T.; Takakuwa, R.; Marchand, S.; Dentz, E.; Bornert, J.-M.; Messadeq, N.; Wendling, O.; Mark, M.; Desvergne, B.; Wahli, W.; et al. PPARγ is required in mature white and brown adipocytes for their survival in the mouse. *Proc. Natl. Acad. Sci. USA* **2004**, *101*, 4543–4547. [CrossRef] [PubMed]
25. Woldt, E.; Terrand, J.; Mlih, M.; Matz, R.L.; Bruban, V.; Foppolo, S.; El Asmar, Z.; Cholet, M.E.; Ninio, E.; Bednarczyk, A.; et al. PPARγ counteracts LRP1-induced vascular calcification by inhibiting a WNT5A signaling pathway. *Nat. Commun.* **2012**, *3*, 1077. [CrossRef]
26. Schuler, M.; Ali, F.; Chambon, C.; Duteil, D.; Bornert, J.-M.; Tardivel, A.; Desvergne, B.; Wahli, W.; Chambon, P.; Metzger, D. PGC1α expression is controlled in skeletal muscles by PPARβ, whose ablation results in fiber-type switching, obesity and type 2 diabetes. *Cell Metab.* **2006**, *4*, 407–414. [CrossRef]
27. Planavila, A.; Rodriguez-Calvo, R.; Jové, M.; Michalik, L.; Wahli, W.; Laguna, J.C.; Vázquez-Carrera, M. Peroxisome proliferator-activated receptor β/δ activation inhibits hypertrophy in neonatal rat cardiomyocytes. *Cardiovasc. Res.* **2005**, *65*, 832–841. [CrossRef]
28. Iglesias, J.; Barg, S.; Yessoufou, A.; Pradevand, S.; McDonald, A.; Bonal, C.; Debril, M.; Metzger, D.; Chambon, P.; Herrera, P.; et al. Modulation of pancreatic β-cell mass and insulin release by PPARβ/δ in mice. *J. Clin. Investig.* **2012**, *122*, 4105–4117. [CrossRef]
29. Leuenberger, N.; Pradervand, S.; Wahli, W. Sumoylated PPARα mediates gender-specific gene repression and protects the liver from estrogen-induced toxicity. *J. Clin. Investig.* **2009**, *119*, 3138–3148. [CrossRef]
30. Rando, G.; Tan, C.K.; Khaled, N.; Montagner, A.; Leuenberger, N.; Bertrand-Michel, J.; Paramalingam, E.; Guillou, H.; Wahli, W. Glucocorticoid receptor–PPARα axis in fetal mouse liver prepares neonates for milk lipid catabolism. *eLife* **2016**, *5*, e11853. [CrossRef]
31. Iroz, A.; Montagner, A.; Benhamed, F.; Levavasseur, F.; Polizzi, A.; Anthony, E.; Régnier, M.; Fouché, E.; Lukowicz, C.; Cauzac, M.; et al. A specific ChREBP and PPARα cross-talk is required for the glucose-mediated FGF21 response. *Cell Rep.* **2017**, *21*, 403–416. [CrossRef] [PubMed]
32. Montagner, A.; Polizzi, A.; Fouché, E.; Ducheix, S.; Lippi, Y.; Lasserre, F.; Barquissau, V.; Régnier, M.; Lukowicz, C.; Benhamed, F.; et al. Liver PPARα is crucial for whole body fatty acid homeostasis and protects from NAFLD. *Gut* **2016**, *65*, 1202–1214. [CrossRef] [PubMed]
33. Tomas, J.; Mulet, C.; Saffarian, A.; Cavin, J.-B.; Ducroc, R.; Regnault, B.; Tan, C.K.; Duszka, K.; Burcelin, R.; Wahli, W.; et al. High-fat diet modifies the PPAR-γ pathway leading to disruption of microbial and physiological ecosystem in murine small intestine. *Proc. Natl. Acad. Sci. USA* **2016**, *113*, E5934–E5943. [CrossRef] [PubMed]
34. Duszka, K.; Ellero-Simatos, S.; Ow, G.S.; Defernez, M.; Tett, A.; Shi, Y.; König, J.; Narbad, A.; Kuznetsov, V.A.; Guillou, H.; et al. Complementary intestinal and microbiome responses to caloric restriction. *Sci. Rep.* **2018**, *8*, 11338. [CrossRef]
35. Lahiri, S.; Kim, H.; Garcia-Perez, I.; Reza, M.M.; Martin, K.A.; Kundu, P.; Cox, L.M.; Selkrig, J.; Posma, J.M.; Zhang, H.; et al. The gut microbiota influences skeletal muscle mass and function in mice. *Sci. Transl. Med.* **2019**, *11*, eaan5662. [CrossRef]
36. Fougerat, A.; Schoiswohl, G.; Polizzi, A.; Régnier, M.; Wagner, C.; Smati, S.; Fougeray, T.; Lippi, Y.; Lasserre, F.; Raho, I.; et al. ATGL-dependent white adipose tissue lipolysis controls hepatocyte PPARα activity. *Cell Rep.* **2022**, *39*, 110910. [CrossRef]
37. Paumelle, R.; Haas, J.T.; Hennuyer, N.; Baugé, E.; Deleye, Y.; Mesotten, D.; Langouche, L.; Vanhoutte, J.; Cudejko, C.; Wouters, K.; et al. Hepatic PPARα is critical in the metabolic adaptation to sepsis. *J. Hepatol.* **2019**, *70*, 963–973. [CrossRef]
38. Aguilar-Recarte, D.; Barroso, E.; Gumà, A.; Pizarro-Delgado, J.; Peña, L.; Ruart, M.; Palomer, X.; Wahli, W.; Vázquez-Carrera, M. GDF15 mediates the metabolic effects of PPARβ/δ by activating AMPK. *Cell Rep.* **2021**, *36*, 109501. [CrossRef]
39. Rostami, A.; Palomer, X.; Pizarro-Delgado, J.; Barroso, E.; Valenzuela-Alcaraz, B.; Crispi, F.; Nistal, J.F.; Hurlé, M.A.; García, R.; Wahli, W.; et al. PPARβ/δ prevents inflammation and fibrosis during diabetic cardiomyopathy. *Pharmacol. Res.* **2024**, *210*, 107515. [CrossRef]

biomolecules

MDPI

Review

PPARs in Clinical Experimental Medicine after 35 Years of Worldwide Scientific Investigations and Medical Experiments

Anna Skoczyńska [1,†], Monika Ołdakowska [2], Agnieszka Dobosz [3,*], Rajmund Adamiec [4,5,†], Sofya Gritskevich [2,‡], Anna Jonkisz [2], Arleta Lebioda [2], Joanna Adamiec-Mroczek [6], Małgorzata Małodobra-Mazur [2] and Tadeusz Dobosz [2,†]

1 Department of Internal and Occupational Medicine and Hypertension, Wroclaw Medical University, Borowska 213, 50-556 Wroclaw, Poland; anna.skoczynska@umw.edu.pl
2 Department of Forensic Medicine, Division of Molecular Techniques, Wroclaw Medical University, M. Sklodowskiej-Curie 52, 50-369 Wroclaw, Poland; monika.oldakowska@umw.edu.pl (M.O.); anna.jonkisz@umw.edu.pl (A.J.); arleta.lebioda@umw.edu.pl (A.L.); malgorzata.malodobra-mazur@umw.edu.pl (M.M.-M.); tadeusz.dobosz@umw.edu.pl (T.D.)
3 Department of Basic Medical Sciences and Immunology, Division of Basic Medical Sciences, Wroclaw Medical University, Borowska 211, 50-556 Wrocław, Poland
4 Department of Diabetology and Internal Medicine, Wroclaw Medical University, Borowska 213, 50-556 Wroclaw, Poland; rajmund.adamiec@umw.edu.pl
5 Department of Internal Medicine, Faculty of Medical and Technical Sciences, Karkonosze University of Applied Sciences, Lwówiecka 18, 58-506 Jelenia Góra, Poland
6 Department of Ophthalmology, Wroclaw Medical University, Borowska 213, 50-556 Wroclaw, Poland; joanna.adamiec-mroczek@umw.edu.pl
* Correspondence: agnieszka.dobosz@umw.edu.pl; Tel.: +48-71-784-04-82
† Retired Author.
‡ Biology Student at Wroclaw University, Political Refugee from Belarus, Volunteer in Department of Forensic Medicine, Wroclaw Medical University.

Abstract: This year marks the 35th anniversary of Professor Walter Wahli's discovery of the PPARs (Peroxisome Proliferator-Activated Receptors) family of nuclear hormone receptors. To mark the occasion, the editors of the scientific periodical Biomolecules decided to publish a special issue in his honor. This paper summarizes what is known about PPARs and shows how trends have changed and how research on PPARs has evolved. The article also highlights the importance of PPARs and what role they play in various diseases and ailments. The paper is in a mixed form; essentially it is a review article, but it has been enriched with the results of our experiments. The selection of works was subjective, as there are more than 200,000 publications in the PubMed database alone. First, all papers done on an animal model were discarded at the outset. What remained was still far too large to describe directly. Therefore, only papers that were outstanding, groundbreaking, or simply interesting were described and briefly commented on.

Keywords: PPARs; peroxisome proliferator-activated receptor; clinical application; experimental therapy

1. Introduction

It is widely believed that the Peroxisome Proliferator-Activated Receptor (PPAR) was described around 1974 by Prof. Walter Wahli. Unfortunately, among his 350 or so published papers, the fundamental, foundational, and pioneering work could not be found (and as there are at least two other contenders for priority [1,2]), we asked Prof. Walter Wahli for a statement on this matter. In response, Prof. W. Wahli identified one of his papers as the first (starting) paper on this topic, indirectly confirming his priority and providing its biographical data [3]. The paper cited is from March 1992 and concerns not a human but the amphibian *Xenopus laevis*, which, incidentally, explains our difficulty in finding it, since PubMed lists, at the time of writing these words, over 300,000 abstracts and full-text papers dealing with PPARs.

To finish this review on time, despite the challenging work of all authors, almost all work done on animal models (including *Xenopus*) was discarded at the outset, as it was assumed that they are usually a preliminary step before work on human material, as animal models are generally very different from humans [4]. Notably, the 'PPARs fad' is slowly passing relative to the publication peak of 2011–2013, when there were 5–6 new publications daily. In the last few years, it has been 4-3 papers per day, following a decreasing trend. New papers are located on the right, descending branch of the Gauss curve. Assuming its complete symmetry, the last single documents in this field are expected around 2040. Tahri-Joutey et al. [5] report (following [6]) that PPAR-alpha, the first of the known receptors, arose at the time of the evolutionary breakthrough of fish and mammals, about 200 million years ago, and evolved three times faster than the other receptors beta/delta and gamma. In 2006, a concise but comprehensive and complete description of the mechanism of action of PPARs appeared in print with co-authorship by Lathion [7]. The paper explained the mechanism of action using PPAR-alpha as an example; however, the activation of all PPAR is very similar. The authors describe the following sequence of events: first, a PPAR heterodimer is formed with the retinoid X receptor (RXR), and then the newly formed complex binds a ligand that defines the end effector. The ligand in question (a substance from a long list of natural or artificial compounds) can be an agonist (increasing activity) or an antagonist (decreasing activity). The described complex (PPAR plus RXR plus ligand) binds to a specific DNA sequence in the gene promoter (PPRE) and triggers transcription (Figure 1).

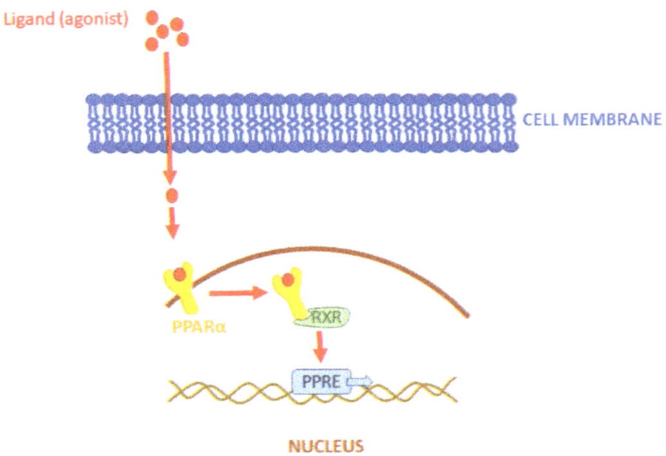

Figure 1. Pattern of PPARα activity. RXR—retinoid X receptor, PPRE—gene promoter.

Figure 2 shows basic information about PPAR isoforms. However, it should be noted that the number of PPAR types is not fully established. For the first few years, it was thought that there were four [3]. Then, their number was reduced to three, but later molecular studies show that there are four (named in 1992: alpha, beta, gamma, and delta) [8].

1.1. Peroxisome Proliferator-Activated Receptor Alpha (PPARα)

PPARα is a ligand-activated transcriptional factor that belongs to the nuclear receptor superfamily. It is also known as N1C1 (nuclear receptor subfamily 1, group C, member 1). PPARα is strongly expressed in tissues with elevated fatty acid catabolism, such as the liver, heart, skeletal muscles, brown adipose tissue, and intestine. It regulates the transcription of multiple genes involved in the intracellular metabolism of lipids and is a major regulator of fatty acid and triglyceride homeostasis [9–11]. PPAR-alpha performs the direct transcriptional control of genes implicated in mitochondrial and peroxisomal β-oxidation and microsomal ω-oxidation also. Modulating the activities of all three fatty acid oxidation systems plays a crucial role in energy expenditure. It also regulates the expression of proteins involved in the transport and β-oxidation of free fatty acids (FFAs) and lipoprotein metabolism genes [12].

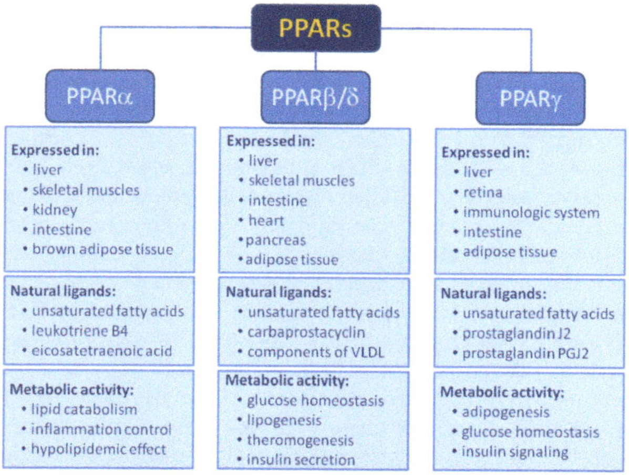

Figure 2. The characteristic of the three isoforms of PPARs: PPARα, PPARβ/δ, and PPARγ (expression, function, and the list of main ligands activating the particular peroxisome proliferator).

The human PPAR-alpha receptor gene (*PPARA*) is located on chromosome 22 at the position *22q12-q13.1*. PPAR-alpha receptor shares a common characteristic with the other members of the PPAR family (PPAR-beta/delta and PPAR-gamma) and has five domains [13]. Among these are the ligand-binding domain, dimerization domain (a dimer forms during the reaction with the PPAR-responsive element), and DNA-binding domain, which binds to the promoter of different genes via the zinc finger. There are two stages in PPAR-alpha transcriptional activation. In the first stage, PPAR-alpha is activated by the natural ligand or agonist, and the second stage is started by the formation of heterodimer PPAR-alpha with a retinoid X receptor (RXR). Then the PPAR-alpha-RXR complex recognizes many genes, binding to PPRE via the zinc finger of the DNA-binding domain. The PPRE is a sequence of nucleotides: AGGTCA, AGGTCA. This sequence is always the same on every gene to which the PPAR-alpha-RXR complex binds [14]. For example, there may be up to 80 different genes regulated by PPAR-alpha in hepatocytes. This action includes the stimulation of the apolipoprotein AI genes, the apolipoprotein AII genes, and the lipoprotein lipase (LPL) gene, and the inhibition of apolipoprotein CIII [14].

1.2. Peroxisome Proliferator-Activated Receptor Beta/Delta (PPARβ/δ)

This receptor type, perhaps because it has numerous links to tumorigenesis, is quite popular; there are about 1200 abstracts and full papers on this topic in the PubMed database. In the late 1980s, it was widely believed that there were two separate receptors, PPAR-beta and PPAR-delta. However, closer examination showed that they were the same, so some researchers felt it was necessary to eliminate one of them. Unfortunately, a big mess in naming ensued, and from Google data, it can be calculated that as many as 83% of researchers use the name "beta", 13% use the name "delta", and 4% refuse to declare their choice and use both names, e.g., "beta/delta".

1.3. Peroxisome Proliferator-Activated Receptor Gamma (PPARγ)

Cardiovascular diseases such as diabetes, hypertension, obesity, and dyslipidemia are the most common causes of death and severe disability. Their treatment consists primarily of modifying environmental factors to normalize metabolic changes, but genetic information is also essential in developing these diseases. The primary gene whose mutation can cause multistep disorders leading to the development of metabolic syndrome and atherosclerosis is the peroxisome proliferator-activated receptor gene γ (*PPARG*). This gene is located on chromosome *3p25*. Its expression can start from different promoters, depending on the site of origin, which results in the possibility of 3 isoforms of this receptor protein: PPARγ-1 (adipose tissue, muscle, heart, liver), PPARγ-2 (mainly adipose tissue), and PPARγ-3 (adipose tissue and colon). The expression of the PPARγ gene depends on, among other things, insulin and food availability, especially free fatty acids. PPARγ modulates insulin sensitivity and immune system activity by controlling the transcription of many proteins essential for carbohydrate and lipid metabolism. PPAR-gamma agonists in type 2 diabetes resulted in lower serum levels of pro-inflammatory markers, such as CRP, white blood cells, soluble CD40, MMP-9, amyloid A, and TNFα [15].

1.4. Pan-PPAR (Alpha, Beta/Delta, Gamma)

The phenomenon of the combined treatment of several seemingly separate (if only partially) phenomena bears the prefix "pan-" in the case of the receptors described; it is about pan-ligands (usually agonists) and pan-PPARs. This phenomenon has been discovered and described as many as three times, yet it is little known and rarely used. The term "pan-PPAR" was noted for the first time in a paper by Cullingford et al. from 2002 [16]. It appeared for the second time in the literature reviewed in a 2007 paper by Rudolph et al. [17]. The authors treated all three PPAR isoforms (alpha, beta/delta, and gamma) together, indirect evidence of which is the new term "pan agonist". Notwithstanding the above, three years later, Huang et al. [18] stated for the third time that separate testing of individual alpha, beta/delta, and gamma activities is incorrect, as the sum of the activities of all three isoforms should be considered. As an argument, they cited that using one of the experimental ligands initially significantly reduces, by decreasing the PPARα activity, the rate of cancer cell proliferation. Still, the subsequent increase in activity within the pan-PPAR phenomenon, specifically PPAR-gamma activity, spoils the whole effect (Table 1).

Table 1. Pan-PPAR effect of PPARs: when one type of receptor is activated, the rest is inhibited. ↑—means receptor activation, while ↓ means receptor inhibition.

Possibilities	PPARα	PPARβ/δ	PPARγ
1	↑ activity	↓ activity	↓ activity
2	↓ activity	↓ activity	↑ activity
3	↓ activity	↑ activity	↓ activity

2. Aim of the Work

Of particular note is that about 300,000 papers have been published on PPARs. This is undoubtedly an important topic that should be studied in depth. The published works go in different directions in learning about PPARs and sometimes the research results duplicate each other, thus supporting the validity and correctness of the results obtained by other research teams. This article was written to systematize and collect the existing knowledge about PPARs. It used the following databases: PubMed, EBSCO, Scopus, OMIM, and Google Scholar. By demonstrating the current state of knowledge, it is hoped that this work will contribute to better research planning, reduce unnecessary repetition, accelerate the growth of knowledge, and produce more interesting results.

3. Results

3.1. Introduction

The avalanche of papers dealing with PPARs moved very slowly. Quoting from Vamecq et al. [19], possibly the first pioneering paper on PPARs appeared in the second part of 1965, and from the beginning of 1966 until the end of 1991, on average, a new paper appeared every few months. For example, in 1988, two papers by Reddy et al. [20,21] appeared, writing about PPARs in liver cancer. Another author wrote about a "steroid...hormone receptor" the same year [2]. In contrast, Prof. W. Wahli gave us the date 1992 as the beginning of research on PPARs [8]. Independently, we found the work of Wahli and Martinez in 1991, in which the authors wrote about PPAR and its positive and negative regulations [22].

Since 1992, new works have appeared steadily, following an increasing trend. At the peak of 2011–2012, up to 5 to 6 new works appeared daily. This rate is now lower. In 2023, they came in at about 3 per day (Figure 3). In light of the above, it is astonishing that, for example, typing "PPARs and 1960" into a search engine yielded as many as 9 hits instead of the expected zero, though none related to the year entered.

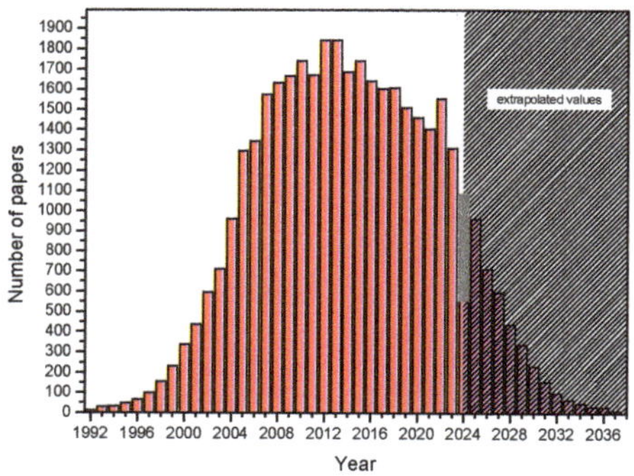

Figure 3. Number of papers published in 1992–2024 (till May 20th) regarding PPARs clinical applications in humans. The number of publications for subsequent years was extrapolated (dashed area).

3.2. Electronic Search Strategy

Data for publication were collected from the Ovid MEDLINE, Embase.com, National Library of Medicine PubMed, and Web of Science databases. Predefined eligibility criteria excluded studies involving animal models. Peer-reviewed journal articles published from 1961 to 20 May 2024 were searched, limited to each year. A detailed search strategy was developed by best practices, as shown in Table 2.

Table 2. Data to be extracted from databases.

Category	Description
Study ID	Authors Year Journal
Model	Human
Intervention	PPAR isotypes: - PPARα - PPARβ/δ - PPARγ - pan-PPAR
Other	The term "PPAR" reported in the title or abstract—yes/no Criteria for inclusion and/or exclusion of data provided—yes/no
Search term type	Free-text

Phrases were searched for in the title, abstract, and keywords provided by the author using Boolean Logic ('OR' and "AND"). Records retrieved in the Embase.com search were limited to only those absent from MEDLINE. The search filter was also adapted to retrieve studies involving human models. No restrictions were placed on language.

Discoveries about PPARs were searched using the terms: "pharmacological PPAR targeting", "PPAR discoveries", "medical PPAR investigations", "peroxisome proliferator-activated receptor", "PPAR AND Human", PPARs AND disease".

Free-text searches for all PPARs (α, β/δ, and γ) and pan-PPAR activity were linked to the terms agonist, modulator, stimulator, and activator using a proximity search so that both terms (e.g., "PPAR" and "agonist") had to occur within five words of each other.

Reference lists of included original studies and relevant review articles obtained from the systematic search were manually reviewed to look for papers scientifically important or interesting to the authors. When a paper was deemed potentially relevant after reviewing the title and abstract, the full text of the study was downloaded. The literature search successfully identified relevant studies, some of which (285 papers) were included in this full-text review. The selection of cited papers was strictly subjective.

3.3. *Calendar with Commentary*

In 1989, 10 new papers arrived. Of these, we were particularly interested in the publication of Liu et al. [23], in which the authors demonstrate that replacing brown marrow with yellow marrow takes place under the control of PPARγ.

The years 1990–1991 did not bring much work regarding the study of PPARs in humans. Animal models (mice, horses) were mainly studied. Anyway, it is worth noticing the paper of Issemann & Green [2], where authors activated PPARs with growth factors, and Alvares et al. [24], who identified the heat shock protein HSP70 as a new possible member of the PPARs superfamily.

The year 1992: in this year, only 10 papers were published, among which, in our opinion, the work of Maksymowich et al. [25] should be singled out. They detected in all members of the steroid receptor superfamily (which form heterodimers with PPARs) a common conserved motif (the DNA sequence), which gives us hope for the future generation of pan-agonist or pan-antagonist PPARs. Then Gottlichet et al. found that chimeric ligands also work [26].

The year 1993 was one in which 29 new papers appeared in the PubMed database, among which it is worth mentioning the following: The paper by Keller and Wahli [27], whose authors see the links between endocrinology and dietetics on the grounds of PPARs. The works of Keller et al. [28] and Issemann et al. [29] in which the role of 9-cis retinoic acid in the activation and regulation of PPARs was demonstrated, and it was established that the mechanism leads through the binding of the said ligand to RXR. The work of Bardot et al. [30], whose authors studied PPAR-RXR heterodimers and acyl CoA oxidase and cytochrome P450 IVA6. The work of Boie et al. [31], whose authors tested the activity of enantiomers of known ligands. The work of Motojima [32], in which the author describes the various functions of PPARs and their activation mechanism. In the paper by Sher et al. [33], the authors describe the results of molecular studies, including cloning. Also interesting is the work of Bass [34], in which the author considers whether the functions performed by PPARs are more general or more specialized, and Gibson [35], in which he considers the future of PPARs. However, we were most impressed by the paper by Chen et al. [36], in which the authors claim to have evidence that there are not three PPARs but as many as five. If their argument is widely accepted, we are in for another nomenclature revolution, in which case the problems arising from the withdrawal of one receptor (beta/delta) will turn out to be a trifle indeed.

In 1994, 33 new papers were published, of which we highlighted the work of Kainu et al. [37], who demonstrated the brain receptor PPARα. In 2009, Aleshin et al. [38] confirmed this in rat brains. In addition, they revealed the presence of other isoforms in the brain, as did the work of Lemberger et al. [39], which showed that stress hormones could be ligands of PPARs.

In 1995, 49 new papers appeared, of which the following caught our attention: The work of Wahli et al. [40], in which the authors extensively discuss the panorama of lipid metabolism and its relationship with PPARs. The work of Gustafsson [41] in which it was pointed out that the toxicity of ligands may be a result of their binding to the receptor. The work of Bocos et al. [42] was repeated the following year by Lemberger et al. [43] (co-authored with W. Wahli), in which the authors found that fatty acids alone can activate PPARs. The paper by Yu et al. [44] described a new activator, HETEs, and its unusual properties. The paper by Keller et al. [45] (co-authored by W. Wahli), whose authors detail the role of RXRs in PPARs activation. The work of Juge-Aubry (co-author Wahli) [46], which considers the effect of thyroid hormone receptors, and the work of Baes et al. [47], who studied not an agonist but a PPAR-alpha antagonist.

The year 1996 brought 66 interesting publications, of which we would like to mention the following: The papers by Formann et al. [48] and Wilson et al. [49] in which the authors describe ligands and activators of PPARs, and the paper by Gibson, in which the author tries to anticipate the future of PPARs. In the Aubert et al. [50] paper, the authors describe the first activator from a new family (thiazolidinediones). The paper by Patel et al. [51] is one in which the authors show that thiazolidinediones brilliantly activate PPARγ but, on the other hand, promote atherosclerosis. Then there is the paper in which Brun et al. [52] analyze the differences between the response to the same stimulus of different isoforms of the receptors studied.

In 1997, 100 new papers arrived, and here are the most important among them: The paper by Guan et al. [53] reporting the presence of PPAR receptors in the human ureter, potentially bringing PPARs into the orbit of urology. In addition to these is the work of Bernlohr et al. [54], describing extracellular proteins and lipid transporters; the work of Clarke and Jump [55] with a description of an independent novel mechanism of PPAR activation by polyunsaturated fatty acids (at this point, one can make a loose hypothesis that this could be one of the reasons for the beneficial effects of PPARs on human health); and the work of Rove [56] on retinoid X receptors.

The year 1998 enriched our knowledge about PPARs with the results of 156 works, the most important of which, in our opinion, worked on the role of PPARs in the functioning of blood vessel walls [57] and skeletal muscles [58] as well as the impact of PPARs on lipid metabolism at the receptor and gene expression levels [59–63]. This has expanded our knowledge about the role of PPARs in metabolic syndrome and the development of atherosclerosis in humans. The inhibition of cell growth and induction of apoptosis by PPAR agonists was first demonstrated in human breast cancer cells [64]. In the same year, Sorensen et al. leaned into the problem of precisely controlling the activity of circumstantial PPARs, which takes on great importance in light of the planned preclinical studies [65], and Goodman found three independent connections between retinoids and schizophrenia [66].

The year 1999 brought us 233 new publications to read, many of which concerned the effects of PPARγ agonists on the growth and differentiation of human cancer cells in cancer tissues (ureter, bladder, prostate, liposarcoma) or cell lines such as colon cancer cells, T-cell lymphoma, or breast cancer cell lines [67–71]. The human T-cell leukemia retrovirus tax protein is a repressor of nuclear receptor signaling [72]. A possible role of reduced PPARγ activation by 15-HETE in the development or progression of prostate cancer has been suspected [73].

In 2000, there were 339 new interesting papers to review and evaluate. Lazennec G. and Wahli W. et al. observed a protein kinase A-dependent enhancement of the activity of all PPARα, β, and γ isotypes on two distinct promoters. Moreover, PKA inhibitors repress the PPAR ligands' effect, suggesting that these ligands act partly by recruiting the PKA pathway. These findings highlight the involvement of the PKA pathway in PPAR action and indicate that it is essential for regulating gene activation by PPAR ligands [74]. Kawahito and co-authors demonstrated that the synovial lining layer, fibroblasts, and endothelial cells show expression of PPARγ in patients with rheumatoid arthritis (RA) and suggested that PPARγ may be an essential immunoinflammatory mediator and its ligands, especially 15d-PGJ2, may be helpful in the treatment of RA [75]. Much attention has been paid to the role of PPARs in oncogenesis, including PPARδ in human endometrial adenocarcinoma and colorectal cancer [76–78] and PPARγ in gastric or prostate cancer [79,80].

In exciting publications, Combs et al. showed that PPARγ agonists inhibit the β-amyloid-stimulated expression of the cytokine genes interleukin-6 and tumor necrosis factor α, which brings us closer to the mechanisms of Alzheimer's disease [81]. In contrast, Zhou et al. drew attention to the over-representation of PPARγ sequence variants in sporadic cases of glioblastoma multiforme [82].

The year 2001 meant 437 new papers for us. Among them, Debrik et al. [83] found that PPAR-gamma has a pleiotropic function in the cell, while Chinetti et al. [84] detected functions performed by PPARs in blood vessel walls, which have serious health implications. Delerive et al. described PPARs in the context of modulation of the inflammatory response, which may have potential therapeutic applications in chronic inflammatory diseases [85]. The authors reported that PPARs activators participate in anti-inflammatory activation in various cell types by inhibiting the expression of pro-inflammatory genes such as cytokines, metalloproteases, and acute phase proteins. In addition, PPARs play an important role in negatively regulating the transcription of inflammatory response genes by antagonizing AP-1, nuclear factor-kappaB (NF-kappaB), a signal transducer and activator of transcription, and the nuclear factor signaling pathways of activated T cells. Duez et al. emphasize that PPARs play an essential role in modulating the development of atherosclerosis through actions at both the metabolic and vascular levels. PPAR-alpha and PPAR-gamma seem to be of the most significant importance in developing this disease, with high levels of their expression. Moreover, PPAR-alpha and PPAR-gamma are involved in immunoregulation, vasculitis, and thrombosis associated with atherosclerosis [86]. Also, Elangbam et al. presented that PPAR-alpha and PPAR-gamma receptors have a beneficial effect on inflammatory diseases, including atherosclerosis, by regulating the production of cytokines, the expression of adhesion molecules on endothelial cells, fibrinolysis, and the modulation of monocyte-derived macrophages [87].

2002: After reading 596 new papers, it can be concluded that, for example, in the review by Stumvoll and Haring, we can learn that the occurrence of the PPAR-gamma2—Pro12Ala polymorphism has a significant impact on the risk of common type 2 diabetes (T2D). The authors emphasized that a thorough understanding of the mechanism of this disease may allow for the prevention or improvement of T2D treatment [88]. Roberts et al. demonstrated that p38 MAP kinase activation is mediated by PPARα transcription and detectable changes in survival and proliferation. Moreover, they presented the problem of how MAP kinases interact with nuclear receptors such as PPARα. This process causes coordinate suppression of apoptosis and cell cycle progression [89]. On the other hand, Berger and Moller presented condensed knowledge about the relationship of PPARs to chronic diseases [90]. According to them, the strong therapeutic effects of PPARα and PPARγ agonists benefit systemic lipid levels, glucose homeostasis, and atherosclerosis. It was emphasized that PPARα activation can effectively alleviate dyslipidemia and that PPARγ agonists reduce insulin resistance. However, in the case of PPARδ, the authors suggest that it may also be an important therapeutic target for selected patients' disorders, including cancer, infertility, and dyslipidemia. Fauconet et al. conducted studies in two different human bladder cancer cell lines, RT4 (derived from a stage I tumor) and T24 (derived from a stage III tumor). VEGF (mRNA and protein) is deferentially increased by the three PPAR isotypes. Additionally, the authors reported that only the MEK inhibitor PD98059 downregulated PPAR ligand-induced VEGF expression [91]. Moreover, Haydon et al. [92] showed that PPARγ agonists are excellent therapy for human osteosarcoma, the biggest metastasis problem in many late carcinomas, while Koumanis et al. [93] looked at correlations between overweight groups and mortality in association with PPAR G2 Pro12Ala genotypes. In contrast, Berger & Wagner [94] described both the health-promoting and pathogenic properties of PPARs.

In the following year, 2003, out of a group of 711, we considered the following to be particularly important. One is an article in which Francis et al. conducted clinical trials in which it was shown that only PPARα agonists improved the prognosis of atherosclerotic heart disease [95]. Another noteworthy work concerns PPAR-gamma ligands and their role as modulators in future cancer treatment strategies. Margeli et al. demonstrated the importance of PPAR-gamma in cell proliferation and cancer, establishing their anticancer properties against a wide range of cancer cells [96]. The following paper outlines the problem of using PPAR-gamma agonists thiazolidinediones (TZDs) to treat type 2 diabetes. Despite the use of TZDs and their ability to reduce and slow down the progression of insulin resistance, including lowering plasma glucose levels and significantly reducing triglycerides and free fatty acids in plasma, patients experienced increased body fat. However, it is suggested that fat redistribution takes place in a favorable direction, from visceral to subcutaneous stores [97].

In 2004, 959 fresh scientific reports were published. One of them is an article by Jiang et al. that pointed out that PPARγ may play an important role as one of the regulators of prostate cancer (PCa) cell differentiation and proliferation. Therefore, PPARγ agonists may find use as adjuncts to watchful waiting in selected subpopulations of patients with localized PCa [98].

In 2005, the number of new works exceeded 1000 when 1297 new papers were published. A big part of them seemed to be useful because they focused on the problem of the targets. For example, Bedu et al. [99] (co-author W. Wahli) described PPARs β/δ as a therapeutic target, and Tenenbaum et al. [100] used a ligand (H-LDL) to achieve a pan-PPAR effect in therapy. Buzzetti et al. examined the association of the PPAR-gamma2 Pro12Ala polymorphism with measurements of insulin sensitivity in a population of obese Italian children (mean age 10.38 +/− 2.8 years). The authors showed that the X12Ala variant (Pro12Ala or Ala12Ala) is significantly associated with greater insulin sensitivity in childhood obesity, which may, therefore, be protected against cardiovascular disease, obesity, and diabetes due to the phenotypic effect of the Ala 12 allele on insulin resistance [101].

The year 2005 also brought several studies of PPARs-specific ligands. A fascinating review paper by Tyagi et al. [102] describes the importance of homocysteine in endocardial function, structure, and remodeling in conjunction with nitric oxide metabolism in diabetic patients with impaired myocyte diastolic relaxation. According to the authors, homocysteine antagonizes PPARγ receptors, increasing oxidative stress and decreasing endothelial nitric oxide levels. Then nitrotyrosine content and matrix metalloproteinase activity in the diabetic heart increase, resulting in myocyte-endothelium disconnection. Harris et al. found that PPAR-gamma activation reduces head and neck cancer proliferation through the mechanism of action of 4-hydroxyphenylretinamide (4-HPR, fenretidine) [103]. In turn, Lowe et al. investigated oxazole-substituted indanylacetic acids in the character of PPAR-gamma and alpha activator ligand [104]. At the same time, LoVerme et al. showed that palmitoylethanolamide (PEA) is an activator of PPAR-alpha [105].

In 2006, the increase in the literature was even more significant, with 1343 new papers, of which Michalik et al. [106] (co-author W. Wahli) was selected. They wrote a valuable, well-done review in which the authors describe the selective effect of PPAR in vivo, which results from the interaction at a given time point between the expression levels of each of the three PPAR and RXR isotypes, the affinity for a specific PPRE promoter, and the availability of ligands and cofactors. Aung et al. investigated the role of peroxisome proliferator-activated receptors (PPARs) alpha and beta in differentiating HT-29 colon cancer cells and MCF-7 breast cancer cells [107]. The authors concluded that PPAR expression depends on the differentiation method and time. These studies have shown that changes in PPAR-alpha levels are not required to differentiate colon cancer cell lines, but changes in PPAR-beta are more closely related to differentiation. Trivedi et al. presented a different role of PPARs, showing that PPARs regulate sebum production. Moreover, this study's authors believe that selectively modulating PPARs activity may constitute a new therapeutic strategy in treating acne [108]. Burdick et al.'s review consolidates information on the role of PPAR-beta in epithelial tissues. It presents critical discrepancies and how to direct research to fully understand how this receptor modulates epithelial homeostasis [109].

In 2007, there were 1577 new publications, some of which have vigorously stirred our imagination, among them a paper by Michalik and Wahli [110] on the differences of PPARs in healthy and diseased (acne, psoriasis, warts, and benign tumors) skin. Akhmetov et al. [111] claim to have found a link between the athletic performance of athletes and several PPAR polymorphisms. It is also worth mentioning the paper of Piqueras et al. [112]; they induced endothelial cell proliferation and angiogenesis using the activation of PPARβ/δ by various experimental ligands. In Grarup et al. [113], the authors report that in a sample of approximately 7500 white, elderly individuals, 12 polymorphic variants of the PPARβ/δ gene correlated with insulin resistance but not with diabetes type. Conversely, Hollingshead et al. conducted studies on human cancer cell lines (HT29, HCT116, LS-174T, HepG2, and HuH7) cultured in the presence or absence of serum and compared in vitro analysis with in vivo analysis. Based on the results obtained (the PPAR-beta/delta ligand did not increase cell growth or Akt phosphorylation, nor did it increase the expression of VEGF or COX2 in any cell line), the authors of this study concluded that PPAR-beta/delta ligands do not enhance tumor formation [114].

The year 2008 brought 1633 news items, and research about the relationship between PPARs and lung cancer is hidden among them. In one such study, Pedchenko et al. demonstrated that PPAR-beta/delta is highly expressed in most lung cancers. Furthermore, PPAR activation induces a proliferative response and survival in non-small cell lung cancer [115]. Borland et al. demonstrated that PPAR-beta/delta ligand activation inhibits keratinocyte proliferation through a PPAR-beta/delta-independent mechanism. Instead, retinoic acid's inhibition of keratinocyte cell proliferation is mediated through a source of PPAR-beta/delta, contradicting the theory that retinoic acid would enhance cell proliferation through activation of PPAR-beta/delta [116]. One interesting study was conducted by Ahmed et al., who examined the expression of PPAR-beta in terminal ovaries and different histological grades of ovarian tumors. The authors of this study found that, unlike other cancers such as colon cancer, endometrial cancer, and head cancer, overexpression of PPAR-beta does not occur in ovarian tumors [117]. In the same year, Billin wrote a review paper in which he summarized the general knowledge available at that time about PPAR-beta/delta agonists and properties in the treatment of dyslipidemia and type 2 diabetes, as well as emerging clinical data on various PPAR-beta/delta agonists. Autor stated that at that time, their use in the treatment of dyslipidemia or type 2 diabetes could not be justified [118]. On the other hand, Gallardo-Soler et al. demonstrated that PPAR-γ/δ-mediated uptake of modified lipoproteins by monocytes, linking lipid metabolism and immunity, is an essential early event in the development of atherosclerosis [119]. In all cells involved in the atherosclerotic process, PPARs can modulate immune pathways through at least three different mechanisms: by directly binding to PPREs of anti-inflammatory cytokine genes, by transrepression of transcription factors such as NF-κB and AP-1; or by corepression [120]. Interestingly, in vitro studies demonstrated similar changes in AngII-treated macrophages: PPARδ activation increased total and free Bcl-6 levels and inhibited AngII activation of MAP kinases, p38, and ERK1/2. These studies uncover crucial proinflammatory mechanisms of AngII and highlight actions of PPARδ activation to inhibit AngII signaling, which is atheroprotective [121].

In 2009, out of 1666 new papers, we decided to give the following list of outstanding, groundbreaking work: Cavalieri et al. [122] tried to look at the role of PPARs (alpha) from the point of view of nutrigenomics, which brought the discussion closer to epigenetics. A theoretical review paper by Porcuna et al. [123] was finally published in 2021, in which the authors point out the great potential of epigenetic tools. In contrast, our first practical work in this field dates back to 2022. These studies demonstrated that the efficiency of PPARγ was increased in adipocytes obtained from insulin and isolated dogwood extracts. These results confirm the involvement of PPARγ in the regulation of cellular metabolism [124]. An interesting study was carried out by Danesi et al., in which green tea extract was added to the cardiomyocyte medium from the first seeding, as a result of which this extract selectively activated the PPAR-beta/delta isoform. Thus, the authors of this study observed that with the increase accompanied by the activation of PPAR-beta/delta, there was a decrease in the production of nitric oxide (NO) synthase and an increase in total antioxidant activity. Therefore, it has been suggested that the activation of PPAR-beta/delta may play a key role in reducing NO production [125]. It is worth looking at two publications of the same year on the role of PPAR-beta in oxidative stress-induced apoptosis in human umbilical vein endothelial cells (HUVEC). In Jiang's publication, it was shown that H_2O_2 decreased the expression and activation of PPAR-beta, which played an essential role in H_2O_2-induced apoptosis in HUVEC [126]. A few months later, the same authors reported that repeated low-level H_2O_2 stress protected HUVECs from subsequent oxidative stress-induced apoptosis by increasing PPARβ expression and activity [127]. In turn, Bishop-Bailey and Bystrom summarized the current knowledge about the critical role PPAR-beta/delta plays in regulating inflammatory processes and immunity [128]. Ramanan et al. conducted research on patients after whole-brain irradiation (WBI). The authors demonstrated that PPARα ligands inhibited or slowed down radiation-induced pro-inflammatory responses in microglia in vitro. Moreover, they prevented the harmful effects of WBI on hippocampal neurogenesis in vivo, and the PPARγ ligand Pio alleviated WBI-induced cognitive impairment [129].

In 2010, the scope of knowledge expanded with the results of 1741 new works, including the work of Genovese et al. [130], who tested propenoic acid derivatives as PPAR-beta/delta agonist ligands. Grimaldi described the regulatory functions of peroxisome proliferator-activated receptor beta in terms of metabolism, inflammation, and cellular stress [131]. In his review, the author emphasized that activation of PPARβ using its strong and specific agonists prevented and reversed abnormalities associated with the metabolic syndrome. Therefore, both genomic and non-genomic modes of action can modulate responses to metabolic, inflammatory, and oxidative stress in several tissues. An interesting study was conducted by McKinnon et al., and it was performed on patients with endometriosis. The study showed that the expression of PPARγ in peritoneal endometriotic lesions was correlated with pelvic pain, dysmenorrhea, and dyspareunia experienced by patients [132]. This year also brought an exciting paper by Mistry and Cresci, who studied PPARs alpha, beta, and gamma to individualize heart failure therapy [133]. It is also important to mention the work of Bassaganya-Riera et al., who studied the effect of PPAR-gamma on respiratory infections with viruses, which offers potential hope for rescue in case of failed viral vector therapies [134].

In 2011, 1672 papers were published, mainly on the role of PPARs in lipid metabolism, inflammation, cell proliferation, and apoptosis. Clinically significant papers were precious. The role of T-cell PPARγ in lymphopenia-associated autoimmunity was determined for the first time. Housley et al. examined the role of PPARγ in CD4+ T cells and found that PPARγ expression in CD4+ CD25− T cells (Teff) is required to develop autoimmunity under lymphopenia. However, an unexpected function for PPARγ in Teff, in contrast to the documented immunosuppressive role of PPARγ, was found in increased Teff proliferation and survival [135]. A clinically valuable study of rheumatoid arthritis patients treated with pioglitazone showed a significant reduction in serum oxidative and inflammatory parameters and a significant improvement in DAS28 compared to the placebo group [136]. In turn, Sertznig and Reichrath presented an exciting summary of the effects of PPAR agonists in dermatology. This review summarizes the current knowledge of PPAR functions in various skin disorders, particularly inflammation and epidermal hyperproliferation (i.e., psoriasis, atopic dermatitis, acne, scleroderma, skin malignancies) [137].

In 2012, 1845 relatively fresh results arrived; the selected works are described below. First, Laschke and Menger [138] had developed a strategy for the use of PPARs in antiangiogenic therapy of endometriosis, while Nickkho-Amiry et al. addressed the modulation of PPARs, which reduces the growth of endometrial cancer cells [139]. Moreover, Knapp et al. [140] showed higher expression of PPARα and PPARβ and lower expression of PPARγ in patients with endometrial cancer than in patients with healthy endometrium. Additionally, Montero et al. [141] demonstrated that activation of PPARα and PPARγ modulates the expression of the human equilibrium nucleoside transporter hENT1, which plays a vital role in chemotherapy. Then Abbas et al. [142] found that a PPARγ agonist can both reduce and increase the risk of cardiovascular events, while Chen et al. [143] studied PPARγ receptors in neurodegenerative diseases, finding them to have a significant effect through NF-kappa beta. Additionally, Greene et al. [144] proved that the level of PPARδ increased during intense strength training and was negatively correlated with the concentration of total cholesterol and LDL, especially in the untrained state. In contrast, the level of PPARα increased intensively after exercise training.

Wadosky and Willis [145] showed in their paper that post-translational regulation of PPARs undergoes ubiquitination and sumoylation, and Peyrou et al. have dealt with PPARs in liver disease and found that they are essential in metabolic dysregulation [146]. Costa and Ciccodicola, on the other hand, investigated whether PPARγ is essential to diabetic retinopathy, finding, however, a beneficial role [147]. Then Balakumar and Mahadevan [148] argued in their paper that PPARs may be involved in the action of statins by showing anti-inflammatory, antioxidant, and antifibrotic effects. Wahli and Michalik [149], on the other hand, consider PPARs to be the main regulator of whole body metabolism. Furthermore, Reichenbach et al. [150] showed that the PPARα agonist Wy14643 inhibits cathepsin B protein expression in endothelial cells. Then, Taiileux et al. [151] claimed that PPARs have potential clinical applications in non-alcoholic fatty liver disease, as evidenced by their participation in modulating triglyceride accumulation. Vidella and Petinelli also mentioned the role of PPARs in NAFLD [152]. In this work, the authors showed that the reduction of PPARα levels is associated with the depletion of the n-3 polyunsaturated fatty acid chain, which may play a role in increasing the DNA binding capacity of the pro-inflammatory factors NFκB and AP-1, thus constituting one of the main mechanisms of progression of steatosis to steatohepatitis. PPARs also play a role in neurological diseases, as demonstrated by Bedendusi et al. [153]. The authors of this study proved that the progression of amyotrophic lateral sclerosis (ALS) leads to the activation of PPARγ in motor neurons, which contributes to the increase in the activity of lipid detoxifying enzymes, such as lipoprotein lipase and glutathione S-transferase α-2. PPARs may participate in the regulation of viral expression, as evidenced by the publication of Hu et al. [154] regarding the HBV virus. This experiment showed that PPARα regulates HBV gene expression through interactions with regulatory elements of the HBV promoter. Interestingly, Zuo et al. demonstrated that the induction of heat injury in fibroblasts activates the action of PPARβ, which has a protective effect on the structure and proliferation of these cells [155].

The same was true in 2013, with 1845 new papers, including the selected ones, such as the paper by Le Foll et al. [156], which analyzed the potential use of peroxisome proliferator-activated receptor (PPAR) agonists as promising new drugs in the treatment of psychoactive substance dependence based on preclinical studies. Aleshin and Reiser [157] described the roles of all PPAR isoforms. They addressed the regulation of oxygen signaling in the brain through pan-PPARs. The paper of Contreras et al. describes PPAR-alpha in the context of the key to metabolic and environmental adaptations of disease substrates [158], and the paper of Akyurek et al. describes PPAR-gamma in overweight children. These researchers found that PPAR-gamma levels in obese children were low [159].

In 2014 we can find 1688 research papers focused on PPAR, some showing the importance of PPARs in anticancer treatment. One such study was conducted by Yao et al. on human breast cancer cell lines (MDA-MB-231—estrogen receptor-negative and MCF7—estrogen receptor positive), where they showed that ligand activation and/or overexpression of PPARβ/δ inhibits the relative carcinogenicity of breast cancer and provides further support for the development of PPARβ/δ ligands to specifically inhibit breast carcinogenesis [160]. In addition, research into the treatment of HCV virus has been carried out. Researchers have shown that using fluoxetine, an HCV titer-lowering drug, also increases the activity of the peroxisome proliferator-activated receptor (PPAR) response element in HCV infection [161].

In 2015—1742 papers were published. One was Montagner et al.'s paper, which reviews peroxisome proliferator-activated receptor PPARβ/δ function in skin wound healing and cancer. The work highlights the significant role of PPARβ/δ in inhibiting keratinocyte apoptosis at the wound edges through activation of the PI3K/PKBα/Akt1 pathway and its role during re-epithelialization in regulating keratinocyte adhesion and migration. The observations and analyses suggest the need to evaluate modulators of PPARβ/δ that attenuate or enhance its activity, depending on the therapeutic target [162].

In 2016, 1643 works were published. An exciting publication concerns the use of nanoparticles to enhance the anti-inflammatory potential of one of the PPARγ agonists—15-deoxy-Δ12,14-prostaglandin J2 (15d-PGJ2). Using solid lipid nanoparticles (SLN) and 15d-PGJ2 at low concentrations, 15d-PGJ2-SLN was developed and investigated for its immunomodulatory potential. In three inflammation models, 15d-PGJ2-SLN reduced neutrophil migration, increased IL-10 levels, and reduced IL-1β and IL-17 in peritoneal fluid. Therefore, using SLN may improve the therapeutic properties of the 15d-PGJ2 [163]. Also, in 2016, Xue et al. constructed the peptide-functionalized nanoparticle platform to deliver PPARγ activator rosiglitazone to adipose tissue vasculature [164]. This system promotes the transformation of white adipose tissue into brown-like adipose tissue and angiogenesis, which facilitates the homing of targeted NPs to adipose angiogenic vessels, thereby amplifying their delivery. In a diet-induced obese mouse model, compared with the control group, these angiogenesis-targeted NPs inhibited body weight gain and modulated several serological markers, including lipids and insulin. These findings suggest that angiogenesis-targeting moieties with angiogenic stimulator-loaded NPs could be incorporated into the clinical treatment of obesity and other metabolic diseases. A different type of carrier, i.e., diamond nanoparticles (DN) was used by Strojna et al. concerning the delivery of curcumin to cells [165]. The major problem with curcumin is its poor bioavailability, which can be improved by adding carriers. DN are large surfaces that are non-toxic, have antiangiogenic properties, and create bio-complexes through a fast and straightforward process of self-organization. The authors investigated the cytotoxicity of complexes of curcumin with DN against liver cancer cells and normal fibroblasts. Preliminary results confirmed the applicability of DN as an efficient carrier of curcumin, which improves its performance against cancer cells in vitro.

In 2017, 1606 papers were published. This year, Fidoamoere et al. [166] elucidated the role of PPAR-alpha in metabolic alterations of the oxygen-deficient microenvironment in newly formed tumors. The same problem was revisited 6 years later by Rolver et al. [167], who looked at the issue of creating, with the help of PPAR-alpha, an acidic microenvironment that favors young tumors. In the same year, Kado et al. [168] proposed to reduce ovarian cancer progression using the exact mechanism. Wright et al. researched human oral cancer cell lines CA-9-22 and NA, which were treated with the PPAR activators eicosatetraenoic acid (ETYA), 15-deoxy-δ-12,14-prostaglandin J2 (PG-J2), thiazolidinedione, and ciglitazone, and then tested on their ability to functionally activate PPARγ luciferase reporter gene constructs. The authors found a significant reduction in cell proliferation and clonogenic potential and concluded that with treatment, there was a decrease in cell proliferation and clonogenic potential [169]. A fascinating paper by Ivanova et al. describes the activation of PPARs in fatty acid and energy metabolism, stating that PPAR-gamma activation is related to atherosclerosis, cardiovascular disorders, and prognosis of cardiovascular surgery [170]. Gross et al. reported that PPAR ligands reduce comorbid disease by affecting WAT combustion storage capacity and fat combustion in BAT and/or peripheral possessions, thereby reducing ectopic fat load. Moreover, a well-prepared clinical study showed that the use of PPARγ has a beneficial effect in patients with non-alcoholic steatomyelitis [171].

The year 2018 brought us 1608 research papers. For example, Borland et al. showed that PPARs have a significant impact on carcinogenicity in squamous cell carcinoma [172]. The authors presented their results on the A431 human squamous cell carcinoma cell line, which showed that stable expression and activation of PPARβ/δ or PPARγ led to reduced carcinogenicity. Furthermore, Leiguez et al. found that in macrophages, the compound MT-III, snake venom phospholipase A2, activated PPARγ and PPARβ/δ and increased protein levels of both transcription factors and CD36 [173]. According to studies conducted by Sun et al. on the human hepatocellular carcinoma cell line HepG2, it is suggested that using 2,4-dichlorophenoxyacetic acid interferes with glucose metabolism in the liver by activating PPARβ [174]. Therefore, it leads to hepatocyte glycogen accumulation by increasing glucose uptake and improving gluconeogenesis (via FoxO1 and CREB).

2019 was particularly abundant in new (1509) and especially in critical reports, of which we have selected a few for presentation. One of the exciting and valuable publications is a review in which the authors present the issues related to chemotherapy-induced neuropathy and the effects of PPARγ activation in treating chemotherapy-induced neuropathic pain [175]. Vallée, et al. demonstrated that non-steroidal anti-inflammatory drugs acting as PPARγ agonists help regulate the WNT/β-catenin pathway and thus control tumor growth through cell cycle arrest, cell differentiation, and apoptosis and may reduce inflammation, oxidative stress, proliferation, invasion, and cell migration [176].

The year 2020 brought 1460 new papers. We found many publications dated that year on the links between PPARs and various types of cancer. One of them is an interesting work of Toraih et al. [177], who addressed the relationship between genotype and clinical prognosis of thyroid cancer patients, finding in a material analysis of 174 samples that the simultaneous presence of micro RNA 27a (miRNA 27a) and PPAR-alpha and RXR-alpha dimer are associated with adverse clinical prognosis. Hirao-Suzuki et al. demonstrated in the MDA-MB-23 human breast cancer cell model that COX-2 expression is positively modulated by PPARβ/δ-mediated signaling [178]. Moreover, Elie-Caille et al., using RTqPCR and Western blot analyses, showed that the peroxisome proliferator-activated receptor β/δ (PPARβ/δ) agonist GW501516 significantly reduced the expression of N-cadherin in the transitional cell carcinoma of the urinary epithelium of the bladder [179]. Moosavi et al. reported that in studies of Cao2 colorectal adenocarcinoma cells, extracellular vehicles EVs derived from gut microbiota such as *F. prausnitzii* increase the permeability of the intestinal barrier, among others, through the PPARα, PPARγ and PPAR β/δ genes [180]. Additionally, Liu et al. demonstrated that genetic variants in PPAR pathway genes, particularly MED1, PRKCA, and PRKCB, may contribute to susceptibility to pancreatic cancer [181]. Wouters et al. demonstrated that inflammation-induced PPARγ expression promotes myelin-induced foam cell formation in macrophages in multiple sclerosis [182]. PPARs are also associated with liver disease. Xia et al. showed that bergenine may be a promising drug candidate for treating liver fibrosis because it activated PPARγ, inhibited TGF-β, and reduced liver fibrosis by inhibiting hepatocyte necrosis and extracellular matrix formation [183]. In studies conducted in hepatocytes treated with non-esterified fatty acids, Shen et al. showed that choline and methionine regulate the transcriptional activity of PPARα and liver X receptor α (LXR-α) through phosphorylation of AMPK-α and regulate SREBP-1c independently of AMPK-α to promote lipid oxidation and transport [184]. Yu et al. showed that PPARβ/δ activating ligands increase the in vitro expression of decidualization biomarkers of endometrial stromal cells (ESCs), and PPARβ/δ antagonists impair decidualization markers [185]. In addition, Zhang et al. showed that the expression of PPARβ/δ was significantly increased in cholesteatoma, and the ligand-activated PPARβ/δ promoted the proliferation of cholesteatoma keratinocytes as a result of the positive regulation of the PDK1/PTEN/AKT/GSK3β/cyclin D1 pathway [186]. In the same year, Wójtowicz et al. [187] proposed using the PPARs in treating neurodegenerative disorders. More Manickam et al. [188] studied the effect of microbiota on skeletal muscle status, da Cruz et al. [189] attempted to control inflammatory processes through PPARβ/δ and manipulation of it with diet, and Wagner and Wagner [190] studied PPARβ/δ as a hallmark of cancer. In this news-rich year, additional mention should be made of Phua et al. [191] work (co-author W. Wahli), in which a PPARβ/δ agonist reduces inflammation in skeletal muscle. Additionally, Zhang et al. researched new biomarkers of myocardial damage. In this work, they demonstrated that small molecules transform myocardial energy metabolism by regulating and translating PPARα [192]. Interestingly, Andrade-Souza et al. researched the effects of exercise on nuclear transcription factors. Studies have shown that PPARα and PPARβ/δ gene transcriptions were increased when exercise was performed twice a day compared to once a day [193]. Furthermore, Faulkner et al. showed that in the human endothelial cell line HUVEC, the tubulogenic effects of compound GW0742 were dependent on PPARβ/δ and demonstrated a regulatory role for PPARβ/δ in regulating endothelial cell behavior and promoting tissue maintenance and repair [194]. In addition, Chai et al. showed that microRNA-9-5p inhibits extracellular matrix deposition and proliferation and induces apoptosis by targeting PPARβ in human hypertrophic scar fibroblasts [195]. It also turns out that PPARs have a potential role in treating diabetic retinopathy, as discovered by Capozzi et al. [196]. In this study, the authors showed that the PPARβ/δ inhibitor—GSK0660 inhibits the palmitic acid-stimulated production of inflammatory mediators by Müller cells. However, Contreras-Lopez et al. demonstrated a different property of PPARs. In their studies, the authors showed that PPARβ/δ participates in the regulation of the immunoregulatory potential of mesenchymal stem cells, therefore dictating their metabolic reprogramming and paving the way to enhance the immunoregulatory properties of MSCs and counteract their versatility [197]. Also of note from this year is the work of Petr et al., who studied the effect of genetic variants of PPARs on the performance of athletes [198]. In addition, Tutunchi et al. [199] reviewed the effects of oleoylethanolamide, a PPARα ligand, as a weight control agent.

There were 1404 newcomers in 2021, among which the following papers should be mentioned: Grabacka et al. [200] found that PPAR-alpha plays a key role in the processes of the innate immune response, and Willems et al. [201] highlighted the possibility of modifying PPAR-alpha (and also beta/delta) activity with photohormones. Kharbanda et al. [202] synthesized forty-eight molecules derived from the arylpropionic acid scaffold and evaluated them for their effects on diabetes based on the excellent docking point demonstrated by all structures made to the PPAR-γ receptor site. The authors concluded that arylpropionic acid derivatives may provide a new perspective toward developing antidiabetic agents with fewer side effects.

Several interesting review papers were also published in 2021. Christofides et al. [203] described the role of PPARs in immune responses, Rayner et al. [204] described PPAR-gamma-mediated neuronal regeneration, and Stark et al., in turn, described the role of PPAR-gamma in allergic diseases [205]. This year also saw the work of Hasegawa et al. who looked at liver diseases and described several potential new drugs using PPARs [206].

2022 was one of the peak years for publications on PPARs, with about 1555 of them recorded in the NCBI database. In our opinion, the most noteworthy are the papers regarding the relationships between non-coding RNA and PPARs. Previously, few works on this topic were published, but in 2022, according to NCBI, there were 146 of them. Therefore, review works are precious, especially those that concern common diseases whose pathogenesis is very complex and whose therapeutic goals are challenging to achieve. Such diseases include NAFLD, the most prevalent form of chronic liver disease in the world and a significant risk factor for developing hepatocellular carcinoma, and NASH. Non-coding, such as microRNA (miRNA) and long non-coding RNA (lncRNA), has been proven to play a significant role in the pathogenesis of these disorders. PPARs were found to be one of the major regulators in the progression of NAFLD. Mukherjee et al. presented reliable knowledge about regulating PPARs through ncRNAs and their role in NAFLD [207]. In the review on the clinical relevance of circulating non-coding RNAs in metabolic diseases (such as obesity, diabetes, cardiovascular diseases, and metabolic syndrome), Dandare cites that miRNA-138 suppresses adipogenic differentiation and is implicated in obesity by interaction with a potent inhibitor of adipogenic differentiation that interferes with Src homology region 2-containing protein (SHP2), an endogenous enhancer of adipogenic PPARγ [208]. A fascinating paper concerns miRNA networks, which form networks of systems and gene expression circuits through molecular signaling and cellular interactions and contribute to various health disorders [209]. It appears necessary to establish genomic signatures to determine multifactorial correlations and reveal variability observed in therapeutic effects, e.g., in patients with cardiovascular disease. Clinically validated miRNA biomarkers and relevant SNPs identified in clinical practice can serve this purpose. They should increase the ability to stratify patients, help select innovative therapeutic regimens, and identify innovative drugs and delivery systems. This paper examines miRNA–gene networks and SNP-derived genomic signatures to highlight specific gene signaling circuits as sources of molecular knowledge relevant to cardiovascular diseases.

The year 2023 resulted in 1310 new publications. We think the following should be mentioned: Kaltdorf et al. [210] proposed a new approach to evaluating experiments using PPARs and a new informatics tool called Boolean networks. Tanaka et al. reported [211] on PPARα and its ligand Ror 1, Ibrahim et al. [212] on the activation of SIRT 1 by taurine, and Yang et al. [213] on the effect of ligand PFAS (polyfluorolkyl) on the PPARα/ACOX 1 heteroduplex. Zhou et al. [214] described the impact of the FNDC5/PARGα heterodimer on macrophage pyroptosis. Nevertheless, in our ranking, the champions of 2023 were Pascoa et al. [215], presenting their prototype tool for identifying new ligands detected in cyanobacteria.

In 2024 (data from January to May 20, includes 385 papers), Kim et al. [216] elucidated the mechanism of PPAR-alpha's effect on the intestinal stem cell pool in the process of calcification of blood vessels, while Liu et al. [217] tackled pectolinargenine, a Chinese folk medicine drug, demonstrating its multiple beneficial effects in many disease entities, including ligands and with the Nrf 2 gene. Exciting research was conducted by Cheng et al. [218], in which they led the study of the role of epigenetic regulation, including DNA methyltransferase 3a (DNMT3a)-mediated methylation and PPARγ-activated receptor inhibition, in the development of intervertebral disc degeneration (IVDD). The authors of this study showed that DNMT3a, by modifying the hypermethylation of the PPARγ promoter, activates the NF-κB pathway and thus contributes to apoptosis and degradation of the extracellular matrix.

In reviewing the literature, a trend in research has been noted: Chinese researchers are often concerned with green tea and drugs used in folk traditional Chinese medicine [219–226]. Indians, on the other hand, are keen on traditional medicines used in Ayurveda (mainly curcumin) [227–229]. Europeans, on the other hand, often try to demonstrate the beneficial effects of cannabis [230–235].

Examining this phenomenon in light of the statements about the lack of conflict of interest might be worthwhile.

4. Discussion

In a situation with so much interest in PPARs and such a huge scattering of detailed research topics, it is tough to summarize the results to date, let alone predict the future. Nevertheless, some things can be expected with reasonable probability. First, in pure science, it is possible to foresee a slow filling of the small gaps in knowledge that still exist. Second, in the field of pharmacology, there will be an acceleration of work on the search for new and better ligands, with little harm to human health. It is also most likely that pan-type approaches will become more common. In conclusion, the field is slowly moving from research to implementation, which bodes well for patients.

PPARα is a crucial regulator of lipid metabolism. It was discovered as the first member of the peroxisome proliferator family [236]. PPARα is mainly expressed in the liver and tissues with increased mitochondrial oxidation and fatty acid catabolism, such as brown adipocytes, cardiac muscle, skeletal muscle, and kidney [237]. PPARα activation leads to the upregulation of enzymes involved in fatty acid uptake, transport to mitochondria, and subsequent oxidation. This stimulates the breakdown of fatty acids, especially during fasting or prolonged exercise when glucose availability is limited and energy is needed. In addition, PPARα activation enhances the removal of circulating triglycerides in the blood by increasing the expression of lipoprotein lipase and other lipolytic enzymes, thereby lowering plasma lipid levels.

PPARγ is expressed in adipose tissue, the colon, the immune system, and the retina. Alternative splicing results in seven different mRNA transcript variants named PPARγ1 to PPARγ7 [238]. PPARγ influences the development, body localization, and adipose tissue metabolism [239]. In addition, it is a significant insulin sensitizer and regulates metabolism and cellular sensitivity to glucose and insulin. Moreover, PPARγ plays a vital role in cell differentiation and the regulation of apoptosis. Another function of PPARγ is to inhibit inflammatory processes, have an anti-atherosclerotic effect, and generally improve cardiac performance. PPARγ also plays an important role in human procreation (in women, it stimulates ovulation, while in men, it regulates sperm biology). When activated by specific ligands, PPARγ binds to the RXR receptor to form a heterodimer, and together, they regulate the expression of multiple genes [15,239]. The same can be observed in PPARα. Ligands for PPARγ can be both natural compounds (docosahexaenoic acid and eicosapentaenoic acid or other polyunsaturated fatty acids and some monounsaturated fatty acids such as oleic acid) and synthetic compounds (thiazolidinediones such as troglitazone, rosiglitazone, and pioglitazone) [240]. Non-steroidal anti-inflammatory drugs are also included in the PPARγ agonist group.

Numerous studies published to date point to the critical role of DNA methylation in regulating the expression and normal function of PPARγ in health and homeostasis by regulating many important vital processes [241–243]. PPARγ is also believed to play a massive role in the pathogenesis of many diseases in which dysregulation of DNA methylation leading to impaired expression disrupts the normal function of the transcription factor. Such disorders include insulin resistance and type 2 diabetes, and the epigenetic component plays a significant role in their development. Epigenetics are those changes in gene function inherited by mitotic or meiotic cells, which are not associated with changes in DNA sequence [244]. Such changes can increase or decrease the expression of a target gene [245]. DNA methylation profiles are greatly influenced by environmental factors. They also affect histone modification, leading to dysregulation of the expression of many genes, including the insulin signaling gene and lipid metabolism genes. In addition to metabolic disorders, epigenetics has been linked to cancer and neurodegenerative diseases [244].

PPARγ and PPARα homologs are subject to epigenetic regulation, mainly DNA methylation [123]. The PPARα promoter is hypermethylated, mainly in patients with non-alcoholic fatty liver disease (NAFLD) and patients with type 2 diabetes [246,247]. PPARα undergoes hydroxymethylation, which affects PPARα expression, predisposing a person to NAFLD [248]. Patients with metabolic syndrome have been found to have significant hyperlipidemia dependent on a different DNA methylation of PPARα [249]. Castellano-Castillo et al. demonstrated hypermethylation of multiple genes that correlated with metabolic dysregulation, including genes essential for regulating adipogenesis, lipid metabolism, and inflammation. This suggests that DNA methylation of the nuclear receptors PPARγ and PPARα is correlated with metabolic deregulation, the pathogenesis of metabolic syndrome, and the induction of inflammation [250].

Considering diabetes treatment, it is known that the use of an adjuvant PPAR agonist influences the control of glycemia, the reduction of HbA1c levels, and the improvement of insulin resistance compared to metformin monotherapy. Moreover, beneficial changes were observed in patients using PPAR agonists, such as a reduction in blood pressure and a decrease in the concentration of inflammatory markers and dyslipidemia. It was mentioned above that in the treatment of diabetes, PPARy agonists can be used, mainly thiazolidinediones (TZD), whose task is to prevent the effects of diabetes by increasing patients' sensitivity to insulin, leading to a reduction in the concentration of both glucose and insulin in the plasma. The most commonly used TZDs are rosiglitazone, pioglitazone, and lobeglitazone sulfate in combination with metformin and other drugs used to treat diabetes [251]. Despite the positive effects of rosiglizatone, such as lowering fasting plasma glucose (FPG) and postprandial serum glucose, lowering HbA1c, and lowering insulin and C-peptide levels, this drug has clinically significant side effects such as edema, anemia, and weight gain [252]. Additionally, an increased risk of cardiovascular disease (CVD) has been observed in patients with unstable heart failure (HF) [253]. In turn, pioglitazone has antidiabetic effects, improves HbA1c, and has a positive effect on serum lipids; however, increasing the duration of pioglitazone use resulted in an increased risk of bladder cancer [254]. Nevertheless, unlike other TZDs, it has been shown that lobeglitazone can be a therapeutic substance that combines the features of both effectiveness and safety [255]. No side or carcinogenic effects were observed in patients taking it, but its antidiabetic effects have been proven [256].

The effect of PPAR agonists in treating diabetes may depend on individual factors such as single nucleotide polymorphism. One of the most frequently studied and analyzed polymorphisms is Pro12Ala, in which there is a nucleotide change from CCA-to-GCA in the codon12 of exon B of the PPARG gene [257]. Multiple studies have shown that the occurrence of this polymorphism was associated with a lower risk of type 2 diabetes as well as improved insulin sensitivity in humans, suggesting that this amino acid change enhances the action of insulin [258,259]. Additionally, it was shown that the occurrence of the Pro12Ala polymorphism was associated with a decrease in triglyceride content in white adipose tissue, skeletal muscles, and the liver, which was the result of increased leptin expression and a decrease in the lipogenesis process, which led to increased insulin sensitivity [257]. The reduction in immune function may result from several mechanisms. As mentioned earlier, pioglitazone is used to reduce insulin sensitivity in diabetes. Numerous studies also show that patients with the Pro12Ala polymorphism showed a better therapeutic response than participants with the Pro12Pro genotype [257]. Additionally, many studies show that the Pro12Ala polymorphism significantly impacts changes in HbA1C, FPG, and TG levels in T2D patients treated with TZD4 [260].

Particular attention should be paid to the importance of pharmacogenetics, which will make individual treatment based on the patient's genetic profile possible in the future.

PPARs, playing an essential role in regulating the inflammatory response, have a significant impact on the course of diseases such as rheumatoid arthritis [75,136,261–263] and inflammatory bowel diseases [264–268]. In the synovium of rheumatoid arthritis (RA) patients, the abnormal migration, proliferation, and activation of fibroblast-like synoviocytes (FLSs) is observed in the joints. In vitro studies demonstrated that the expression of PPAR-γ in the synoviocytes of RA patients and adjuvant arthritis (AA) in experimental animals was significantly lower than that of normal FLSs [263,267]. Synthetic non-fibrate PPARγ ligands (thiazolidinedione family) and fibrates (bezafibrate and fenofibrate), as well as natural PPAR agonists (unsaturated and nitrated fatty acids, eicosanoids), had various effects on the severity and course of RA. In the synoviocytes of RA patients, the induction of inflammatory cytokine mRNA expressions such as TNF-α and IL-1β was significantly inhibited by the 15d-PGJ2, natural PPARγ agonist [261,269]. Using solid lipid nanoparticles may enhance this anti-inflammatory effect of 15d-PGJ2 and rosiglitazone [163]. A recent study has shown that PPARγ alleviates the inflammatory response in TNF-α-induced FLSs by binding to p53 in RA patients [270]. Another study indicated that PPARγ may induce the activation of Wnt/β-catenin signaling during FLSs activation [176,263]; perhaps in this way, PPARγ ligands induce synovial cell apoptosis.

Study results support the use of PPARγ agonists in the treatment of RA. Their beneficial effects, especially in combination with methotrexate, are documented. The mechanisms of this action include the inhibition of ROS and inflammatory cytokines generation, as well as the attenuation of the migration and proliferation of FLSs.

PPARγ is highly expressed in the colon (in epithelial cells and lamina propria mononuclear cells such as macrophages as well as T and B cells), being a key regulatory factor of bacteria-induced mucosal inflammation. A bacteria-induced excess of TLR4 under the influence of lipopolysaccharide (LPS) triggers the NFκB and mitogen-activated kinases (MAPKs) pathways, leading to increased production of inflammatory mediators. In patients diagnosed with inflammatory bowel disease (IBD); ulcerative colitis (UC), or Crohn's disease (CD), as well as in experimentally induced colitis, the depressed PPARγ expression in colon epithelial cells has been shown. Simultaneously, the PPARγ gene has been described as a susceptibility gene for IBD [265]. One of the oldest anti-inflammatory agents used for the treatment of IBD is 5-ASA, the functional synthetic ligand for PPARγ in the colon. PPARγ is the critical receptor mediating the 5-ASA activity by trans-repressing several essential target genes such as NFκB, signal transducers, and activators of transcription [264]. Synthetic (thiazolidinediones, glitazars, non-steroidal anti-inflammatory drugs, aspirin) and natural (conjugated linoleic acid, barley leaf (BL), inosine) PPARγ ligands were used in various models of IBD, with different effectiveness [264–268]. Rosiglitazone was also used in patients suffering from UC, and the endoscopic and clinical results were promising [271,272]. As UC is closely associated with gut microbiota dysbiosis, Li et al., using 16S rRNA gene-based microbiota analysis, found that dietary supplementation of BL ameliorated dextran sulfate sodium (DSS)-induced gut microbiota dysbiosis and protects against DSS-induced colitis. The mechanism of this protective BL action resulted from improved intestinal mucosal barrier functions via the activation of PPARγ signaling. Similarly to the effect of exogenous treatment with inosine, BL protects against DSS-induced colitis by improving adenosine 2A receptor/PPARγ-dependent mucosal barrier functions [268]. The conjugated linoleic acid (CLA) induced PPARγ and δ and repressed tumor necrosis factor α (TNF-α) expression and NFκB activation while inducing the immunoregulatory cytokine transforming growth factor β 1 (TGF-β1). Clinically, CLA also ameliorated DSS- and CD4+-induced colitis through a PPAR γ-dependent mechanism [273]. Additionally, PPARγ receptors are directly involved in the mechanism of action of mesalazine, which is primarily used and effective in UC.

The impact of PPARγ on the course of IBD seems to pertain not solely to UC but also CD. Numerous animal studies have concluded that PPARγ agonists may have higher efficacy in maintaining rather than inducing IBD remission and that the therapeutic effect of PPARγ is mainly dependent on its abundance in target tissues.

It is also essential to emphasize the role of PPAR-alpha gene polymorphisms on cardiovascular risk.

The risk of coronary heart disease (CHD) has been suggested to be associated with polymorphisms of peroxisome proliferator-activated receptors. The results were controversial. In the case-control study, T allele carriers of C161T polymorphism were not significantly associated with CHD, while T allele carriers showed a higher risk of acute coronary syndrome (ACS). The meta-analysis indicated that compared with CC homozygous, T allele carriers had lower CHD but higher ACS risk. Other polymorphisms were also significantly associated with CHD risk under the dominant model: PPAR-alpha intron 7G/C polymorphism and L162V polymorphism [274]. The association of PPAR-alpha L162V polymorphism with cardiovascular risk may result from genetically conditioned changes in lipid metabolism. The serum total cholesterol and LDL-cholesterol levels were higher in PPAR-alpha V162 allele carriers in non-diabetic CHD. The increasing effect of the PPAR-alpha V162 allele on serum cholesterol levels was weakened with the presence of the PPAR-gamma 161T allele in the non-diabetic CHD patients. The ApoE4-PPAR-alpha V162 allelic combination of the ApoE/PPARA genes was found to be more frequent in diabetic CHD patients, independent of serum lipids. It was suggested that the PPAR-alpha L162V polymorphism may have diverse effects on serum lipids, and CHD risk depends on the presence of diabetes [275].

In our study, compared to healthy men, the frequency of the V allele of the L162V polymorphism was four times higher in men with confirmed coronary atherosclerosis. We concluded that L162V polymorphism in the gene for PPAR-alpha seems to be associated with atherosclerosis through a mechanism including regulation of the interleukin-6 level [276]. On the other hand, in patients with type 2 diabetes, there was a trend towards a lower prevalence of atherosclerosis and lower CHD prevalence in carriers versus noncarriers of the V allele. These data suggest that the PPAR-alpha polymorphism L162V might protect against the development of atherosclerosis or CHD in patients with DM-2 [277].

Recently, an increasing number of studies indicate that various PPAR gene polymorphisms influence the effects of PPAR agonists on lipids and cardiovascular risk. Moreover, much evidence shows that lipid metabolism is influenced by gene-gene and gene-environment interactions involving PPAR-alpha genes [278].

The data presented above indicate that PPAR-alpha agonists, by reducing metabolic disorders associated with atherogenic dyslipidemia, reduce the so-called residual risk of cardiovascular events. Currently, scientific societies recommend the use of PPAR-alpha agonists in people with hypertriglyceridemia after the LDL cholesterol level has been reduced to the target values. An attempt to reduce the non-HDL cholesterol to the target values can be considered (a) combination therapy using omega-3 fatty acids (PUFA, at 2–4 g/day) and statin in patients with triglyceride concentrations above 2.3 mmol/L (200 mg/dL) despite statin treatment; (b) combination therapy using choline fenofibrate and statin can be considered as a part of primary prevention in patients with triglyceride concentrations above 2.3 mmol/L (200 mg/dL) whose LDL cholesterol levels have been reduced to the target values, especially where the HDL cholesterol levels are low; (c) combination therapy using choline fenofibrate and statin should be considered in high-risk patients with triglyceride concentrations above 2.3 mol/L (200 mg/dL) whose LDL cholesterol levels have been reduced to the target values, especially where the HDL cholesterol levels are low [279,280]. The anti-arthrogenic mechanism of fibrates is shown in Figure 4.

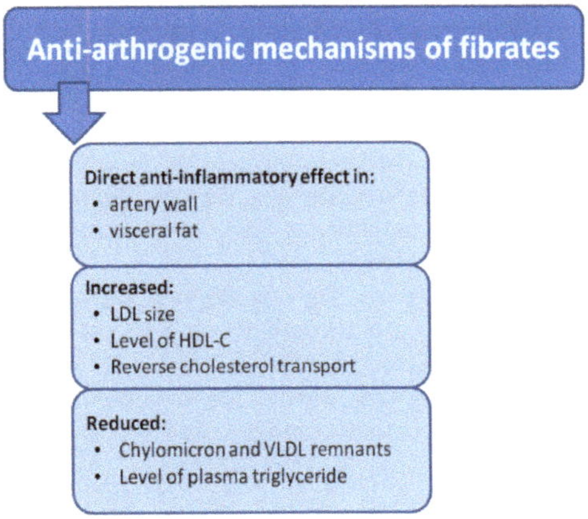

Figure 4. Anti-arthrogenic mechanism of fibrates.

PPAR agonists already have and may have many more medical applications now and in the future (Figure 5).

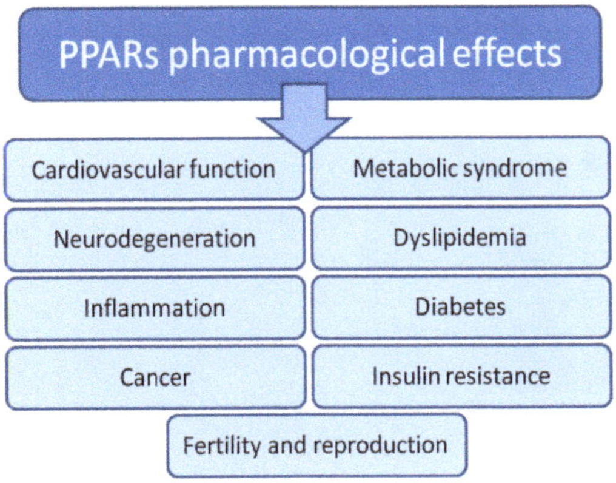

Figure 5. PPARs pharmacological effect.

This problem was widely discussed by Cheng et al. in their fascinating review exploring PPAR modulators in medical research, showing their application and utility. The authors collected and described existing knowledge about different classes of PPAR agonists, which are under various phases of clinical investigations or have already been approved for medical use [281]. Anyway, it is also worth mentioning the paper by Colapitro et al. [282], which describes new therapeutic fibrates and PPAR agonists such as bezafibrate and fenofibrate. In addition, the paper describes other PPAR ligands, such as Seladerpar, Elafibranor, and Saroglitazar, tested in patients with PBC (Primary Biliary Cholangitis). Bezafibrate has the most promising effect on liver disease, improving symptoms in patients with PBC. However, despite its broad spectrum of activity, it should be used with caution in patients with chronic kidney disease and those with cirrhosis, in whom it increases bilirubin levels and impairs liver function. However, in patients who do not yet have cirrhosis it could be used successfully. Seladelpar is a PPAR-δ agonist. It has choleretic properties, resulting from altered bile acid synthesis, and anti-inflammatory properties. Elafibranor, a dual PPARα/δ agonist, is a promising drug due to the lack of adverse effects of PPAR-γ ligands on the heart. Saroglitazar, on the other hand, is mainly a PPAR-α activator and exerts agonist effects on PPAR-γ; that is, it has dual PPAR agonist activity. It is used to treat NASH, diabetic dyslipidemia, and diabetes uncontrolled by statins. Its effects are also being studied in PBC. These drugs show improvement in ALP but have only been tested in the short term and on a small group of patients. Further studies are needed to determine the dose and test the drug's safety. Ideally, these patients should be selected for individual double or triple combination therapy.

Unfortunately, side effects are the main problem in designing and developing new synthetic PPAR ligands. Therefore, future PPAR research should focus on minimizing the side effects of PPAR agonists and the potential of PPAR antagonists. Using such compounds will open a new chapter in using PPAR ligands in medicine.

Generally, ligands of PPARs can be agonists (full, partial, or inverse) or natural antagonists. There are 84 types of PPAR synthetic ligands used as drugs to treat various diseases. For example, PPARα agonists (fibrates), PPARγ agonists (thiazolidinediones -TZDs) and combined PPARα/γ agonists (glitazars) are therapeutic agents for CVD prevention, T2DM and dyslipidemia. Unfortunately, PPAR ligands are reported to have multiple organ toxicity (Figure 6) [283].

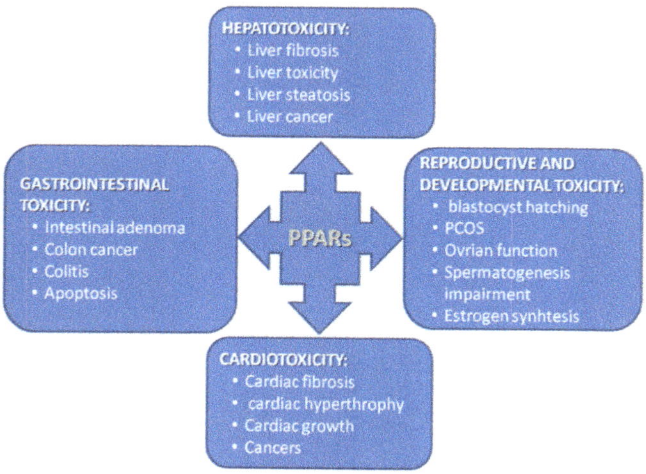

Figure 6. Systemic toxicities of PPARs.

TZDs can cause hepatotoxicity, heart failure, or fluid retention. Glitazone's side effects include body weight gain, peripheral edema, and congestive heart failure [283]. Fenofibrate therapy causes pulmonary embolism, pancreatitis, an evaluated plasma creatinine level (reversible after therapy ends), and an increased rate of noncardiovascular mortality. However, fibrate risk should be treated individually [284].

5. Concluding Remarks

For 35 years, knowledge about PPARs has significantly expanded. Over the years, thousands of studies have been performed. PPARs are an exciting group of transcription factors due to their connection with various chronic diseases, such as cancer, diabetes, obesity, arteriosclerosis, and neurodegenerative diseases. PPAR agonists are used to treat these diseases. For example, PPARα agonists treat hyperlipidemia or modulation of inflammatory response, while PPARγ agonists have potent hypoglycemic potential in patients with insulin resistance and cancer [285]. However, time and many studies are still needed to confirm the safety and effectiveness of the long-term use of PPAR agonists. In addition, particular attention should be paid to individual variations, such as polymorphisms that influence the function of PPARs. However, more research is still needed to fully understand the functions of PPARs.

Author Contributions: Conceptualization, T.D., A.D., M.O., A.S. and R.A.; resources, T.D., M.M.-M., A.S. and R.A.; data curation, A.D., S.G., A.S., A.J. and A.L.; writing—original draft preparation, T.D., A.D., A.S. and M.O.; writing—review and editing, A.D., A.S., T.D. and M.O.; visualization, M.O., A.D., J.A.-M., M.M.-M. and A.S. All authors have read and agreed to the published version of the manuscript.

Funding: This research received no external funding.

Institutional Review Board Statement: Not applicable.

Informed Consent Statement: Not applicable.

Data Availability Statement: Data are contained within the article.

Conflicts of Interest: The authors declare no conflicts of interest.

References

1. Vamecq, J.; Latruffe, N. Medical significance of peroxisome proliferator-activated receptors. *Lancet* **1999**, *354*, 141–148. [CrossRef]
2. Issemann, I.; Green, S. Activation of a member of the steroid hormone receptor superfamily by peroxisome proliferators. *Nature* **1990**, *347*, 645–650. [CrossRef]
3. Dreyer, C.; Krey, G.; Keller, H.; Givel, F.; Helftenbein, G.; Wahli, W. Control of the peroxisomal β-oxidation pathway by a novel family of nuclear hormone receptors. *Cell* **1992**, *68*, 879–887. [CrossRef]
4. Reilly, M.M.; Rossor, A.M. Humans: The Ultimate Animal Models. *J. Neurol. Neurosurg. Psychiatry* **2020**, *91*, 1132–1136. [CrossRef]
5. Tahri-Joutey, M.; Andreoletti, P.; Surapureddi, S.; Nasser, B.; Cherkaoui-Malki, M.; Latruffe, N. Mechanisms Mediating the Regulation of Peroxisomal Fatty Acid Beta-Oxidation by PPARα. *Int. J. Mol. Sci.* **2021**, *22*, 8969. [CrossRef] [PubMed]
6. Zhou, T.; Yan, X.; Wang, G.; Liu, H.; Gan, X.; Zhang, T.; Wang, J.; Li, L. Evolutionary Pattern and Regulation Analysis to Support Why Diversity Functions Existed within PPAR Gene Family Members. *Biomed. Res. Int.* **2015**, *2015*, 613910. [CrossRef] [PubMed]
7. Lathion, C.; Michalik, L.; Wahli, W. Physiological Ligands of PPARs in Inflammation and Lipid Homeostasis. *Future Lipidol.* **2006**, *1*, 191–201. [CrossRef]
8. Willson, T.M.; Brown, P.J.; Sternbach, D.D.; Henke, B.R. The PPARs: From orphan receptors to drug discovery. *J. Med. Chem.* **2000**, *43*, 527–550. [CrossRef]
9. Chinetti-Gbaguidi, G.; Fruchart, J.C.; Staels, B. Role of the PPAR family of nuclear receptors in the regulation of metabolic and cardiovascular homeostasis: New approaches to therapy. *Curr. Opin. Pharmacol.* **2005**, *5*, 177–183. [CrossRef] [PubMed]
10. Seok, H.; Cha, B.S. Refocusing Peroxisome Proliferator Activated Receptor-α: A New Insight for Therapeutic Roles in Diabetes. *Diabetes Metab. J.* **2013**, *37*, 326–332. [CrossRef]
11. Waterham, H.R.; Ferdinandusse, S.; Wanders, R.J. Human disorders of peroxisome metabolism and biogenesis. *Biochim. Biophys. Acta.* **2016**, *1863*, 922–933. [CrossRef] [PubMed]
12. Staels, B.; Fruchart, J.C. Therapeutic Roles of Peroxisome Proliferator–Activated Receptor Agonists. *Diabetes* **2005**, *54*, 2460–2470. [CrossRef]
13. Pyper, S.R.; Viswakarma, N.; Jia, Y.; Zhu, Y.J.; Fondell, J.D.; Reddy, J.K. PRIC295, a Nuclear Receptor Coactivator, Identified from PPARα-Interacting Cofactor Complex. *PPAR Res.* **2010**, *2010*, 173907. [CrossRef] [PubMed]
14. Lefebvre, P.; Chinetti, G.; Fruchart, J.-C.; Staels, B. Sorting out the roles of PPARα in energy metabolism and vascular homeostasis. *J. Clin. Investig.* **2006**, *116*, 571–580. [CrossRef] [PubMed]

15. Gacka, M.; Adamiec, R. Mutacje genu receptora aktywowanego przez proliferatory peroksysomów γ (PPARγ)—Implikacje kliniczne. *Postępy Hig. I Med. Doświadczalnej* **2004**, *58*, 483–489.
16. Cullingford, T.E.; Dolphin, C.T.; Sato, H. The peroxisome proliferator-activated receptor alpha-selective activator ciprofibrate upregulates expression of genes encoding fatty acid oxidation and ketogenesis enzymes in rat brain. *Neuropharmacology* **2002**, *42*, 724–730. [CrossRef] [PubMed]
17. Rudolph, J.; Chen, L.; Majumdar, D.; Bullock, W.H.; Burns, M.; Claus, T.; Dela Cruz, F.E.; Daly, M.; Ehrgott, F.J.; Johnson, J.S.; et al. Indanylacetic acid derivatives carrying 4-thiazolyl-phenoxy tail groups, a new class of potent PPAR alpha/gamma/delta pan agonists: Synthesis, structure-activity relationship, and in vivo efficacy. *J. Med. Chem.* **2007**, *50*, 984–1000. [CrossRef] [PubMed]
18. Huang-Jin, H.; Kuei-Jen, L.; Hsin, W.Y.; Hsin-Yi, C.; Fuu-Jen, T.; Calvin, Y.-C.C. A novel strategy for designing the selective PPAR agonist by the "sum of activity" model. *J. Biomol. Struct. Dyn.* **2010**, *28*, 187–200. [CrossRef] [PubMed]
19. Vamecq, J.; Cherkaoui-Malki, M.; Andreoletti, P.; Latruffe, N. The human peroxisome in health and disease: The story of an oddity becoming a vital organelle. *Biochimie* **2014**, *98*, 4–15. [CrossRef]
20. Reddy, J.K.; Lalvvai, N.D.; Farber, E. Carcinogenesis by peroxisome proliferators: Evaluation of the risk of hypolipidemic drugs and industrial plasticizers to humans. *Crit. Rev. Toxicol.* **1988**, *18*, 1–58. [CrossRef]
21. Reddy, J.K.; Rao, M.S. Peroxisome proliferators and cancer: Mechanisms and implications. *Adv. Exp. Med. Biol.* **1988**, *283*, 131–148.
22. Martinez, E.; Givel, F.; Wahli, W.; Keller, H. The peroxisome proliferator-activated receptor (PPAR) is a transcriptional regulator of the peroxisomal beta-oxidation pathway. *J. Cell Sci.* **1991**, *100*, 711–717.
23. Liu, J.; Ormö, M.; Nyström, A.C.; Claesson, J.; Giordanetto, F. Transient expression, purification and characterisation of human full-length PPARγ2 in HEK293 cells. *Protein Expr. Purif.* **2013**, *89*, 189–195. [CrossRef] [PubMed]
24. Alvares, K.; Carrillo, A.; Yuan, P.M.; Kawano, H.; Morimoto, R.I.; Reddy, J.K. Identification of cytosolic peroxisome proliferator binding protein as a member of the heat shock protein HSP70 family. *Proc. Natl. Acad. Sci. USA* **1990**, *87*, 5293–5297. [CrossRef] [PubMed]
25. Maksymowych, A.B.; Hsu, T.C.; Litwack, G. A novel, highly conserved structural motif is present in all members of the steroid receptor superfamily. *Receptor* **1992**, *2*, 225–240. [PubMed]
26. Göttlicher, M.; Demoz, A.; Svensson, D.; Tollet, P.; Berge, R.K.; Gustafsson, J.A. Structural and metabolic requirements for activators of the peroxisome proliferator-activated receptor. *Biochem. Pharmacol.* **1993**, *46*, 2177–2184. [CrossRef]
27. Keller, H.; Wahli, W. Peroxisome proliferator-activated receptors A link between endocrinology and nutrition? *Trends Endocrinol. Metab.* **1993**, *4*, 291–296. [CrossRef]
28. Keller, H.; Mahfoudi, A.; Dreyer, C.; Hihi, A.K.; Medin, J.; Ozato, K.; Wahli, W. Peroxisome proliferator-activated receptors and lipid metabolism. *Ann. N. Y. Acad. Sci.* **1993**, *684*, 157–173. [CrossRef]
29. Issemann, I.; Prince, R.A.; Tugwood, J.D.; Green, S. The retinoid X receptor enhances the function of the peroxisome proliferator activated receptor. *Biochimie* **1993**, *75*, 251–256. [CrossRef] [PubMed]
30. Bardot, O.; Aldridge, T.C.; Latruffe, N.; Green, S. PPAR-RXR heterodimer activates a peroxisome proliferator response element upstream of the bifunctional enzyme gene. *Biochem. Biophys. Res. Commun.* **1993**, *192*, 37–45. [CrossRef]
31. Boie, Y.; Adam, M.; Rushmore, T.H.; Kennedy, B.P. Enantioselective activation of the peroxisome proliferator-activated receptor. *J. Biol. Chem.* **1993**, *268*, 5530–5534. [CrossRef] [PubMed]
32. Motojima, K. Proliferator-Activated Receptor (PPAR): Structure, Mechanisms of Activation and Diverse Functions. *Cell Struct. Funct.* **1993**, *18*, 267–277. [CrossRef] [PubMed]
33. Sher, T.; Yi, H.F.; McBride, O.W.; Gonzalez, F.J. cDNA cloning, chromosomal mapping, and functional characterization of the human peroxisome proliferator activated receptor. *Biochemistry* **1993**, *32*, 5598–5604. [CrossRef] [PubMed]
34. Bass, N.M. Cellular binding proteins for fatty acids and retinoids: Similar or specialized functions? *Mol. Cell Biochem.* **1993**, *123*, 191–202. [CrossRef] [PubMed]
35. Gibson, G.G. Peroxisome proliferators: Paradigms and prospects. In Proceedings of the 1992 Conference on Toxicology: Application of Advances in Toxicology to Risk Assessment; Dodd, D.E., Clewell, H.J., III, Mattie, D.R., Eds.; ManTech Environmental Technology, Inc.: Dayton, OH, USA, 1993; pp. 194–202.
36. Chen, F.; Law, S.W.; O'Malley, B.W. Identification of two mPPAR related receptors and evidence for the existence of five subfamily members. *Biochem. Biophys. Res. Commun.* **1993**, *196*, 671–677. [CrossRef] [PubMed]
37. Kainu, T.; Wikström, A.C.; Gustafsson, J.A.; Pelto-Huikko, M. Localization of the peroxisome proliferator-activated receptor in the brain. *Neuroreport* **1994**, *5*, 2481–2485. [CrossRef] [PubMed]
38. Aleshin, S.; Grabeklis, S.; Hanck, T.; Sergeeva, M.; Reiser, G. Peroxisome proliferator-activated receptor (PPAR)-gamma positively controls and PPARalpha negatively controls cyclooxygenase-2 expression in rat brain astrocytes through a convergence on PPARbeta/delta via mutual control of PPAR expression levels. *Mol Pharmacol.* **2009**, *76*, 414–424. [CrossRef] [PubMed]
39. Lemberger, T.; Saladin, R.; Vázquez, M.; Assimacopoulos, F.; Staels, B.; Desvergne, B.; Wahli, W.; Auwerx, J. Expression of the peroxisome proliferator-activated receptor alpha gene is stimulated by stress and follows a diurnal rhythm. *J. Biol. Chem.* **1996**, *271*, 1764–1769. [CrossRef] [PubMed]
40. Wahli, W.; Braissant, O.; Desvergne, B. Peroxisome proliferator activated receptors: Transcriptional regulators of adipogenesis, lipid metabolism and more.... *Chem. Biol.* **1995**, *2*, 261–266. [CrossRef]
41. Gustafsson, J.A. Receptor-mediated toxicity. *Toxicol. Lett.* **1995**, *82–83*, 465–470. [CrossRef]

42. Bocos, C.; Göttlicher, M.; Gearing, K.; Banner, C.; Enmark, E.; Teboul, M.; Crickmore, A.; Gustafsson, J.A. Fatty acid activation of peroxisome proliferator-activated receptor (PPAR). *J. Steroid. Biochem. Mol. Biol.* **1995**, *53*, 467–473. [CrossRef] [PubMed]
43. Lemberger, T.; Desvergne, B.; Wahli, W. Peroxisome proliferator-activated receptors: A nuclear receptor signaling pathway in lipid physiology. *Annu. Rev. Cell Dev. Biol.* **1996**, *12*, 335–363. [CrossRef] [PubMed]
44. Yu, K.; Bayona, W.; Kallen, C.B.; Harding, H.P.; Ravera, C.P.; McMahon, G.; Brown, M.; Lazar, M.A. Differential activation of peroxisome proliferator-activated receptors by eicosanoids. *J. Biol. Chem.* **1995**, *270*, 23975–23983. [CrossRef]
45. Keller, H.; Givel, F.; Perroud, M.; Wahli, W. Signaling cross-talk between peroxisome proliferator-activated receptor/retinoid X receptor and estrogen receptor through estrogen response elements. *Mol. Endocrinol.* **1995**, *9*, 794–804. [CrossRef] [PubMed]
46. Juge-Aubry, C.E.; Gorla-Bajszczak, A.; Pernin, A.; Lemberger, T.; Wahli, W.; Burger, A.G.; Meier, C.A. Peroxisome proliferator-activated receptor mediates cross-talk with thyroid hormone receptor by competition for retinoid X receptor. Possible role of a leucine zipper-like heptad repeat. *J. Biol. Chem.* **1995**, *270*, 18117–18122. [CrossRef]
47. Baes, M.; Castelein, H.; Desmet, L.; Declercq, P.E. Antagonism of COUP-TF and PPAR alpha/RXR alpha on the activation of the malic enzyme gene promoter: Modulation by 9-cis RA. *Biochem. Biophys. Res. Commun.* **1995**, *215*, 338–345. [CrossRef]
48. Forman, B.M.; Chen, J.; Evans, R.M. The peroxisome proliferator-activated receptors: Ligands and activators. *Ann. N. Y. Acad. Sci.* **1996**, *804*, 266–275. [CrossRef]
49. Willson, T.M.; Cobb, J.E.; Cowan, D.J.; Wiethe, R.W.; Correa, I.D.; Prakash, S.R.; Beck, K.D.; Moore, L.B.; Kliewer, S.A.; Lehmann, J.M. The structure-activity relationship between peroxisome proliferator- activated receptor gamma agonism and the antihyperglycemic activity of thiazolidinediones. *J. Med. Chem.* **1996**, *39*, 665–668. [CrossRef]
50. Aubert, J.; Ailhaud, G.; Negrel, R. Evidence for a novel regulatory pathway activated by (carba)prostacyclin in preadipose and adipose cells. *FEBS Lett.* **1996**, *397*, 117–121. [CrossRef]
51. Patel, C.B.; De Lemos, J.A.; Wyne, K.L.; McGuire, D.K. Thiazolidinediones and risk for atherosclerosis: Pleiotropic effects of PPar gamma agonism. *Diab. Vasc. Dis. Res.* **2006**, *3*, 65–71. [CrossRef]
52. Brun, R.P.; Tontonoz, P.; Forman, B.M.; Ellis, R.; Chen, J.; Evans, R.M.; Spiegelman, B.M. Differential activation of adipogenesis by multiple PPAR isoforms. *Genes Dev.* **1996**, *10*, 974–984. [CrossRef] [PubMed]
53. Guan, Y.; Zhang, Y.; Davis, L.; Breyer, M.D. Expression of peroxisome proliferator-activated receptors in urinary tract of rabbits and humans. *Am. J. Physiol.* **1997**, *273*, F1013–F1022. [CrossRef] [PubMed]
54. Bernlohr, D.A.; Simpson, M.A.; Hertzel, A.V.; Banaszak, L.J. Intracellular lipid-binding proteins and their genes. *Annu. Rev. Nutr.* **1997**, *17*, 277–303. [CrossRef] [PubMed]
55. Clarke, S.D.; Jump, D. Polyunsaturated fatty acids regulate lipogenic and peroxisomal gene expression by independent mechanisms. *Prostaglandins Leukot Essent Fat. Acids.* **1997**, *57*, 65–69. [CrossRef] [PubMed]
56. Rowe, A. Retinoid X receptors. *Int. J. Biochem. Cell Biol.* **1997**, *29*, 275–278. [CrossRef] [PubMed]
57. Marx, N.; Schönbeck, U.; Lazar, M.A.; Libby, P.; Plutzky, J. Peroxisome Proliferator-Activated Receptor Gamma Activators Inhibit Gene Expression and Migra-tion in Human Vascular Smooth Muscle Cells. *Circ Res.* **1998**, *83*, 1097–1103. [CrossRef] [PubMed]
58. Kruszynska, Y.T.; Mukherjee, R.; Jow, L.; Dana, S.; Paterniti, J.R.; Olefsky, J.M. Skeletal muscle peroxisome proliferator- activated receptor-gamma expression in obesity and non- insulin-dependent diabetes mellitus. *J. Clin. Investig.* **1998**, *101*, 543–548. [CrossRef] [PubMed]
59. Schulman, I.G.; Shao, G.; Heyman, R.A. Transactivation by Retinoid X Receptor–Peroxisome Proliferator-Activated Receptor gamma (PPARgamma) Heterodimers: Intermolecular Synergy Requires Only the PPAR? Hormone-Dependent Activation Function. *Mol. Cell Biol.* **1998**, *18*, 3483–3494. [CrossRef]
60. Ricote, M.; Huang, J.; Fajas, L.; Li, A.; Welch, J.; Najib, J.; Witztum, J.L.; Auwerx, J.; Palinski, W.; Glass, C.K. Expression of the peroxisome proliferator-activated receptor gamma (PPARgamma) in human atherosclerosis and regulation in macrophages by colony stimulating factors and oxidized low density lipoprotein. *Proc. Natl. Acad. Sci. USA* **1998**, *95*, 7614–7619. [CrossRef]
61. Zhou, Y.-T.; Shimabukuro, M.; Wang, M.-Y.; Lee, Y.; Higa, M.; Milburn, J.L.; Newgard, C.B.; Unger, R.H. Role of peroxisome proliferator-activated receptor alpha in disease of pancreatic ß cells. *Proc. Natl. Acad. Sci. USA* **1998**, *95*, 8898–8903. [CrossRef]
62. Ribon, V.; Johnson, J.H.; Camp, H.S.; Saltiel, A.R. Thiazolidinediones and insulin resistance: Peroxisome proliferator activated receptor gamma activation stimulates expression of the CAP gene. *Proc. Natl. Acad. Sci. USA* **1998**, *95*, 14751–14756. [CrossRef]
63. Shimomura, I.; Hammer, R.E.; Richardson, J.A.; Ikemoto, S.; Bashmakov, Y.; Goldstein, J.L.; Brown, M.S. Insulin resistance and diabetes mellitus in transgenic mice expressing nuclear SREBP-1c in adi-pose tissue: Model for congenital generalized lipodystrophy. *Genes Dev.* **1998**, *12*, 3182–3194. [CrossRef] [PubMed]
64. Elstner, E.; Müller, C.; Koshizuka, K.; Williamson, E.A.; Park, D.; Asou, H.; Shintaku, P.; Said, J.W.; Heber, D.; Koeffler, H.P. Ligands for peroxisome proliferator-activated receptor gamma and retinoic acid receptor inhibit growth and induce apoptosis of human breast cancer cells in vitro and in BNX mice. *Proc. Natl. Acad. Sci. USA* **1998**, *95*, 8806–8811. [CrossRef] [PubMed]
65. Sørensen, H.N.; Treuter, E.; Gustafsson, J.A. Regulation of peroxisome proliferator-activated receptors. *Vitam. Horm.* **1998**, *54*, 121–166. [CrossRef] [PubMed]
66. Goodman, A.B. Three independent lines of evidence suggest retinoids as causal to schizophrenia. *Proc. Natl. Acad. Sci. USA* **1998**, *95*, 7240–7244. [CrossRef]
67. Guan, Y.-F.; Zhang, Y.-H.; Breyer, R.M.; Davis, L.; Breyer, M.D. Expression of Peroxisome Proliferator-Activated Receptor Gamma (PPARgamma) in Human Transition-al Bladder Cancer and its Role in Inducing Cell Death. *Neoplasia* **1999**, *1*, 330–339. [CrossRef]

68. Demetri, G.D.; Fletcher, C.D.M.; Mueller, E.; Sarraf, P.; Naujoks, R.; Campbell, N.; Spiegelman, B.M.; Singer, S. Induction of solid tumor differentiation by the peroxisome proliferator-activated receptor-gamma ligand troglitazone in patients with liposarcoma. *Proc. Natl. Acad. Sci. USA* **1999**, *96*, 3951–3956. [CrossRef] [PubMed]
69. Kitamura, S.; Miyazaki, M.; Shinomura, Y.; Kondo, S.; Kanayama, S.; Matsuzawa, Y. Peroxisome Proliferator-activated Receptor Gamma Induces Growth Arrest and Differentiation Markers of Human Colon Cancer Cells. *Jpn J. Cancer Res.* **1999**, *90*, 75–80. [CrossRef]
70. Sarraf, P.; Mueller, E.; Smith, W.M.; Wright, H.M.; Kum, J.B.; Aaltonen, L.A.; de la Chapelle, A.; Spiegelman, B.M.; Eng, C. Loss-of-function mutations in PPAR gamma associated with human colon cancer. *Mol. Cell.* **1999**, *3*, 799–804. [CrossRef]
71. Zhu, Y.; Qi, C.; Jain, S.; Le Beau, M.M.; Espinosa, R., 3rd; Atkins, G.B.; Lazar, M.A.; Yeldandi, A.V.; Rao, M.S.; Reddy, J.K. Amplification and overexpression of peroxisome proliferator activated receptor binding protein (PBP/PPARBP) gene in breast cancer. *Proc. Natl. Acad. Sci. USA* **1999**, *96*, 10848–10853. [CrossRef]
72. Doucas, V.; Evans, R.M. The human T-cell leukemia virus type 1 tax oncoprotein represses nuclear receptor signaling. *Proc. Natl. Acad. Sci. USA* **1999**, *96*, 2633–2638. [CrossRef] [PubMed]
73. Shappell, S.B.; Boeglin, W.E.; Olson, S.J.; Kasper, S.; Brash, A.R. 15-Lipoxygenase-2 (15-LOX-2) Is Expressed in Benign Prostatic Epithelium and Reduced in Prostate Adenocarcinoma. *Am. J. Pathol.* **1999**, *155*, 235–245. [CrossRef] [PubMed]
74. Lazennec, G.; Canaple, L.; Saugy, D.; Wahli, W. Activation of peroxisome proliferator-activated receptors (PPARs) by their ligands and protein kinase A activators. *Mol. Endocrinol.* **2000**, *14*, 1962–1975. [CrossRef] [PubMed] [PubMed Central]
75. Kawahito, Y.; Kondo, M.; Tsubouchi, Y.; Hashiramoto, A.; Bishop-Bailey, D.; Inoue, K.; Kohno, M.; Yamada, R.; Hla, T.; Sano, H. 15-deoxy-delta(12,14)-PGJ(2) induces synoviocyte apoptosis and sup-presses adjuvant-induced arthritis in rats. *J. Clin. Investig.* **2000**, *106*, 189–197. [CrossRef] [PubMed] [PubMed Central]
76. Tong, B.J.; Tan, J.; Tajeda, T.; Das, S.K.; Chapman, J.A.; DuBoist, R.N.; Dey, S.K. Heightened Expression of Cyclooxygenase-2 and Peroxisome Proliferator-Activated Recep-tor-δ in Human Endometrial Adenocarcinoma. *Neoplasia* **2000**, *2*, 483–490. [CrossRef] [PubMed]
77. Gupta, R.A.; Tan, J.; Krause, W.F.; Geraci, M.W.; Willson, T.M.; Dey, S.K.; DuBois, R.N. Prostacyclin-mediated activation of peroxisome proliferator-activated receptor δ in colorectal cancer. *Proc. Natl. Acad. Sci. USA* **2000**, *97*, 13275–13280. [CrossRef] [PubMed]
78. Jehl-Pietri, C.; Bastie, C.; Gillot, I.; Luquet, S.; Grimaldi, P.A. Peroxisome-proliferator-activated recep-tor delta mediates the effects of long-chain fatty acids on post-confluent cell proliferation. *Biochem. J.* **2000**, *350*, 93–98. [CrossRef] [PubMed]
79. Sato, H.; Ishihara, S.; Kawashima, K.; Moriyama, N.; Suetsugu, H.; Kazumori, H.; Okuyama, T.; Rumi, M.A.; Fukuda, R.; Nagasue, N.; et al. Expression of peroxisome proliferator-activated receptor (PPAR)gamma in gastric cancer and inhibitory effects of PPARgamma agonists. *Br. J. Cancer.* **2000**, *83*, 1394–1400. [CrossRef] [PubMed]
80. Mueller, E.; Smith, M.; Sarraf, P.; Kroll, T.; Aiyer, A.; Kaufman, D.S.; Oh, W.; Demetri, G.; Figg, W.D.; Zhou, X.P.; et al. Effects of ligand activation of peroxisome proliferator-activated receptor gamma in human prostate cancer. *Proc. Natl. Acad. Sci. USA* **2000**, *97*, 10990–10995. [CrossRef]
81. Combs, C.K.; Johnson, D.E.; Karlo, J.C.; Cannady, S.B.; Landreth, G.E. Inflammatory mechanisms in Alz-heimer's disease: Inhibition of beta-amyloid-stimulated proinflammatory responses and neurotoxi-city by PPARgamma agonists. *J. Neurosci.* **2000**, *20*, 558–567. [CrossRef]
82. Zhou, X.P.; Smith, W.M.; Gimm, O.; Mueller, E.; Gao, X.; Sarraf, P.; Prior, T.W.; Plass, C.; von Deimling, A.; Black, P.M.; et al. Over-representation of PPARgamma sequence variants in sporadic cases of glioblastoma multiforme: Preliminary evidence for common low penetrance modifiers for brain tumour risk in the general population. *J. Med. Genet.* **2000**, *37*, 410–414. [CrossRef] [PubMed]
83. Debril, M.B.; Renaud, J.P.; Fajas, L.; Auwerx, J.; Girard, J. Transcription factors and nuclear receptors interact with the SWI/SNF complex through the BAF60c subunit. *J. Biol. Chem.* **1999**, *274*, 8945–8951. [CrossRef]
84. Chinetti, G.; Fruchart, J.C.; Staels, B. Peroxisome proliferator-activated receptors (PPARs): Nuclear receptors with functions in the vascular wall. *Z Kardiol.* **2001**, *90* (Suppl. S3), 125–132. [CrossRef]
85. Delerive, P.; Fruchart, J.C.; Staels, B. Peroxisome Proliferator-Activated Receptors in Inflammation Control. *J. Endocrinol.* **2001**, *169*, 453–459. [CrossRef]
86. Duez, H.; Fruchart, J.C.; Staels, B. PPARS in Inflammation, Atherosclerosis and Thrombosis. *J. Cardiovasc. Risk* **2001**, *8*, 187–194. [CrossRef]
87. Elangbam, C.S.; Tyler, R.D.; Lightfoot, R.M. Peroxisome Proliferator-Activated Receptors in Atherosclerosis and Inflammation--an Update. *Toxicol. Pathol.* **2001**, *29*, 224–231. [CrossRef]
88. Stumvoll, M.; Häring, H. The Peroxisome Proliferator-Activated Receptor-Gamma2 Pro12Ala Polymorphism. *Diabetes* **2002**, *51*, 2341–2347. [CrossRef]
89. Roberts, R.A.; Chevalier, S.; Hasmall, S.C.; James, N.H.; Cosulich, S.C.; Macdonald, N. PPARα and the Regulation of Cell Division and Apoptosis. *Toxicology* **2002**, *181–182*, 167–170. [CrossRef] [PubMed]
90. Berger, J.; Moller, D.E. The Mechanisms of Action of PPARs. *Annu. Rev. Med.* **2002**, *53*, 409–435. [CrossRef] [PubMed]
91. Fauconnet, S.; Lascombe, I.; Chabannes, E.; Adessi, G.-L.; Desvergne, B.; Wahli, W.; Bittard, H. Differential Regulation of Vascular Endothelial Growth Factor Expression by Peroxisome Proliferator-Activated Receptors in Bladder Cancer Cells. *J. Biol. Chem.* **2002**, *277*, 23534–23543. [CrossRef]

92. Haydon, R.C.; Zhou, L.; Feng, T.; Breyer, B.; Cheng, H.; Jiang, W.; Ishikawa, A.; Peabody, T.; Montag, A.; Simon, M.A.; et al. Nuclear receptor agonists as potential differentiation therapy agents for human osteosarcoma. *Clin. Cancer Res.* **2002**, *8*, 1288–1294. [PubMed]
93. Koumanis, D.J.; Christou, N.V.; Wang, X.L.; Gilfix, B.M. Pilot study examining the frequency of several gene polymorphisms in a morbidly obese population. *Obes. Surg.* **2002**, *12*, 759–764. [CrossRef] [PubMed]
94. Berger, J.; Wagner, J.A. Physiological and therapeutic roles of peroxisome proliferator-activated receptors. *Diabetes Technol. Ther.* **2002**, *4*, 163–174. [CrossRef] [PubMed]
95. Francis, G.A.; Annicotte, J.-S.; Auwerx, J. PPAR Agonists in the Treatment of Atherosclerosis. *Curr. Opin. Pharmacol.* **2003**, *3*, 186–191. [CrossRef] [PubMed]
96. Margeli, A.; Kouraklis, G.; Theocharis, S. Peroxisome Proliferator Activated Receptor-γ (PPAR-γ) Ligands and Angiogenesis. *Angiogenesis* **2003**, *6*, 165–169. [CrossRef] [PubMed]
97. Larsen, T.M.; Toubro, S.; Astrup, A. PPARgamma Agonists in the Treatment of Type II Diabetes: Is Increased Fatness Commensurate with Long-Term Efficacy? *Int. J. Obes. Relat. Metab. Disord. J. Int. Assoc. Study Obes.* **2003**, *27*, 147–161. [CrossRef] [PubMed]
98. Jiang, M.; Shappell, S.B.; Hayward, S.W. Approaches to Understanding the Importance and Clinical Implications of Peroxisome Proliferator-Activated Receptor Gamma (PPARgamma) Signaling in Prostate Cancer. *J. Cell. Biochem.* **2004**, *91*, 513–527. [CrossRef] [PubMed]
99. Bedu, E.; Wahli, W.; Desvergne, B. Peroxisome proliferator-activated receptor beta/delta as a therapeutic target for metabolic diseases. *Expert Opin. Ther. Targets.* **2005**, *9*, 861–873. [CrossRef] [PubMed]
100. Tenenbaum, A.; Motro, M.; Fisman, E.Z. Dual and pan-peroxisome proliferator-activated receptors (PPAR) co-agonism: The bezafibrate lessons. *Cardiovasc. Diabetol.* **2005**, *4*, 14. [CrossRef]
101. Buzzetti, R.; Petrone, A.; Caiazzo, A.M.; Alemanno, I.; Zavarella, S.; Capizzi, M.; Mein, C.A.; Osborn, J.A.; Vania, A.; di Mario, U. PPAR-Gamma2 Pro12Ala Variant Is Associated with Greater Insulin Sensitivity in Childhood Obesity. *Pediatr. Res.* **2005**, *57*, 138–140. [CrossRef]
102. Tyagi, S.C.; Rodriguez, W.; Patel, A.M.; Roberts, A.M.; Falcone, J.C.; Passmore, J.C.; Fleming, J.T.; Joshua, I.G. Hyperhomocysteinemic diabetic cardiomyopathy: Oxidative stress, remodeling, and endothelial-myocyte uncoupling. *J. Cardiovasc. Pharmacol. Ther.* **2005**, *10*, 1–10. [CrossRef] [PubMed]
103. Harris, G.; Ghazallah, R.A.; Nascene, D.; Wuertz, B.; Ondrey, F.G. PPAR activation and decreased proliferation in oral carcinoma cells with 4-HPR. *Otolaryngol. Head Neck Surg.* **2005**, *133*, 695–701. [CrossRef] [PubMed]
104. Lowe, D.B.; Bifulco, N.; Bullock, W.H.; Claus, T.; Coish, P.; Dai, M.; Dela Cruz, F.E.; Dickson, D.; Fan, D.; Hoover-Litty, H.; et al. Substituted indanylacetic acids as PPAR-alpha-gamma activators. *Bioorg. Med. Chem. Lett.* **2006**, *16*, 297–301. [CrossRef] [PubMed]
105. LoVerme, J.; La Rana, G.; Russo, R.; Calignano, A.; Piomelli, D. The search for the palmitoylethanolamide receptor. *Life Sci.* **2005**, *77*, 1685–1698. [CrossRef] [PubMed]
106. Michalik, L.; Auwerx, J.; Berger, J.P.; Chatterjee, V.K.; Glass, C.K.; Gonzalez, F.J.; Grimaldi, P.A.; Kadowaki, T.; Lazar, M.A.; O'Rahilly, S.; et al. International Union of Pharmacology. LXI. Peroxisome proliferator-activated receptors. *Pharmacol. Rev.* **2006**, *58*, 726–741. [CrossRef] [PubMed]
107. Aung, C.S.; Faddy, H.M.; Lister, E.J.; Monteith, G.R.; Roberts-Thomson, S.J. Isoform Specific Changes in PPAR Alpha and Beta in Colon and Breast Cancer with Differentiation. *Biochem. Biophys. Res. Commun.* **2006**, *340*, 656–660. [CrossRef]
108. Trivedi, N.R.; Cong, Z.; Nelson, A.M.; Albert, A.J.; Rosamilia, L.L.; Sivarajah, S.; Gilliland, K.L.; Liu, W.; Mauger, D.T.; Gabbay, R.A.; et al. Peroxisome Proliferator-Activated Receptors Increase Human Sebum Production. *J. Investig. Dermatol.* **2006**, *126*, 2002–2009. [CrossRef]
109. Burdick, A.D.; Kim, D.J.; Peraza, M.A.; Gonzalez, F.J.; Peters, J.M. The Role of Peroxisome Proliferator-Activated Receptor-Beta/Delta in Epithelial Cell Growth and Differentiation. *Cell. Signal.* **2006**, *18*, 9–20. [CrossRef]
110. Michalik, L.; Wahli, W. Peroxisome proliferator-activated receptors (PPARs) in skin health, repair and disease. *Biochim. Biophys. Acta.* **2007**, *1771*, 991–998. [CrossRef]
111. Akhmetov, I.I.; Astranenkova, I.V.; Rogozkin, V.A. Association of PPARD gene polymorphism with human physical performance. *Mol. Biol.* **2007**, *41*, 852–857. (In Russian)
112. Piqueras, L.; Reynolds, A.R.; Hodivala-Dilke, K.M.; Alfranca, A.; Redondo, J.M.; Hatae, T.; Tanabe, T.; Warner, T.D.; Bishop-Bailey, D. Activation of PPARbeta/delta induces endothelial cell proliferation and angiogenesis. *Arterioscler. Thromb. Vasc. Biol.* **2007**, *27*, 63–69. [CrossRef] [PubMed]
113. Grarup, N.; Albrechtsen, A.; Ek, J.; Borch-Johnsen, K.; Jørgensen, T.; Schmitz, O.; Hansen, T.; Pedersen, O. Variation in the peroxisome proliferator-activated receptor delta gene in relation to common metabolic traits in 7495 middle-aged white people. *Diabetologia* **2007**, *50*, 1201–1208. [CrossRef] [PubMed]
114. Hollingshead, H.E.; Killins, R.L.; Borland, M.G.; Girroir, E.E.; Billin, A.N.; Willson, T.M.; Sharma, A.K.; Amin, S.; Gonzalez, F.J.; Peters, J.M. Peroxisome Proliferator-Activated Receptor-Beta/Delta (PPARbeta/Delta) Ligands Do Not Potentiate Growth of Human Cancer Cell Lines. *Carcinogenesis* **2007**, *28*, 2641–2649. [CrossRef] [PubMed]
115. Pedchenko, T.V.; Gonzalez, A.L.; Wang, D.; DuBois, R.N.; Massion, P.P. Peroxisome Proliferator-Activated Receptor Beta/Delta Expression and Activation in Lung Cancer. *Am. J. Respir. Cell Mol. Biol.* **2008**, *39*, 689–696. [CrossRef] [PubMed]

116. Borland, M.G.; Foreman, J.E.; Girroir, E.E.; Zolfaghari, R.; Sharma, A.K.; Amin, S.; Gonzalez, F.J.; Ross, A.C.; Peters, J.M. Ligand Activation of Peroxisome Proliferator-Activated Receptor-Beta/Delta Inhibits Cell Proliferation in Human HaCaT Keratinocytes. *Mol. Pharmacol.* **2008**, *74*, 1429–1442. [CrossRef] [PubMed]
117. Ahmed, N.; Riley, C.; Quinn, M.A. An Immunohistochemical Perspective of PPAR Beta and One of Its Putative Targets PDK1 in Normal Ovaries, Benign and Malignant Ovarian Tumours. *Br. J. Cancer* **2008**, *98*, 1415–1424. [CrossRef] [PubMed]
118. Billin, A.N. PPAR-Beta/Delta Agonists for Type 2 Diabetes and Dyslipidemia: An Adopted Orphan Still Looking for a Home. *Expert Opin. Investig. Drugs* **2008**, *17*, 1465–1471. [CrossRef] [PubMed]
119. Gallardo-Soler, A.; Gómez-Nieto, C.; Campo, M.L.; Marathe, C.; Tontonoz, P.; Castrillo, A.; Corraliza, I. Arginase I induction by modified lipoproteins in macrophages: A peroxisome proliferator-activated receptor-gamma/delta-mediated effect that links lipid metabolism and immunity. *Mol. Endocrinol.* **2008**, *22*, 1394–1402. [CrossRef] [PubMed]
120. Fernandez, A.Z. Peroxisome proliferator-activated receptors in the modulation of the im-mune/inflammatory response in atherosclerosis. *PPAR Res.* **2008**, *2008*, 285842. [CrossRef]
121. Takata, Y.; Liu, J.; Yin, F.; Collins, A.R.; Lyon, C.J.; Lee, C.H.; Atkins, A.R.; Downes, M.; Barish, G.D.; Evans, R.M.; et al. PPARdelta-mediated antiinflammatory mechanisms inhibit angiotensin II-accelerated atherosclerosis. *Proc. Natl. Acad. Sci. USA* **2008**, *105*, 4277–4282. [CrossRef]
122. Cavalieri, D.; Calura, E.; Romualdi, C.; Marchi, E.; Radonjic, M.; Van Ommen, B.; Müller, M. Filling gaps in PPAR-alpha signaling through comparative nutrigenomics analysis. *BMC Genom.* **2009**, *10*, 596. [CrossRef] [PubMed]
123. Porcuna, J.; Mínguez-Martínez, J.; Ricote, M. The PPARα and PPARγ Epigenetic Landscape in Cancer and Immune and Metabolic Disorders. *Int. J. Mol. Sci.* **2021**, *22*, 10573. [CrossRef] [PubMed]
124. Małodobra-Mazur, M.; Cierzniak, A.; Ryba, M.; Sozański, T.; Piórecki, N.; Kucharska, A.Z. *Cornus mas* L. Increases Glucose Uptake and the Expression of *PPARG* in Insulin-Resistant Adipocytes. *Nutrients* **2022**, *14*, 2307. [CrossRef] [PubMed]
125. Danesi, F.; Di Nunzio, M.; Boschetti, E.; Bordoni, A. Green Tea Extract Selectively Activates Peroxisome Proliferator-Activated Receptor Beta/Delta in Cultured Cardiomyocytes. *Br. J. Nutr.* **2009**, *101*, 1736–1739. [CrossRef] [PubMed]
126. Jiang, B.; Liang, P.; Zhang, B.; Song, J.; Huang, X.; Xiao, X. Role of PPAR-Beta in Hydrogen Peroxide-Induced Apoptosis in Human Umbilical Vein Endothelial Cells. *Atherosclerosis* **2009**, *204*, 353–358. [CrossRef] [PubMed]
127. Jiang, B.; Liang, P.; Zhang, B.; Huang, X.; Xiao, X. Enhancement of PPAR-β Activity by Repetitive Low-Grade H_2O_2 Stress Protects Human Umbilical Vein Endothelial Cells from Subsequent Oxidative Stress-Induced Apoptosis. *Free Radic. Biol. Med.* **2009**, *46*, 555–563. [CrossRef]
128. Bishop-Bailey, D.; Bystrom, J. Emerging Roles of Peroxisome Proliferator-Activated Receptor-Beta/Delta in Inflammation. *Pharmacol. Ther.* **2009**, *124*, 141–150. [CrossRef] [PubMed]
129. Ramanan, S.; Zhao, W.; Riddle, D.R.; Robbins, M.E. Role of PPARs in Radiation-Induced Brain Injury. *PPAR Res.* **2010**, *2010*, 234975. [CrossRef] [PubMed]
130. Genovese, S.; Foreman, J.E.; Borland, M.G.; Epifano, F.; Gonzalez, F.J.; Curini, M.; Peters, J.M. A natural propenoic acid derivative activates peroxisome proliferator-activated receptor-beta/delta (PPARbeta/delta). *Life Sci.* **2010**, *86*, 493–498. [CrossRef]
131. Grimaldi, P.A. Metabolic and Nonmetabolic Regulatory Functions of Peroxisome Proliferator-Activated Receptor Beta. *Curr. Opin. Lipidol.* **2010**, *21*, 186–191. [CrossRef]
132. McKinnon, B.; Bersinger, N.A.; Huber, A.W.; Kuhn, A.; Mueller, M.D. PPAR-γ Expression in Peritoneal Endometriotic Lesions Correlates with Pain Experienced by Patients. *Fertil. Steril.* **2010**, *93*, 293–296. [CrossRef] [PubMed]
133. Mistry, N.F.; Cresci, S. PPAR transcriptional activator complex polymorphisms and the promise of individualized therapy for heart failure. *Heart Fail Rev.* **2010**, *15*, 197–207. [CrossRef] [PubMed]
134. Bassaganya-Riera, J.; Song, R.; Roberts, P.C.; Hontecillas, R. PPAR-gamma activation as an anti-inflammatory therapy for respiratory virus infections. *Viral Immunol.* **2010**, *23*, 343–352. [CrossRef] [PubMed]
135. Housley, W.J.; Adams, C.O.; Vang, A.G.; Brocke, S.; Nichols, F.C.; LaCombe, M.; Rajan, T.V.; Clark, R.B. Peroxisome proliferator-activated receptor gamma is required for CD4+ T cell-mediated lymphopenia-associated autoimmunity. *J. Immunol.* **2011**, *187*, 4161–4169. [CrossRef]
136. Shahin, D.; Toraby, E.E.; Abdel-Malek, H.; Boshra, V.; Elsamanoudy, A.Z.; Shaheen, D. Effect of peroxisome proliferator-activated receptor gamma agonist (pioglitazone) and methotrexate on disease activity in rheumatoid arthritis (experimental and clinical study). *Clin. Med. Insights Arthritis Musculoskelet Disord.* **2011**, *4*, 1–10. [CrossRef] [PubMed]
137. Sertznig, P.; Reichrath, J. Peroxisome proliferator-activated receptors (PPARs) in dermatology: Challenge and promise. *Dermatoendocrinol* **2011**, *3*, 130–135. [CrossRef] [PubMed]
138. Laschke, M.W.; Menger, M.D. Anti-angiogenic treatment strategies for the therapy of endometriosis. *Hum. Reprod. Update.* **2012**, *18*, 682–702. [CrossRef] [PubMed]
139. Nickkho-Amiry, M.; McVey, R.; Holland, C. Peroxisome proliferator-activated receptors modulate proliferation and angiogenesis in human endometrial carcinoma. *Mol. Cancer Res.* **2012**, *10*, 441–453. [CrossRef]
140. Knapp, P.; Chabowski, A.; Błachnio-Zabielska, A.; Jarząbek, K.; Wołczyński, S. Altered peroxisome-proliferator activated receptors expression in human endometrial cancer. *PPAR Res.* **2011**, *2012*, 471524. [CrossRef]
141. Montero, T.D.; Racordon, D.; Bravo, L.; Owen, G.I.; Bronfman, M.L.; Leisewitz, A.V. PPARα and PPARγ regulate the nucleoside transporter hENT1. *Biochem. Biophys. Res. Commun.* **2012**, *419*, 405–411. [CrossRef]

142. Abbas, A.; Blandon, J.; Rude, J.; Elfar, A.; Mukherjee, D. PPAR- γ agonist in treatment of diabetes: Cardiovascular safety considerations. *Cardiovasc. Hematol. Agents Med. Chem.* **2012**, *10*, 124–134. [CrossRef] [PubMed]
143. Chen, Y.C.; Wu, J.S.; Tsai, H.D.; Huang, C.Y.; Chen, J.J.; Sun, G.Y.; Lin, T.N. Peroxisome proliferator-activated receptor gamma (PPAR-γ) and neurodegenerative disorders. *Mol. Neurobiol.* **2012**, *46*, 114–124. [CrossRef] [PubMed]
144. Greene, N.P.; Fluckey, J.D.; Lambert, B.S.; Greene, E.S.; Riechman, S.E.; Crouse, S.F. Regulators of blood lipids and lipoproteins? PPARδ and AMPK, induced by exercise, are correlated with lipids and lipoproteins in overweight/obese men and women. *Am. J. Physiol. Endocrinol. Metab.* **2012**, *303*, E1212–E1221. [CrossRef] [PubMed]
145. Wadosky, K.M.; Willis, M.S. The story so far: Post-translational regulation of peroxisome proliferator-activated receptors by ubiquitination and SUMOylation. *Am. J. Physiol. Heart Circ. Physiol.* **2012**, *302*, H515–H526. [CrossRef] [PubMed]
146. Peyrou, M.; Ramadori, P.; Bourgoin, L.; Foti, M. PPARs in Liver Diseases and Cancer: Epigenetic Regulation by MicroRNAs. *PPAR Res.* **2012**, *2012*, 757803. [CrossRef] [PubMed]
147. Costa, V.; Ciccodicola, A. Is PPARG the key gene in diabetic retinopathy? *Br. J. Pharmacol.* **2012**, *165*, 1–3. [CrossRef]
148. Balakumar, P.; Mahadevan, N. Interplay between statins and PPARs in improving cardiovascular outcomes: A double-edged sword? *Br. J. Pharmacol.* **2012**, *165*, 373–379. [CrossRef] [PubMed]
149. Wahli, W.; Michalik, L. PPARs at the crossroads of lipid signaling and inflammation. *Trends. Endocrinol. Metab.* **2012**, *23*, 351–363. [CrossRef]
150. Reichenbach, G.; Starzinski-Powitz, A.; Sloane, B.F.; Doll, M.; Kippenberger, S.; Bernd, A.; Kaufmann, R.; Meissner, M. PPARα agonist Wy14643 suppresses cathepsin B in human endothelial cells via transcriptional, post-transcriptional and post-translational mechanisms. *Angiogenesis* **2013**, *16*, 223–233. [CrossRef]
151. Tailleux, A.; Wouters, K.; Staels, B. Roles of PPARs in NAFLD: Potential therapeutic targets. *Biochim. Biophys. Acta* **2012**, *1821*, 809–818. [CrossRef]
152. Videla, L.A.; Pettinelli, P. Misregulation of PPAR Functioning and Its Pathogenic Consequences Associated with Nonalcoholic Fatty Liver Disease in Human Obesity. *PPAR Res.* **2012**, *2012*, 107434. [CrossRef] [PubMed]
153. Benedusi, V.; Martorana, F.; Brambilla, L.; Maggi, A.; Rossi, D. The Peroxisome Proliferator-activated Receptor γ (PPARγ) Controls Natural Protective Mechanisms against Lipid Peroxidation in Amyotrophic Lateral Sclerosis*. *J. Biol. Chem.* **2012**, *287*, 35899–35911. [CrossRef] [PubMed]
154. Hu, W.; Wang, X.; Ding, X.; Li, Y.; Zhang, X.; Xie, P.; Yang, J.; Wang, S. MicroRNA-141 represses HBV replication by targeting PPARA. *PLoS ONE* **2012**, *7*, e34165. [CrossRef] [PubMed]
155. Zuo, C.; Liang, P.; Huang, X. Role of PPARbeta in fibroblast response to heat injury. *Indian J. Biochem. Biophys.* **2012**, *49*, 219–227. [PubMed]
156. Le Foll, B.; Di Ciano, P.; Panlilio, L.V.; Goldberg, S.R.; Ciccocioppo, R. Peroxisome proliferator-activated receptor (PPAR) agonists as promising new medications for drug addiction: Preclinical evidence. *Curr. Drug Targets* **2013**, *14*, 768–776. [CrossRef] [PubMed]
157. Aleshin, S.; Reiser, G. Role of the peroxisome proliferator-activated receptors (PPAR)-α, β/δ and γ triad in regulation of reactive oxygen species signaling in brain. *Biol. Chem.* **2013**, *394*, 1553–1570. [CrossRef] [PubMed]
158. Contreras, A.V.; Torres, N.; Tovar, A.R. PPAR-α as a key nutritional and environmental sensor for metabolic adaptation. *Adv. Nutr.* **2013**, *4*, 439–452. [CrossRef] [PubMed]
159. Akyürek, N.; Aycan, Z.; Çetinkaya, S.; Akyürek, Ö.; Yilmaz Ağladioğlu, S.; Ertan, Ü. Peroxisome proliferator activated receptor (PPAR)-gamma concentrations in childhood obesity. *Scand J. Clin. Lab. Investig.* **2013**, *73*, 355–360. [CrossRef] [PubMed]
160. Yao, P.L.; Morales, J.L.; Zhu, B.; Kang, B.H.; Gonzalez, F.J.; Peters, J.M. Activation of Peroxisome Proliferator-Activated Receptor-β/δ (PPAR-β/δ) Inhibits Human Breast Cancer Cell Line Tumorigenicity. *Mol. Cancer Ther.* **2014**, *13*, 1008–1017. [CrossRef]
161. Young, K.C.; Bai, C.H.; Su, H.C.; Tsai, P.J.; Pu, C.Y.; Liao, C.S.; Tsao, C.W. Fluoxetine a novel anti-hepatitis C virus agent via ROS-, JNK-, and PPARβ/γ-dependent pathways. *Antivir. Res.* **2014**, *110*, 158–167. [CrossRef]
162. Montagner, A.; Wahli, W.; Tan, N.S. Nuclear receptor peroxisome proliferator activated receptor (PPAR) β/δ in skin wound healing and cancer. *Eur. J. Dermatol.* **2015**, *Suppl. 1*, 4–11. [CrossRef]
163. de Melo, N.F.; de Macedo, C.G.; Bonfante, R.; Abdalla, H.B.; da Silva, C.M.; Pasquoto, T.; de Lima, R.; Fraceto, L.F.; Clemente-Napimoga, J.T.; Napimoga, M.H. 15d-PGJ2-Loaded Solid Lipid Nanopar-ticles: Physicochemical Characterization and Evaluation of Pharmacological Effects on In-flammation. *PLoS ONE* **2016**, *11*, e0161796. [CrossRef] [PubMed]
164. Xue, Y.; Xu, X.; Zhang, X.Q.; Farokhzad, O.C.; Langer, R. Preventing diet-induced obesity in mice by adipose tissue transformation and angiogenesis using targeted nanoparticles. *Proc. Natl. Acad. Sci. USA.* **2016**, *113*, 5552–5557. [CrossRef]
165. Strojny, B.; Grodzik, M.; Sawosz, E.; Winnicka, A.; Kurantowicz, N.; Jaworski, S.; Kutwin, M.; Urbańska, K.; Hotowy, A.; Wierzbicki, M.; et al. Diamond Nanoparticles Modify Curcumin Activity: In Vitro Studies on Cancer and Normal Cells and In Ovo Studies on Chicken Embryo Model. *PLoS ONE* **2016**, *11*, e0164637. [CrossRef] [PubMed]
166. Fidoamore, A.; Cristiano, L.; Laezza, C.; Galzio, R.; Benedetti, E.; Cinque, B.; d'Angelo, M.; Castelli, V.; Cifone, M.G.; Ippoliti, R.; et al. Annamaria Cimini Energy metabolism in glioblastoma stem cells: PPARα a metabolic adaptor to intratumoral microenvironment. *Oncotarget* **2017**, *8*, 108430–108450. [CrossRef] [PubMed]
167. Rolver, M.G.; Holland, L.K.K.; Ponniah, M.; Prasad, N.S.; Yao, J.; Schnipper, J.; Kramer, S.; Elingaard-Larsen, L.; Pedraz-Cuesta, E.; Liu, B.; et al. Chronic acidosis rewires cancer cell metabolism through PPARα signaling. *Int. J. Cancer.* **2023**, *152*, 1668–1684. [CrossRef] [PubMed]

168. Kado, T.; Kusakari, N.; Tamaki, T.; Murota, K.; Tsujiuchi, T.; Fukushima, N. Oleic acid stimulates cell proliferation and BRD4-L-MYC-dependent glucose transporter transcription through PPARα activation in ovarian cancer cells. *Biochem. Biophys. Res. Commun.* **2023**, *657*, 24–34. [CrossRef] [PubMed]
169. Wright, S.K.; Wuertz, B.R.; Harris, G.; Abu Ghazallah, R.; Miller, W.A.; Gaffney, P.M.; Ondrey, F.G. Functional Activation of PPARγ in Human Upper Aerodigestive Cancer Cell Lines. *Mol. Carcinog.* **2017**, *56*, 149–162. [CrossRef] [PubMed]
170. Ivanova, E.A.; Myasoedova, V.A.; Melnichenko, A.A.; Orekhov, A.N. Peroxisome Proliferator-Activated Receptor (PPAR) Gamma Agonists as Therapeutic Agents for Cardiovascular Disorders: Focus on Atherosclerosis. *Curr. Pharm. Des.* **2017**, *23*, 1119–1124. [CrossRef] [PubMed]
171. Gross, B.; Pawlak, M.; Lefebvre, P.; Staels, B. PPARs in Obesity-Induced T2DM, Dyslipidaemia and NAFLD. *Nat. Rev. Endocrinol.* **2017**, *13*, 36–49. [CrossRef]
172. Borland, M.G.; Kehres, E.M.; Lee, C.; Wagner, A.L.; Shannon, B.E.; Albrecht, P.P.; Zhu, B.; Gonzalez, F.J.; Peters, J.M. Inhibition of tumorigenesis by peroxisome proliferator-activated receptor (PPAR)-dependent cell cycle blocks in human skin carcinoma cells. *Toxicology* **2018**, *404–405*, 25–32. [CrossRef] [PubMed]
173. Leiguez, E.; Giannotti, K.C.; Viana, M.D.N.; Matsubara, M.H.; Fernandes, C.M.; Gutiérrez, J.M.; Lomonte, B.; Teixeira, C. A Snake Venom-Secreted Phospholipase A2 Induces Foam Cell Formation Depending on the Activation of Factors Involved in Lipid Homeostasis. *Mediat. Inflamm.* **2018**, *2018*, 2547918. [CrossRef] [PubMed]
174. Sun, H.; Shao, W.; Liu, H.; Jiang, Z. Exposure to 2,4-dichlorophenoxyacetic acid induced PPARβ-dependent disruption of glucose metabolism in HepG2 cells. *Environ. Sci. Pollut. Res.* **2018**, *25*, 17050–17057. [CrossRef] [PubMed]
175. Quintão, N.L.M.; Santin, J.R.; Stoeberl, L.C.; Corrêa, T.P.; Melato, J.; Costa, R. Pharmacological Treatment of Chemotherapy-Induced Neuropathic Pain: PPAR Agonists as a Promising Tool. *Front. Neurosci.* **2019**, *13*, 907. [CrossRef] [PubMed]
176. Vallée, A.; Lecarpentier, Y.; Vallée, J.N. Targeting the Canonical WNT/β-Catenin Pathway in Cancer Treatment Using Non-Steroidal Anti-Inflammatory Drugs. *Cells* **2019**, *8*, 726. [CrossRef] [PubMed]
177. Toraih, E.A.; Fawzy, M.S.; Abushouk, A.I.; Shaheen, S.; Hobani, Y.H.; Alruwetei, A.M.; Badran, D.I. Prognostic Value of the miRNA-27a and PPAR/RXRα Signaling Axis in Patients with Thyroid Carcinoma. *Epigenomics* **2020**, *12*, 1825–1843. [CrossRef] [PubMed]
178. Hirao-Suzuki, M.; Takeda, S.; Koga, T.; Takiguchi, M.; Toda, A. Cannabidiolic acid dampens the expression of cyclooxygenase-2 in MDA-MB-231 breast cancer cells: Possible implication of the peroxisome proliferator-activated receptor β/δ abrogation. *J. Toxicol. Sci.* **2020**, *45*, 227–236. [CrossRef] [PubMed]
179. Elie-Caille, C.; Lascombe, I.; Péchery, A.; Bittard, H.; Fauconnet, S. Molecular and nanoscale evaluation of N-cadherin expression in invasive bladder cancer cells under control conditions or GW501516 exposure. *Mol. Cell Biochem.* **2020**, *471*, 113–127. [CrossRef]
180. Moosavi, S.M.; Akhavan Sepahi, A.; Mousavi, S.F.; Vaziri, F.; Siadat, S.D. The effect of *Faecalibacterium prausnitzii* and its extracellular vesicles on the permeability of intestinal epithelial cells and expression of PPARs and ANGPTL4 in the Caco-2 cell culture model. *J. Diabetes Metab. Disord.* **2020**, *19*, 1061–1069. [CrossRef]
181. Liu, X.; Qian, D.; Liu, H.; Abbruzzese, J.L.; Luo, S.; Walsh, K.M.; Wei, Q. Genetic variants of the peroxisome proliferator-activated receptor (PPAR) signaling pathway genes and risk of pancreatic cancer. *Mol Carcinog.* **2020**, *59*, 930–939. [CrossRef]
182. Wouters, E.; Grajchen, E.; Jorissen, W.; Dierckx, T.; Wetzels, S.; Loix, M.; Tulleners, M.P.; Staels, B.; Stinissen, P.; Haidar, M.; et al. Altered PPARγ Expression Promotes Myelin-Induced Foam Cell Formation in Macrophages in Multiple Sclerosis. *Int. J. Mol. Sci.* **2020**, *21*, 9329. [CrossRef] [PubMed]
183. Xia, Y.; Li, J.; Chen, K.; Feng, J.; Guo, C. Bergenin Attenuates Hepatic Fibrosis by Regulating Autophagy Mediated by the PPAR-γ/TGF-β Pathway. *PPAR Res.* **2020**, *2020*, e6694214. [CrossRef] [PubMed]
184. Shen, J.; Sun, B.; Yu, C.; Cao, Y.; Cai, C.; Yao, J. Choline and methionine regulate lipid metabolism via the AMPK signaling pathway in hepatocytes exposed to high concentrations of nonesterified fatty acids. *J. Cell Biochem.* **2020**, *121*, 3667–3678. [CrossRef] [PubMed]
185. Yu, J.; Berga, S.L.; Zou, W.; Rajakumar, A.; Man, M.; Sidell, N.; Taylor, R.N. Human Endometrial Stromal Cell Differentiation is Stimulated by PPARβ/δ Activation: New Targets for Infertility? *J. Clin. Endocrinol. Metab.* **2020**, *105*, 2983–2995. [CrossRef] [PubMed]
186. Zhang, C.; Liu, Y.W.; Chi, Z.; Chen, B. Ligand-Activated Peroxisome Proliferator-Activated Receptor β/δ Facilitates Cell Proliferation in Human Cholesteatoma Keratinocytes. *PPAR Res.* **2020**, *2020*, e8864813. [CrossRef] [PubMed]
187. Wójtowicz, S.; Strosznajder, A.K.; Jeżyna, M.; Strosznajder, J.B. The Novel Role of PPAR Alpha in the Brain: Promising Target in Therapy of Alzheimer's Disease and Other Neurodegenerative Disorders. *Neurochem. Res.* **2020**, *45*, 972–988. [CrossRef] [PubMed]
188. Manickam, R.; Duszka, K.; Wahli, W. PPARs and Microbiota in Skeletal Muscle Health and Wasting. *Int. J. Mol. Sci.* **2020**, *21*, 8056. [CrossRef]
189. da Cruz, B.O.; Cardozo, L.F.M.F.; Magliano, D.C.; Stockler-Pinto, M.B. Nutritional strategies to modulate inflammation pathways via regulation of peroxisome proliferator-activated receptor β/δ. *Nutr. Rev.* **2020**, *78*, 207–214. [CrossRef] [PubMed]
190. Wagner, K.D.; Wagner, N. PPARs and Myocardial Infarction. *Int. J. Mol. Sci.* **2020**, *21*, 9436. [CrossRef]
191. Phua, W.W.T.; Tan, W.R.; Yip, Y.S.; Hew, I.D.; Wee, J.W.K.; Cheng, H.S.; Leow, M.K.S.; Wahli, W.; Tan, N.S. PPARβ/δ Agonism Upregulates Forkhead Box A2 to Reduce Inflammation in C2C12 Myoblasts and in Skeletal Muscle. *Int. J. Mol. Sci.* **2020**, *21*, 1747. [CrossRef]

192. Zhang, Z.; Su, H.; Ahmed, R.Z.; Zheng, Y.; Jin, X. Critical biomarkers for myocardial damage by fine particulate matter: Focused on PPARα-regulated energy metabolism. *Environ. Pollut.* **2020**, *264*, 114659. [CrossRef] [PubMed]
193. Andrade-Souza, V.A.; Ghiarone, T.; Sansonio, A.; Santos Silva, K.A.; Tomazini, F.; Arcoverde, L.; Fyfe, J.; Perri, E.; Saner, N.; Kuang, J.; et al. Exercise twice-a-day potentiates markers of mitochondrial biogenesis in men. *FASEB J.* **2020**, *34*, 1602–1619. [CrossRef]
194. Faulkner, A.; Lynam, E.; Purcell, R.; Jones, C.; Lopez, C.; Board, M.; Wagner, K.-D.; Wagner, N.; Carr, C.; Wheeler-Jones, C. Context-dependent regulation of endothelial cell metabolism: Differential effects of the PPARβ/δ agonist GW0742 and VEGF-A. *Sci. Rep.* **2020**, *10*, 7849. [CrossRef]
195. Chai, C.Y.; Tai, I.C.; Zhou, R.; Song, J.; Zhang, C.; Sun, S. MicroRNA-9-5p inhibits proliferation and induces apoptosis of human hypertrophic scar fibroblasts through targeting peroxisome proliferator-activated receptor β. *Biol. Open.* **2020**, *9*, bio051904. [CrossRef] [PubMed]
196. Capozzi, M.E.; Savage, S.R.; McCollum, G.W.; Hammer, S.S.; Ramos, C.J.; Yang, R.; Bretz, C.A.; Penn, J.S. The peroxisome proliferator-activated receptor-β/δ antagonist GSK0660 mitigates retinal cell inflammation and leukostasis. *Exp. Eye Res.* **2020**, *190*, 107885. [CrossRef] [PubMed]
197. Contreras-Lopez, R.A.; Elizondo-Vega, R.; Torres, M.J.; Vega-Letter, A.M.; Luque-Campos, N.; Paredes-Martinez, M.J.; Pradenas, C.; Tejedor, G.; Oyarce, K.; Salgado, M.; et al. PPARβ/δ-dependent MSC metabolism determines their immunoregulatory properties. *Sci. Rep.* **2020**, *10*, 11423. [CrossRef]
198. Petr, M.; Maciejewska-Skrendo, A.; Zajac, A.; Chycki, J.; Stastny, P. Association of Elite Sports Status with Gene Variants of Peroxisome Proliferator Activated Receptors and Their Transcriptional Coactivator. *Int. J. Mol. Sci.* **2019**, *21*, 162. [CrossRef]
199. Tutunchi, H.; Ostadrahimi, A.; Saghafi-Asl, M.; Hosseinzadeh-Attar, M.J.; Shakeri, A.; Asghari-Jafarabadi, M.; Roshanravan, N.; Farrin, N.; Naemi, M.; Hasankhani, M. Oleoylethanolamide supplementation in obese patients newly diagnosed with non-alcoholic fatty liver disease: Effects on metabolic parameters, anthropometric indices, and expression of PPAR-α, UCP1, and UCP2 genes. *Pharmacol. Res.* **2020**, *156*, 104770. [CrossRef]
200. Grabacka, M.; Pierzchalska, M.; Płonka, P.M.; Pierzchalski, P. The Role of PPAR Alpha in the Modulation of Innate Immunity. *Int. J. Mol. Sci.* **2021**, *22*, 10545. [CrossRef]
201. Willems, S.; Morstein, J.; Hinnah, K.; Trauner, D.; Merk, D. A Photohormone for Light-Dependent Control of PPARα in Live Cells. *J. Med. Chem.* **2021**, *64*, 10393–10402. [CrossRef]
202. Kharbanda, C.; Alam, M.S.; Hamid, H.; Ali, Y.; Nazreen, S.; Dhulap, A.; Alam, P.; Pasha, M.A.Q. In Silico Designing, in Vitro and in Vivo Evaluation of Potential PPAR-γ Agonists Derived from Aryl Propionic Acid Scaffold. *Bioorganic Chem.* **2021**, *106*, 104458. [CrossRef] [PubMed]
203. Christofides, A.; Konstantinidou, E.; Jani, C.; Boussiotis, V.A. The role of peroxisome proliferator-activated receptors (PPAR) in immune responses. *Metabolism* **2021**, *114*, 154338. [CrossRef] [PubMed]
204. Rayner, M.L.D.; Healy, J.; Phillips, J.B. Repurposing Small Molecules to Target PPAR-γ as New Therapies for Peripheral Nerve Injuries. *Biomolecules* **2021**, *11*, 1301. [CrossRef] [PubMed]
205. Stark, J.M.; Coquet, J.M.; Tibbitt, C.A. The Role of PPAR-γ in Allergic Disease. *Curr. Allergy Asthma Rep.* **2021**, *21*, 45. [CrossRef] [PubMed]
206. Hasegawa, S.; Yoneda, M.; Kurita, Y.; Nogami, A.; Honda, Y.; Hosono, K.; Nakajima, A. Cholestatic Liver Disease: Current Treatment Strategies and New Therapeutic Agents. *Drugs* **2021**, *81*, 1181–1192. [CrossRef] [PubMed]
207. Mukherjee, A.G.; Wanjari, U.R.; Gopalakrishnan, A.V.; Katturajan, R.; Kannampuzha, S.; Murali, R.; Namachivayamm, A.; Ganesan, R.; Renu, K.; Dey, A.; et al. Exploring the Regulatory Role of ncRNA in NAFLD: A Particular Focus on PPARs. *Cells* **2022**, *24*, 3959. [CrossRef] [PubMed]
208. Dandare, A.; Khan, M.J.; Naeem, A.; Liaquat, A. Clinical relevance of circulating non-coding RNAs in metabolic diseases: Emphasis on obesity, diabetes, cardiovascular diseases and metabolic syndrome. *Genes Dis.* **2022**, *10*, 2393–2413. [CrossRef] [PubMed]
209. Chatzopoulou, F.; Kyritsis, K.A.; Papagiannopoulos, C.I.; Galatou, E.; Mittas, N.; Theodoroula, N.F.; Papazoglou, A.S.; Karagiannidis, E.; Chatzidimitriou, M.; Papa, A.; et al. Dissecting miRNA-Gene Networks to Map Clinical Utility Roads of Pharmacogenomics-Guided Therapeutic Decisions in Cardiovascular Precision Medicine. *Cells* **2022**, *11*, 607. [CrossRef]
210. Kaltdorf, M.; Breitenbach, T.; Karl, S.; Fuchs, M.; Kessie, D.K.; Psota, E.; Prelog, M.; Sarukhanyan, E.; Ebert, R.; Jakob, F.; et al. Software JimenaE allows efficient dynamic simulations of Boolean networks, centrality and system state analysis. *Sci. Rep.* **2023**, *13*, 1855. [CrossRef]
211. Tanaka, Y.; Minami, Y.; Endo, M. Ror1 promotes PPARα-mediated fatty acid metabolism in astrocytes. *Genes Cells.* **2023**, *28*, 307–318. [CrossRef]
212. Ibrahim, M.A.A.; Abdeljawaad, K.A.A.; Roshdy, E.; Mohamed, D.E.M.; Ali, T.F.S.; Gabr, G.A.; Jaragh-Alhadad, L.A.; Mekhemer, G.A.H.; Shawky, A.M.; Sidhom, P.A.; et al. In Silico Drug Discovery of SIRT2 Inhibitors from Natural Source as Anticancer Agents. *Sci. Rep.* **2023**, *13*, 2146. [CrossRef]
213. Yang, W.; Ling, X.; He, S.; Cui, H.; Yang, Z.; An, H.; Wang, L.; Zou, P.; Chen, Q.; Liu, J.; et al. PPARα/ACOX1 as a novel target for hepatic lipid metabolism disorders induced by per- and polyfluoroalkyl substances: An integrated approach. *Environ. Int.* **2023**, *178*, 108138. [CrossRef] [PubMed]

214. Zhou, B.; Zhao, G.; Li, H.; Zhao, Q.; Liu, D. FNDC5/PPARa Pathway Alleviates THP-1-derived Macrophage Pyroptosis and Its Mechanism. *Altern. Ther. Health Med.* **2023**, *29*, 32–42. [PubMed]
215. Páscoa, I.; Biltes, R.; Sousa, J.; Preto, M.A.C.; Vasconcelos, V.; Castro, L.F.; Ruivo, R.; Cunha, I. A Multiplex Molecular Cell-Based Sensor to Detect Ligands of PPARs: An Optimized Tool for Drug Discovery in Cyanobacteria. *Sensors* **2023**, *23*, 1338. [CrossRef] [PubMed]
216. Kim, G.; Chen, Z.; Li, J.; Luo, J.; Castro-Martinez, F.; Wisniewski, J.; Cui, K.; Wang, Y.; Sun, J.; Ren, X.; et al. Gut-liver axis calibrates intestinal stem cell fitness. *Cell.* **2024**, *187*, 914–930.e20. [CrossRef] [PubMed]
217. Liu, P.; Li, Q.; Zhu, G.; Zhang, T.; Tu, D.; Zhang, F.; Finel, M.; He, Y.; Ge, G. Characterization of the Glucuronidating Pathway of Pectolinarigenin, the Major Active Constituent of the Chinese Medicine Daji, in Humans and Its Influence on Biological Activities. *J. Ethnopharmacol.* **2024**, *319*, 117280. [CrossRef] [PubMed]
218. Cheng, P.; Wei, H.-Z.; Chen, H.-W.; Wang, Z.-Q.; Mao, P.; Zhang, H.-H. DNMT3a-Mediated Methylation of PPARγ Promote Intervertebral Disc Degeneration by Regulating the NF-κB Pathway. *J. Cell. Mol. Med.* **2024**, *28*, e18048. [CrossRef] [PubMed]
219. Shyni, G.L.; Kavitha, S.; Indu, S.; Arya, A.D.; Anusree, S.S.; Vineetha, V.P.; Vandana, S.; Sundaresan, A.; Raghu, K.G. Chebulagic acid from Terminalia chebula enhances insulin mediated glucose uptake in 3T3-L1 adipocytes via PPARγ signaling pathway. *Biofactors* **2014**, *40*, 646–657. [CrossRef]
220. Zhang, D.; Wei, G. Green tea and the prevention of breast cancer: A case-control study in Southeast China. *Carcinogenesis* **2009**, *30*, 1351–1356. [CrossRef]
221. Tang, J.; Zheng, J.S.; Fang, L.; Jin, Y. Catechins and their therapeutic benefits to inflammatory bowel disease. *Molecules* **2009**, *14*, 5004–5015.
222. Chen, Z.P.; Schell, J.B.; Ho, C.T.; Chen, K.Y.; Li, S. Green tea epigallocatechin gallate shows a pronounced growth inhibitory effect on cancerous cells but not on their normal counterparts. *Cancer Lett.* **2020**, *278*, 101–112. [CrossRef]
223. Shao, M.; Guo, D.; Lu, W.; Chen, X.; Ma, L.; Wu, Y.; Zhang, X.; Wang, Q.; Wang, X.; Li, W.; et al. Identification of the active compounds and drug targets of Chinese medicine in heart failure based on the PPARs-RXRα pathway. *J. Ethnopharmacol.* **2020**, *257*, 112859. [CrossRef] [PubMed]
224. Tang, G.; Li, S.; Zhang, C.; Chen, H.; Wang, N.; Feng, Y. Clinical efficacies, underlying mechanisms and molecular targets of Chinese medicines for diabetic nephropathy treatment and management. *Acta Pharm. Sin B.* **2021**, *11*, 2749–2767. [CrossRef]
225. Xu, R.; Zhang, J.; Hu, X.; Xu, P.; Huang, S.; Cui, S.; Guo, Y.; Yang, H.; Chen, X.; Jiang, C. Yi-shen-hua-shi granules modulate immune and inflammatory damage via the ALG3/PPARγ/NF-κB pathway in the treatment of immunoglobulin a nephropathy. *J. Ethnopharmacol.* **2024**, *319*, 117204. [CrossRef] [PubMed]
226. Yuan, Z.; Qiao, H.; Wang, Z.; Wang, H.; Han, M.; Zhang, W.; Zhou, Y.; Hassan, H.M.; Zhao, W.; Qin, T. Taohe Chengqi decoction alleviated metabolic-associated fatty liver disease by boosting branched chain amino acids catabolism in the skeletal muscles of type 2 diabetes mellitus. *Phytomedicine* **2024**, *126*, 155315. [CrossRef]
227. Kuttan, R.; Bhanumathy, P.; Nirmala, K.; George, M.C.; Maliakel, B. Potential anticancer activity of turmeric (Curcuma longa). *Cancer Lett.* **2009**, *136*, 11–16. [CrossRef] [PubMed]
228. Aggarwal, B.B.; Sung, B. Pharmacological basis for the role of curcumin in chronic diseases: An age-old spice with modern targets. *Trends Pharmacol. Sci.* **2009**, *30*, 85–94. [CrossRef] [PubMed]
229. Anandharajan, R.; Jaiganesh, S.; Shankernarayanan, N.P.; Viswakarma, R.A.; Balakrishnan, A. In vitro glucose uptake activity of Aegles marmelos and Syzygium cumini by activation of Glut-4, PI3 kinase and PPARgamma in L6 myotubes. *Phytomedicine* **2006**, *13*, 434–441. [CrossRef] [PubMed]
230. Khosropoor, S.; Alavi, M.S.; Etemad, L.; Roohbakhsh, A. Cannabidiol goes nuclear: The role of PPARγ. *Phytomedicine* **2023**, *114*, 154771. [CrossRef]
231. Burstein, S. PPAR-gamma: A nuclear receptor with affinity for cannabinoids. *Life Sci.* **2005**, *77*, 1674–1684. [CrossRef]
232. Mechoulam, R.; Parker, L.A.; Gallily, R. Cannabidiol: An overview of some pharmacological aspects. *J. Clin. Pharmacol.* **2002**, *42* (Suppl. S11), 11S–19S. [CrossRef] [PubMed]
233. Robson, P. Human studies of cannabinoids and medicinal cannabis. *Handb. Exp. Pharmacol.* **2005**, *168*, 719–756. [CrossRef]
234. Sun, Y.; Alexander, S.P.; Kendall, D.A.; Bennett, A.J. Cannabinoids and PPARalpha signalling. *Biochem. Soc. Trans.* **2006**, *34*, 1095–1097. [CrossRef] [PubMed]
235. Brown, I.; Cascio, M.G.; Rotondo, D.; Pertwee, R.G.; Heys, S.D.; Wahle, K.W. Cannabinoids and omega-3/6 endocannabinoids as cell death and anticancer modulators. *Prog. Lipid Res.* **2013**, *52*, 80–109. [CrossRef] [PubMed]
236. Blüher, M.; Lübben, G.; Paschke, R. Analysis of the relationship between the Pro12Ala variant in the PPAR-gamma2 gene and the response rate to therapy with pioglitazone in patients with type 2 diabetes. *Diabetes Care* **2003**, *26*, 825–831. [CrossRef] [PubMed]
237. Blüher, M.; Klemm, T.; Gerike, T.; Krankenberg, H.; Schuler, G.; Paschke, R. Lack of association between peroxisome proliferator-activated receptor-gamma-2 gene variants and the occurrence of coronary heart disease in patients with diabetes mellitus. *Eur. J. Endocrinol.* **2002**, *146*, 545–551. [CrossRef]
238. Usuda, D.; Kanda, T. Peroxisome proliferator-activated receptors for hypertension. *World J. Cardiol.* **2014**, *6*, 744–754. [CrossRef] [PubMed]
239. Berger, J.; Leibowitz, M.D.; Doebber, T.W.; Elbrecht, A.; Zhang, B.; Zhou, G.; Biswas, C.; Cullinan, C.A.; Hayes, N.S.; Li, Y.; et al. Novel peroxisome proliferator-activated receptor (PPAR) gamma and PPARdelta ligands produce distinct biological effects. *J. Biol. Chem.* **1999**, *274*, 6718–6725. [CrossRef] [PubMed]

240. Bidzińska, B.; Demissie, M.; Tworowska, U.; Milewicz, A. Receptory aktywowane przez proliferatory peroksysomów a gospodarka lipidowa i węglowodanowa—rola fizjologiczna i znaczenie kliniczne. *Diabetol. Pol.* **2000**, *7*, 258–264. (In Polish)
241. Houseknecht, K.L.; Cole, B.M.; Steele, P.J. Peroxisome proliferator-activated receptor gamma (PPARgamma) and its ligands: A review. *Domest. Anim. Endocrinol.* **2002**, *22*, 1–23. [CrossRef]
242. Lindi, V.I.; Uusitupa, M.I.; Lindström, J.; Louheranta, A.; Eriksson, J.G.; Valle, T.T.; Hämäläinen, H.; Ilanne-Parikka, P.; Keinänen-Kiukaanniemi, S.; Laakso, M.; et al. Finnish Diabetes Prevention Study. Association of the Pro12Ala polymorphism in the PPAR-gamma2 gene with 3-year incidence of type 2 diabetes and body weight change in the Finnish Diabetes Prevention Study. *Diabetes* **2002**, *51*, 2581–2586. [CrossRef] [PubMed]
243. Sanchez, J.L.G.; Rios, M.S.; Perez, C.F.; Laakso, M.; Larrad, M.T.M. Effect of the Pro12Ala polymorphism of the peroxisome proliferator-activated receptor g -2 gene on adiposity, insulin sensitivity and lipid profile in the Spanish population. *Eur. J. Endocrinol.* **2000**, *147*, 495–501.
244. Demissie, M. Związek Polimorfizmu Genu Receptora Aktywowanego Proliferatorami Peroksysomów g 2 z Zaburzeniami Gospodarki Węglowodanowej i Lipidowej Oraz Profilem Hormonalnym u Osób z Należną Masą Ciała i Otyłych [Association of Peroxisome Proliferator-Activated Receptor Gene g 2 Polymorphism with Carbohydrate and Lipid Metabolism Disorders and Hormonal Profile in Normal-Weight and Obese Subjects]. Ph.D. Disertation, Wroclaw Medical Univerity, Wrocław, Poland, 2003.
245. Douglas, J.A.; Erdos, M.R.; Watanabe, R.M.; Braun, A.; Johnston, C.L.; Oeth, P.; Mohlke, K.L.; Valle, T.T.; Ehnholm, C.; Buchanan, T.A.; et al. The peroxisome proliferator-activated receptor-gamma2 Pro12A1a variant: Association with type 2 diabetes and trait differences. *Diabetes* **2001**, *50*, 886–890. [CrossRef] [PubMed]
246. Ju, J.; Huang, Q.; Sun, J.; Zhao, X.; Guo, X.; Jin, Y.; Ma, W.; Wang, W. Correlation between PPAR-α methylation level in peripheral blood and atherosclerosis of NAFLD patients with DM. *Exp. Ther. Med.* **2018**, *15*, 2727–2730. [CrossRef] [PubMed]
247. Wang, J.; Zhang, Y.; Zhuo, Q.; Tseng, Y.; Wang, J.; Ma, Y.; Zhang, J.; Liu, J. TET1 promotes fatty acid oxidation and inhibits NAFLD progression by hydroxymethylation of PPARα promoter. *Nutr. Metab.* **2020**, *17*, 46. [CrossRef] [PubMed]
248. Theys, C.; Lauwers, D.; Perez-Novo, C.; Vanden Berghe, W. PPARα in the Epigenetic Driver Seat of NAFLD: New Therapeutic Opportunities for Epigenetic Drugs? *Biomedicines* **2022**, *10*, 3041. [CrossRef] [PubMed]
249. Yideng, J.; Zhihong, L.; Jiantuan, X.; Jun, C.; Guizhong, L.; Shuren, W. Homocysteine-Mediated PPARα,γ DNA Methylation and Its Potential Pathogenic Mechanism in Monocytes. *DNA Cell Biol.* **2008**, *27*, 143–150. [CrossRef] [PubMed]
250. Castellano-Castillo, D.; Moreno-Indias, I.; Sanchez-Alcoholado, L.; Ramos-Molina, B.; Alcaide-Torres, J.; Morcillo, S.; Ocaña-Wilhelmi, L.; Tinahones, F.; Queipo-Ortuño, M.I.; Cardona, F. Altered Adipose Tissue DNA Methylation Status in Metabolic Syndrome: Relationships Between Global DNA Methylation and Specific Methylation at Adipogenic, Lipid Metabolism and Inflammatory Candidate Genes and Metabolic Variables. *J. Clin. Med.* **2019**, *8*, 87. [CrossRef] [PubMed]
251. Hong, F.; Xu, P.; Zhai, Y. The Opportunities and Challenges of Peroxisome Proliferator-Activated Receptors Ligands in Clinical Drug Discovery and Development. *Int. J. Mol. Sci.* **2018**, *19*, 2189. [CrossRef]
252. Kahn, B.B.; McGraw, T.E. Rosiglitazone, PPARγ, and Type 2 Diabetes. *N. Engl. J. Med.* **2010**, *363*, 2667–2669. [CrossRef]
253. Mitka, M. Panel Recommends Easing Restrictions on Rosiglitazone Despite Concerns About Cardiovascular Safety. *JAMA* **2013**, *310*, 246–247. [CrossRef] [PubMed]
254. Levin, D.; Bell, S.; Sund, R.; Hartikainen, S.A.; Tuomilehto, J.; Pukkala, E.; Keskimäki, I.; Badrick, E.; Renehan, A.G.; Buchan, I.E.; et al. Pioglitazone and Bladder Cancer Risk: A Multipopulation Pooled, Cumulative Exposure Analysis. *Diabetologia* **2015**, *58*, 493–504. [CrossRef] [PubMed]
255. Kim, S.H.; Kim, S.G.; Kim, D.M.; Woo, J.-T.; Jang, H.C.; Chung, C.H.; Ko, K.S.; Park, J.H.; Park, Y.S.; Kim, S.J.; et al. Safety and Efficacy of Lobeglitazone Monotherapy in Patients with Type 2 Diabetes Mellitus over 52 Weeks: An Open-Label Extension Study. *Diabetes Res. Clin. Pract.* **2015**, *110*, e27–e30. [CrossRef] [PubMed]
256. Shin, D.; Kim, T.-E.; Yoon, S.H.; Cho, J.-Y.; Shin, S.-G.; Jang, I.-J.; Yu, K.-S. Assessment of the Pharmacokinetics of Co-Administered Metformin and Lobeglitazone, a Thiazolidinedione Antihyperglycemic Agent, in Healthy Subjects. *Curr. Med. Res. Opin.* **2012**, *28*, 1213–1220. [CrossRef]
257. Priya, S.S.; Sankaran, R.; Ramalingam, S.; Sairam, T.; Somasundaram, L.S. Genotype Phenotype Correlation of Genetic Polymorphism of PPAR Gamma Gene and Therapeutic Response to Pioglitazone in Type 2 Diabetes Mellitus- A Pilot Study. *J. Clin. Diagn. Res. JCDR* **2016**, *10*, FC11–FC14. [CrossRef] [PubMed]
258. Stumvoll, M.; Häring, H. Reduced Lipolysis as Possible Cause for Greater Weight Gain in Subjects with the Pro12Ala Polymorphism in PPARgamma2? *Diabetologia* **2002**, *45*, 152–153. [CrossRef] [PubMed]
259. Stumvoll, M.; Wahl, H.G.; Löblein, K.; Becker, R.; Machicao, F.; Jacob, S.; Häring, H. Pro12Ala Polymorphism in the Peroxisome Proliferator-Activated Receptor-Gamma2 Gene Is Associated with Increased Antilipolytic Insulin Sensitivity. *Diabetes* **2001**, *50*, 876–881. [CrossRef] [PubMed]
260. Jang, E.J.; Lee, D.H.; Im, S.-S.; Yee, J.; Gwak, H.S. Correlation between PPARG Pro12Ala Polymorphism and Therapeutic Responses to Thiazolidinediones in Patients with Type 2 Diabetes: A Meta-Analysis. *Pharmaceutics* **2023**, *15*, 1778. [CrossRef] [PubMed]
261. Ji, J.D.; Cheon, H.; Jun, J.B.; Choi, S.J.; Kim, Y.R.; Lee, Y.H.; Kim, T.H.; Chae, I.J.; Song, G.G.; Yoo, G.H.; et al. Effects of Peroxisome Proliferator-activated Receptor-γ (PPAR-γ) on the Expression of Inflammatory Cytokines and Apoptosis Induction in Rheumatoid Synovial Fibroblasts and Monocytes. *J. Autoimmun.* **2001**, *17*, 215–221. [CrossRef]

262. Ganeb, S.S.; El-Brashy, A.E.-W.S.; Baraka, E.A.; Aboelazm, A.A.; Shaza, A.; Abdul Basset, S.A.A. Peroxisome proliferator-activated receptor gamma expression in peripheral monocytes from rheumatoid arthritis patients. *Egypt. Rheumatol.* **2016**, *38*, 141–146. [CrossRef]
263. Li, X.F.; Sun, Y.Y.; Bao, J.; Chen, X.; Li, Y.H.; Yang, Y.; Zhang, L.; Huang, C.; Wu, B.-M.; Meng, X.-M.; et al. Functional role of PPAR-γ on the proliferation and migration of fibroblast-like synoviocytes in rheumatoid arthritis. *Sci. Rep.* **2017**, *7*, 1–13. [CrossRef] [PubMed]
264. Rousseaux, C.; Lefebvre, B.; Dubuquoy, L.; Lefebvre, P.; Romano, O.; Auwerx, J.; Metzger, D.; Wahli, W.; Desvergne, B.; Naccari, G.C.; et al. Intestinal anti-inflammatory effect of 5-aminosalicylic acid is dependent on peroxisome proliferator-activated receptor-γ. *J. Exp. Med.* **2005**, *201*, 1205–1215. [CrossRef]
265. Dubuquoy, L.; Rousseaux, C.; Thuru, X.; Peyrin-Biroulet, L.; Romano, O.; Chavatte, P.; Chamaillard, M.; Desreumaux, P. PPARgamma as a new therapeutic target in inflammatory bowel diseases. *Gut* **2006**, *55*, 1341–1349. [CrossRef]
266. Annese, V.; Rogai, F.; Settesoldi, A.; Bagnoli, S. PPARγ in Inflammatory Bowel Disease. *PPAR Res.* **2011**, *2012*, 620839. [CrossRef]
267. da Rocha, G.H.O.; de Paula-Silva, M.; Broering, M.F.; Scharf, P.R.D.S.; Matsuyama, L.S.A.S.; Maria-Engler, S.S.; Farsky, S.H.P. Pioglitazone-Mediated Attenuation of Experimental Colitis Relies on Cleaving of Annexin A1 Released by Macrophages. *Front. Pharmacol.* **2020**, *11*, 591561. [CrossRef]
268. Li, D.; Feng, Y.; Tian, M.; Ji, J.; Hu, X.; Chen, F. Gut microbiota-derived inosine from dietary barley leaf supplementation attenuates colitis through PPARγ signaling activation. *Microbiome* **2021**, *9*, 83. [CrossRef] [PubMed]
269. Liu, Y.; Wang, J.; Luo, S.; Zhan, Y.; Lu, Q. The roles of PPARγ and its agonists in autoimmune diseases: A comprehensive review. *J. Autoimmun.* **2020**, *113*, 102510. [CrossRef]
270. Li, X.F.; Yin, S.Q.; Li, H.; Yang, Y.L.; Chen, X.; Song, B.; Wu, S.; Wu, Y.Y.; Wang, H.; Li, J. PPAR-γ alleviates the inflammatory response in TNF-α-induced fibroblast-like synoviocytes by binding to p53 in rheumatoid arthritis. *Acta Pharmacol Sin.* **2023**, *44*, 454–464. [CrossRef]
271. Lewis, J.D.; Lichtenstein, G.R.; Stein, R.B.; Deren, J.J.; Judge, T.A.; Fogt, F.; Furth, E.E.; Demissie, E.J.; Hurd, L.B.; Su, C.G.; et al. An open-label trial of the PPAR-gamma ligand rosiglitazone for active ulcerative colitis. *Am. J. Gastroenterol.* **2001**, *96*, 3323–3328. [CrossRef] [PubMed]
272. Liang, H.L.; Ouyang, Q. A clinical trial of rosiglitazone and 5-aminosalicylate combination for ulcerative colitis. *Zhonghua Nei ke Za Zhi* **2006**, *45*, 548–551.
273. Bassaganya-Riera, J.; Rynolds, K.; Martino-Catt, S.; Cui, Y.; Hennighausen, L.; Gonzalez, F.; Rohrer, J.; Benninghoff, A.U.; Hontecillas, R. Activation of PPAR γ and δ by conjugated linoleic acid mediates protection from experimental inflammatory bowel disease. *Gastroenterology* **2004**, *127*, 777–791. [CrossRef] [PubMed]
274. Qian, Y.; Li, P.; Zhang, J.; Shi, Y.; Chen, K.; Yang, J.; Wu, Y.; Ye, X. Association between peroxisome proliferator-activated receptor-alpha, delta, and gamma polymorphisms and risk of coronary heart disease: A case-control study and meta-analysis. *Medicine* **2016**, *95*, e4299. [CrossRef] [PubMed]
275. Yilmaz-Aydogan, H.; Kurnaz, O.; Kucukhuseyin, O.; Akadam-Teker, B.; Kurt, O.; Eronat, A.P.; Tekeli, A.; Bugra, Z.; Ozturk, O. Different effects of PPARA, PPARG and ApoE SNPs on serum lipids in patients with coronary heart disease based on the presence of diabetes. *Gene* **2013**, *523*, 20–26. [CrossRef] [PubMed]
276. Skoczynska, A.; Dobosz, T.; Poreba, R.; Turczyn, B.; Derkacz, A.; Zoledziewska, M.; Jonkisz, A.; Lebioda, A. The dependence of serum interleukin-6 level on PPAR-alpha polymorphism in men with coronary atherosclerosis. *Eur. J. Intern. Med.* **2005**, *16*, 501–506. [CrossRef] [PubMed]
277. Gouni-Berthold, I.; Giannakidou, E.; Müller-Wieland, D.; Faust, M.; Kotzka, J.; Berthold, H.K.; Krone, W. Association between the PPARalpha L162V polymorphism, plasma lipoprotein levels, and atherosclerotic disease in patients with diabetes mellitus type 2 and in nondiabetic controls. *Am. Heart J.* **2004**, *147*, 1117–1124. [CrossRef] [PubMed]
278. Dong, C.; Zhou, H.; Shen, C.; Yu, L.G.; Ding, Y.; Zhang, Y.H.; Guo, Z.R. Role of peroxisome proliferator-activated receptors gene polymorphisms in type 2 diabetes and metabolic syndrome. *World J. Diabetes* **2015**, *6*, 654–661. [CrossRef] [PubMed]
279. Visseren, F.L.J.; Mach, F.; Smulders, Y.M.; Carballo, D.; Koskinas, K.C.; Bäck, M.; Benetos, A.; Biffi, A.; Boavida, J.-M.; Capodanno, D.; et al. ESC National Cardiac Societies, ESC Scientific Document Group. 2021 ESC Guidelines on cardiovascular disease prevention in clinical practice. *Eur. Heart J.* **2021**, *42*, 3227–3237. [CrossRef] [PubMed]
280. Dobrowolski, P.; Prejbisz, A.; Kuryłowicz, A.; Baska, A.; Burchardt, P.; Chlebus, K.; Dzida, G.; Jankowski, P.; Jaroszewicz, J.; Jaworski, P.; et al. Metabolic syndrome—A new definition and management guidelines: A joint position paper by the Polish Society of Hypertension, Polish Society for the Treatment of Obesity, Polish Lipid Association, Polish Association for Study of Liver, Polish Society of Family Medicine, Polish Society of Lifestyle Medicine, Division of Prevention and Epidemiology Polish Cardiac Society, "Club 30" Polish Cardiac Society, and Division of Metabolic and Bariatric Surgery Society of Polish Surgeons. *Arch. Med. Sci.* **2022**, *18*, 1133–1156. [CrossRef]
281. Cheng, H.S.; Tan, W.R.; Low, Z.S.; Marvalim, C.; Lee, J.Y.H.; Tan, N.S. Exploration and Development of PPAR Modulators in Health and Disease: An Update of Clinical Evidence. *Int. J. Mol. Sci.* **2019**, *20*, 5055. [CrossRef]
282. Colapietro, F.; Gershwin, M.E.; Lleo, A. PPAR agonists for the treatment of primary biliary cholangitis: Old and new tales. *J. Transl. Autoimmun.* **2023**, *6*, 100188. [CrossRef]
283. Xi, Y.; Zhang, Y.; Zhu, S.; Luo, Y.; Xu, P.; Huang, Z. PPAR-Mediated Toxicology and Applied Pharmacology. *Cells* **2020**, *9*, 352. [CrossRef] [PubMed]

284. Okopień, B.; Bułdak, Ł.; Bołdys, A. Benefits and risks of the treatment with fibrates—A comprehensive summary. *Expert Rev. Clin. Pharmacol.* **2018**, *11*, 1099–1112. [CrossRef] [PubMed]
285. Kersten, S.; Wahli, W. Peroxisome proliferator-activated receptor agonists. *Exp.-Suppl. Only* **2000**, *89*, 141–152.

biomolecules

Hypothesis

Molecular Thumbprints: Biological Signatures That Measure Loss of Identity

Pallavi R. Devchand *, Michael Dicay and John L. Wallace

Department of Physiology and Pharmacology, University of Calgary, Calgary, AB T2N 4N1, Canada; mdicay@ucalgary.ca (M.D.); wallacej@ucalgary.ca (J.L.W.)
* Correspondence: pallavi.devchand@ucalgary.ca

Abstract: Each life is challenged to adapt to an ever-changing environment with integrity—simply put, to maintain identity. We hypothesize that this mission statement of adaptive homeostasis is particularly poignant in an adaptive response, like inflammation. A maladaptive response of unresolved inflammation can seed chronic disease over a lifetime. We propose the concept of a molecular thumbprint: a biological signature of loss of identity as a measure of incomplete return to homeostasis after an inflammatory response. Over time, personal molecular thumbprints can measure dynamic and precise trajectories to chronic inflammatory diseases and further loss of self to cancer. Why is this important? Because the phenotypes and molecular signatures of established complex inflammatory diseases are a far cry from the root of the complex problem, let alone the initial seed. Understanding the science behind key germinating seeds of disease helps to identify molecular factors of susceptibility, resilience, and early dietary or drug intervention. We pilot this hypothesis in a rat colitis model that is well-established for understanding molecular mechanisms of colonic health, disease, and transition of colitis to cancer.

Keywords: resilience; susceptibility; inflammation; anti-inflammatory drugs; clinical signatures

1. Introduction

The body is resilient: not all inflammatory events germinate to disease. In fact, acute inflammation is an adaptive response that is protective, most often goes unnoticed, and is self-resolving (Figure 1).

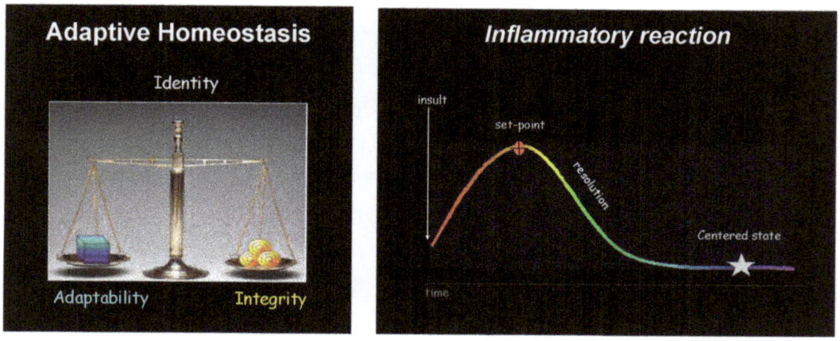

Figure 1. Adaptive homeostasis. Identity is a fine balance between adaptability and integrity. An insult triggers inflammation, which reaches a set point and then resolves successfully to return the organism and tissue back to its centered state (homeostasis).

We provide a preliminary evaluation of our hypothesis that an incomplete response of adaptive homeostasis can be defined by a loss of identity and measured by a biological signature. This biological thumbprint harbors molecular mechanisms that, over time, amplify the further loss of identity. For example, it can seed chronic inflammation and subsequent progression to complex diseases gradually (Figure 2) through successive new identities over time where maladaptation causes I_0 transitions to I_1, I_1 to I_2, and so on, with many iterations and, in some cases, the transition of I_n from chronic disease to cancer.

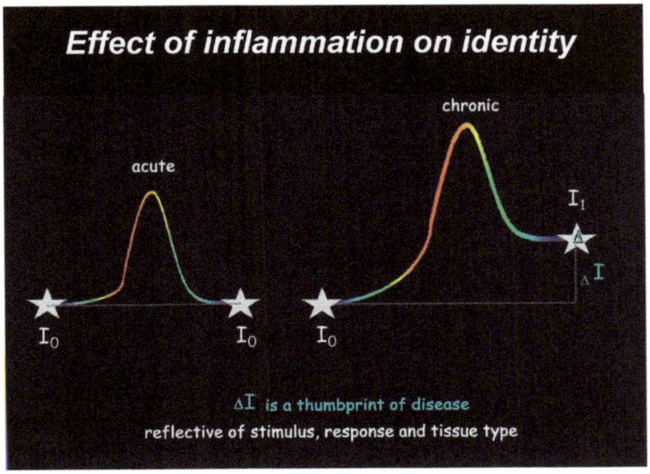

Figure 2. Identity measures a thumbprint of disease. A successful acute inflammation returns the organism back to its original identity $\mathbf{I_0}$. In multifactorial chronic inflammatory diseases, the initial inflammatory response is maladaptive. Incomplete resolution of inflammation returns the organism to a new identity, $\mathbf{I_1}$. The difference in identity, $\mathbf{\Delta I}$, can be measured and provides a thumbprint of disease.

After a tissue heals from an injury or insult, specialized cells must regain identity to perform their function. Over time, successive losses of molecular determinants of identity can progress to susceptibility to chronic disease, loss of function, and loss of resilience. The definitive case of loss of identity is cancer.

Unresolved inflammation can be quantified by several means and measures, including differences in protein expression, altered metabolic or structural profiles, DNA modifications, and gene expression. The ultimate goal would be detailed and comprehensive biological thumbprints of longitudinal data using current multifactorial and genetic bioinformatic systems. The ambition would be to have biological thumbprints as wholistic clinical measurements with utility in wellness, diagnosis, treatment, and patient responsiveness.

Here, we pilot this hypothesis in a simple, tangible, focused, and well-established model. This rat TNBS-induced colitis model has proven utility in drug discovery and also in understanding molecular mechanisms of colonic health, disease, drug impacts, and transition of colitis to cancer.

2. Materials and Methods

Animals: Male Wistar rats (175–200 g) were obtained from Charles River Breeding Farms (Montreal, PQ, Canada). The rats were fed standard laboratory rodent chow and tap water. All experiments were performed as approved by the Animal Care Committee of the University of Calgary (Protocol #M03140) and in accordance with the guidelines of the Canadian Council on Animal Care and NIH.

Induction of colitis using TNBS: Colitis was induced as previously described [1,2]. Briefly, 8 rats were lightly anesthetized with halothane, and an infant feeding tube fitted onto a blunt 18-gauge needle was inserted rectally. The tip of the tube was placed approximately 8 cm into the colon. For treatment, 30 mg of 2,4,6-trinitrobenzene sulphonic acid

(TNBS) in 0.5 mL of 50% ethanol was instilled. Control rats were age-matched and naïve (untreated). After 6 weeks, the rats were sacrificed.

In the present study, the severity of colonic damage was blindly scored on a 0 to 10 scale using criteria that have been previously reported in detail [3]. Briefly, a score of 0 represents normal appearance, a score of 1 is given for focal hyperemia but no ulcers, a score of 2 is given for ulceration without associated inflammation, and a score of 3 or greater is given when ulceration and inflammation are both evident (the score increasing further with the extent of ulceration). In most cases, the colonic damage score in rats sacrificed 6 weeks after TNBS administration is 0 or 1. In rare cases, ulceration of the colon is observed. However, as this study is focused on resolved colitis, any rat with a colonic damage score of greater than 1 was excluded from further analysis. After scoring, samples of the distal colon were frozen for subsequent measurement of MPO activity as an index of granulocyte infiltration and gene expression studies below. The standard *t*-test was used for statistical analyses ($p < 0.05$).

Isolation of Total RNA: Frozen colonic tissue was lysed in TRIzol using a hand homogenizer. Total RNA was isolated by the TRIzol method followed by DNase I digestion, as described previously [4]. Contaminating genomic DNA was removed using the Message Clean kit (Genehunter, Nashville, TN, USA). The purified RNA was resuspended in DEPC-treated ddH_2O, quantitated (OD_{260}), and qualitated ($OD_{260/280}$) by UV-spectroscopy, as described in [5].

FDD-PCR: FDD-PCR was performed using a modified protocol of Liang and Pardee (2003) [6]. Briefly, cDNAs were generated from 0.2 mg of RNA using 0.2 μM anchor primer 5′ACGACTCACTATAGGCTTTTTTTTTTTCA3′ and 20 units of Superscript RT enzyme (GIBCO, Waltham, MA, USA) in a 20 μL reaction containing 1 unit RNasin (Promega, Madison, WI, USA), 25 μM dNTPs, 50 mM Tris pH 8.3, 75 mM KCl, 3 mM $MgCl_2$, and 10 mM DTT. After one hour of incubation at 42 °C, reactions were terminated by heat inactivation of the enzyme (72 °C for 15 min).

Complementary DNA fragments were amplified with 0.2 μM each of FITC-conjugated anchor and 1 of 24 arbitrary primers in a 20 μL reaction consisting of 1.5 mM $MgCl_2$, 20 μM dNTPs, and 1 unit AmpliTaq enzyme (Boehringer Mannheim, Mannheim, Germany) using the following amplification program: 94 °C for 2 min; 4 cycles (94 °C for 15 s, 50 °C for 30 s, 72 °C for 2 min), 25 cycles (94 °C for 15 s, 60 °C for 30 s, 72 °C for 2 min), and an extension cycle of 72 °C for 7 min.

PCR products were resolved on a denaturing 4.5% polyacrylamide gel (Genomyx LR DNA Sequencer, GeneHunter Core Facility, Nashville, TN, USA) and visualized by fluorography. Bands of interest were excised, reamplified, purified and cloned in PCR-Trap vector. The DNA inserts were sequenced in both directions (Harvard Biopolymer Core Facility, Boston, MA, USA). Sequences were identified by BLAST searches.

Semi-quantitative RT-PCR: Gene changes observed in representative animals used in FDD-PCR were confirmed using standard methods of semi-quantitative RT-PCR. All reactions were performed in a thermocycler (*DNAEngine*, MJResearch) using a heated lid. Poly-A containing primer (50 nmol of Poly$T_{1\ 2}$-T7 primer) was annealed to 3 μg of RNA, in a total volume of 12 μL, by heating to 70 °C for 10 min followed by quick cooling to 4 °C. These primer–template complexes were used in a 20 μL reverse-transcription (RT) reaction performed in 50 mM Tris pH8.3, 75 mM KCl, 3 mM $MgCl_2$, 10 mM DTT, and 0.5 mM dNTPs. Prior to the addition of 200 U of Superscript II Reverse Transcriptase (Gibco), the reaction was incubated at 25 °C for 10 min and 42 °C for 2 min. The RT reaction was carried out at 42 °C for 50 min and terminated by heat inactivation of the enzyme (72 °C for 15 min). The cDNA was purified by digestion of RNA with 2 U of RNase H (37 °C for 20 min), followed by purification on a QIAquick gel extraction kit (Qiagen, Hilden, Germany). Finally, the cDNA was eluted in 50 μL of ddH_2O and stored at −20 °C until further use. The PCRs (total volume of 50 μL) utilized the Qiagen Master mix, 0.2 mM of each gene-specific primer and 2 μL of the above cDNAs. Reactions were performed in a thermocycler using the following program: denaturation at 94 °C for 2 min, amplification for 30 cycles (94 °C for 0.5 min, 48 °C for 0.5 min, 72 °C for 2 min), followed by extension cycle of 72 °C for 10 min. Annealing temperatures differed depending on the primer sets used. Products were separated by electrophoresis of 10 µL of PCR on 1% agarose gels and visualized by ethidium bromide staining. Results were photographed, scanned, and analyzed for comparative band intensities using the NIH Imager program, described in [5]. As with this pilot, follow-up experiments will continue to provide confirmation of the DD-PCR results by using the tissue/organs from mice that were used to generate the DD-PCR ($n = 3$ expectation of at least of 3 determinates for each sample) and then also in the other mice ($n = 8$) of the two cohorts in the TNBS-colitis experiment described above. The standard *t*-test will be used for statistical analyses.

3. Results

3.1. Rat Model for a Thumbprint of Unresolved Colitis

Over the past decades, the TNBS rat model has proven useful in understanding the biology of gut health, disease mechanisms, pharmacological intervention, exploration of new investigational drugs, and most notably, in understanding the risks and benefits of over-the-counter medicines, like globally consumed non-steroidal anti-inflammatory drugs. TNBS produces a disease that is histopathologically similar to Crohn's disease [1]. Key roles for T cells and for several cytokines that have been shown to be important in human IBD have also been shown to be important in the rat TNBS-induced colitis model. This experimental system involves only a single treatment of the rodent (no surgery, etc.) and has been well characterized in terms of the resolution phase of colitis, the boundary to colon cancer and the pharmacology of xenobiotics related to eicosanoid function in the GI, particularly COX- 2-related xenobiotics; (for review, see Wallace and Devchand, 2005 [7]).

This single shot of controlled and localized administration of TNBS is in contrast to the DSS model of colitis, in which the animals are given DSS in drinking water for various periods of time. DSS produces inflammation of the intestine but not very much ulceration. Since our hypothesis deals with post-colitis effects, it is imperative to keep the confounding variables, such as the degree of colitis and site of inflammation, to a minimum. Importantly, transmural inflammation or granulomas (features of Crohn's disease) are both induced by TNBS.

In previous studies, we found that colitis had resolved by 6 weeks after TNBS administration; that is, the macroscopic appearance of the colon, colonic myeloperoxidase activity (a marker of granulocyte infiltration), and colonic prostaglandin E_2 synthesis were no longer different from those in healthy controls [1,2].

Here, the aim of our hypothesis was to compare the difference in the identity of resolved inflammation. Therefore, this simple experiment compared two sets of rats: naïve controls versus post-colitis (Figure 3).

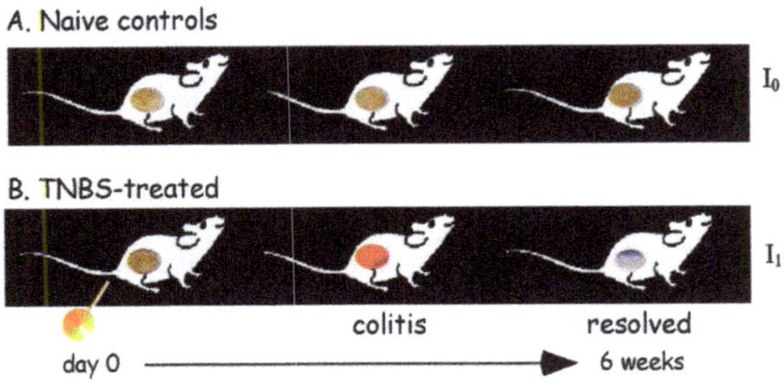

Figure 3. Experimental model. After 6 weeks, TNBS-treated rats (identity $\mathbf{I_1}$) were compared to naïve controls (original identity $\mathbf{I_0}$). Tissues around the precise location of TNBS treatment were harvested for analysis.

Since the single application of rectally administered TNBS was defined as 8 cm into the colon, the experiment allowed for precision tissue harvesting.

3.2. *Visualizing Unresolved Inflammation*

Here, a preliminary test of our hypothesis was performed using gene expression measurements. In our hands, differential-display polymerase chain reaction (DD-PCR) has been particularly useful in the identification of eicosanoid ligand-triggered pathways [4]. The primary advantage is highly reproducible and tangible data sets at every step of the process. This is often apparent in low-copy number transcripts and alternatively spliced transcripts (often in 3′ non-coding regions). The high-resolution separation of cDNAs on a sequencing gel affords advantages over cDNA CHIP arrays, including identification of novel transcripts, novel alternate spliced transcripts, and low-level transcripts such as nuclear cofactors; for review, see Liang and Pardee, 2003 [6]. Inherent in this method is the cloning of products that correspond to differentially expressed fragments of genes and can subsequently be used as probes in Northern blots, for example. These advantages are particularly important for pathway discovery so that the technique can be streamlined and derivative reagents like clones can be multifunctional for future use (Northern blots, in situ hybridization, FISH). The trade-off on these advantages is time and effort—the high copy number of transcripts that differ in comparison groups could most probably be identified by array or RNAseq analyses in a shorter time frame.

In this pilot experiment, twenty-four cDNA subsets were analyzed for reset after colitis. Duplicate samples of rats of each group were used for gene expression analyses by fluorescent differential display polymerase chain reaction (Figure 4). Close examination of all FDD-PCR reaction lanes revealed that ΔI was very small: ~30 bands differed when comparing naïve rats versus rats that reset after colitis. These differences varied in signal character, including intensity (high or low) and direction of expression (up-regulated or down-regulated).

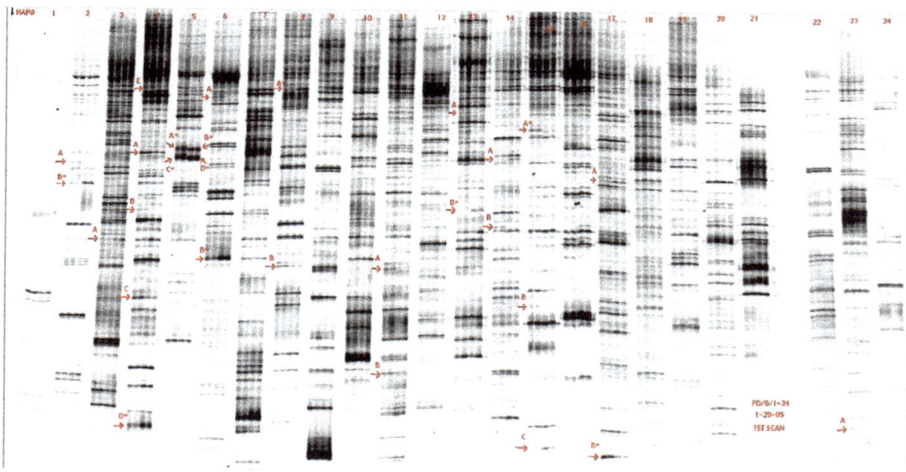

Figure 4. Fluorescent Differential Display Polymerase Chain Reactions. RT-PCR reactions of each treatment group were resolved on a sequencing gel. Lanes derived from each primer pair (1–24) are labeled above, with the left two lanes from naïve controls (original identity $\mathbf{I_0}$) and the right two lanes resolved TNBS-treated rats (identity $\mathbf{I_1}$). The bands of differential expression are indicated and labeled systematically. Original image can be found in Supplementary Materials.

3.3. Tangible Data

Twenty bands were excised for further processing (Figure 5). The fragments were amplified, cloned, screened, and sequenced.

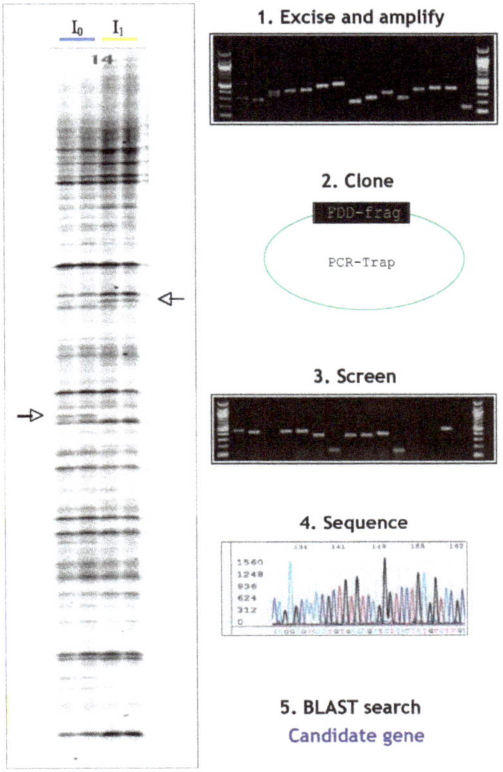

Figure 5. Identifying the candidate genes. This schematic shows the pipeline of analysis. Bands of interest were excised, amplified by PCR, and ligated into the PCR-TRAP vector to derive bacterial clones. After screening, plasmid inserts were sequenced. The candidate genes were identified by BLAST search. Original image can be found in Supplementary Materials.

3.4. BLAST to Gene Identification

Eighteen bands were successfully extracted and cloned. Sequence matches were identified by public database homology using BLAST (Tables 1 and 2). Referencing back to the original bands in the FDD-PCR sequencing gel, 12 were down-regulated in post-colitis and 6 up-regulated. Tables 1 and 2 list the genes identified, including *sv2b*, *fra-2*, *sephs-1*, *pallidin*, *DNA primase*, *ptger4*, *rbbp5*, and *mical 3*. Interestingly, in rats with post-resolved colitis, our analysis indicated a potential switch in isoform expression of synaptic vesicle glycoprotein 2: fragment of *sv2b* transcript 1 was up-regulated (Table 2) whereas that of transcript 2 was down-regulated (Table 1). Of note, both *sv2b* transcripts are of low-signal fragments on the FDD-PCR gel. Some excised bands are mapped to chromosome regions or by homology to other species. And one is coded on the mitochondrial genome.

Table 1. Down-regulated genes. These 12 cDNA fragments showed lower expression in rats with resolved colitis. For each, the upper number indicates location on the FDD-PCR gel. Gene names identified by BLAST are in bold. The protein name, when available, is indicated. Finally, the corresponding regions of the transcript that were excised and amplified from the band are indicated.

Low Signal	High Signal
2A-1 **Sv2b** Synaptic vesicle glycoprotein 2b Transcript 1 5271-5872	**5A-1** **Pallidin** Syntaxin-13 interacting protein 1871-2041 1770-1846
14B-3 Sim to region on Chr4 on mouse 76075-75953	**5A′-1** Aligns to mChr8 BAC clone Possibly a MHC gene
23A-1 **Fra-2** Fos-related antigen DNA 105-234	**5C-2** Rat homologue of mKIAA1930 4091-4386
4B-1 **Sephs 1** Selenophosphate synthetase	**6A-1** **DNA Primase** Subunit 49 1171-1374 1407-1591
15B-2 Homology to region in mChr1	**11A-1** **Ptger** EP4 receptor 2619-2811
	15A-4 Sim to region on Chr11 on mouse 63633-751. 63766-882
	21B-2 **Rbbp5** Retinoblastoma binding protein 5 Similar to mChr1 23480-23132 in rat 766-897;898-970

Table 2. Up-regulated genes. These 6 transcript fragments showed higher expression in rats with resolved colitis. For each, the upper number indicates location on the FDD-PCR gel. Gene names identified by BLAST are in bold. The protein name, when available, is indicated. Finally the corresponding regions of the transcript that were excised and amplified from the band are indicated.

Low Signal	High Signal
2B-w **Sv2b** Synaptic vesicle glycoprotein 2b Transcript 2 5271-5942	**5D-1** Rat homologue of *m.musc* cDNA BC051083 1095-1329
21A-2 Similar to region on Chr 15 on mouse 168293-7953	**8A-1** Rat homologue of **MICAL3** Microtubule-associated monooxygenase calponin Lim-domain containing 3 1546-1694
	13B-1 Receptor for hyaluron-mediated motility Coded on mitochrondrial genome 308-144
	14A-4 Image clone 7123360 2688-2770 2519-2631 78bp homologous to interferon-induced P44

4. Discussion

In our hypothesis, we approach health as a challenge to modulate molecular mechanisms of adaptive homeostasis that define the dynamic border of identity. In humans, chronic inflammation of colitis is associated with a predisposition to colorectal cancer, and the risk of incidence increases with the duration of colonic inflammation [8]. This fact supports the concept that chronic inflammation results in small but significant changes in identity, and over time, the increased inability to maintain homeostasis after successive bouts of inflammation, compounded with other environmental triggers, ultimately manifests as the changed identity of cancer.

Here, our hypothesis focused on the final stages of the resolution phase of inflammation to better understand how, after one bout of substantial inflammation, a specialized tissue like the colon retains identity. The reparative ability of the colon is remarkable. At the transcript level, the measured **ΔI**, the change in identity post-colitis, was very small. Only ~30 differences were observed from the 24 subsets of cDNA that were analyzed by FDD-PCR. Sequenced fragments identified several interesting proteins: transcription factors (the fos protein FRA2), the chromatin modulator Rbbp5, DNA maintenance enzyme (DNA primase), vesicle transport and cellular motility (Palladin, the transmembrane transport Svb2, the enzyme Mical3, the hyaluronan-binding protein HMMR), the eicosanoid receptor PTGER4, and the selenide metabolic enzyme SEPHS2.

In the context of our animal model of defining a biological fingerprint post-resolution of TNBS-induced colitis, the identification of PTGER4 is particularly exciting. Altered expression of this eicosanoid G-coupled receptor has long been associated with Crohn's disease in humans. For example, in a genome-wide association study, Glas et al. (2012) identified a region at Chr 5p13.1 that was associated with both modulating PTGER4 expression and susceptibility to Crohn's disease [9]. Prager et al. (2014) used a subset of German patients to refine the *rs7720838* variant as a determinant of increased susceptibility to disease [10]. Most recently, Drew et al. (2024) deduced that variants in this 5p13.1 region (like *rs350047*) are mechanistically relevant to the chemo-preventative effect of regular aspirin/non-steroidal anti-inflammatory drug (NSAID) use in patients [11]. Interestingly, using an animal model, Na et al. demonstrated that PTGER4 expression in intestinal macrophages is essential for supporting the regeneration of injured epithelium [12].

At the molecular level, the mechanisms that cause the disease remain elusive [13]. Current treatments of colitis, like the NSAID mesalamine, aim to control chronic inflammation to sustain colonic integrity over time [14]. Substantial evidence indicates that one target of NSAIDs, the enzyme cyclooxygenase-2 (COX-2), plays several central roles in modulating molecular and cellular mechanisms of gut mucosal defense (Figure 6 and Wallace and Devchand, 2005 [7]). Under normal healthy conditions, COX-2 couples to a battery of downstream enzymes to generate potent lipid mediators involved in processes of adaptive homeostasis, healing, and innate immunity. These include the regulation of mucus and bicarbonate secretion by the gastric and duodenal epithelium, mucosal blood flow, epithelial cell proliferation, epithelial restitution, and mucosal immunocyte function. In the context of an inflammatory reaction, the COX-2 axis is dynamic: in combination with downstream enzymes, it regulates the amplification of the inflammation and also modulates the resolution program in part via prostaglandin D_2 and E_2 lipid signaling pathways [3].

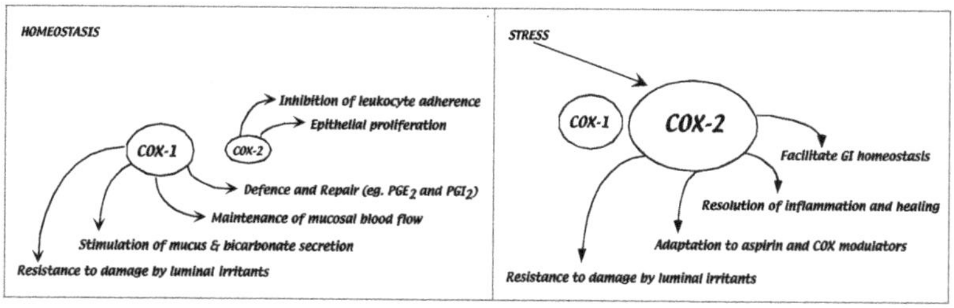

Figure 6. Duality of COX-2 function in dynamics of adaptive homeostasis.

Under normal healthy conditions of the gastrointestinal system, endogenous COX-2-derived lipid mediators are involved in processes of adaptive homeostasis, healing, and innate immunity. The COX-2 gene is rapidly induced by stress, and the downstream products are potent lipid signals that enhance resistance to injury and regulate the dynamics of inflammation and resolution.

To date, there are no polymorphisms in the COX-2 gene itself that have been linked to IBD (http://www.ncbi.nlm.nih.gov/projects/SNP, accessed on 12 October 2020). The functions of prostanoids are dynamic, site-specific, and complex [15]. Given the myriad of potent lipid signals and pathways that connect to COX-2 in health and disease states, the duality of COX-2 function in the dynamics of adaptive homeostasis of the GI tract is an important clue to colitis-related disease susceptibility. For instance, COX-2 protein levels remain elevated after a single bout of colitis, and the increased enzyme activity is a determinant of susceptibility to genotoxic-induced cancer formation [16]. It is clear that dysregulation of COX-2 function can be associated with various disease states, including diminished resistance to injury, ulcers, colitis, and colon cancer, see Dubois review [17]. However, the nature of specific downstream lipid products and their signal transduction pathways that mediate associated homeostatic functions remain under-explored.

The most well-characterized products of the COX-2 reaction are the AA-derived prostanoids. Major advances have been made in understanding signaling by arachidonic acid-derived lipid mediators [7,18–20]. Eicosanoids are generated in small amounts (nano- to mico-molar quantities) and are rapidly inactivated. Even though these autocoids are short-lived, they are potent stimulators of bioactivity. A retrospective view of eicosanoid research emphasizes the importance of a multidisciplinary approach coupled with the use and development of stable mimetics [21]. For example, distinct biosynthetic enzymes downstream of COX-2 dynamically produce the main lipid mediators linked to the resolution of inflammation: PGE_2 and PGD_2 [22,23]. The use of a PGD_2 synthetic antagonist has implicated PGD_2 as a determinant of colitis-induced cancer formation [16,24].

Intracellular targets for eicosanoids have been postulated for many years. Recent studies indicate that eicosanoid signaling molecules can directly interact with nuclear receptors (PPARs) to modulate gene expression (e.g., [25–28]). In the context of PMNs where eicosanoid signals are present intracellularly, these ligand-activated transcription factors have potentially significant roles. Consistent with this notion is the finding that in PPARα knock-out mice, an LTB_4-mediated inflammatory response is attenuated [28]. The role of PPAR β/δ has been a very active area of research in colon cancer models [17] and the mechanisms underlying chemo-preventative effects of NSAIDs [29].

In the context of shared ligands between cell-surface and nuclear receptors, future molecular modeling and docking studies would complement and advance understanding of drug–protein specificity, mechanisms of action, drug dynamics in detailed binding free energy calculations, and underlying biology based on predictions of the strength and stability of the interactions. A companion study in this collection of articles shows a wonderful example of structural studies on the evaluation of PPAR binding to drugs [30].

Many cell surface receptors also modulate gene expression via non-receptor transcription factors in a tightly controlled fashion. One example of an elaborate system is the activation pathway for the nuclear factor kappa B (NFKB) complex containing FOS and JUN transcription factors [31]. In its active form, NFKB is considered a pro-inflammatory transcription factor, as exemplified by its ability to mediate TNFα-induced transcription of the pro-inflammatory cytokine interleukin1b [32]. In this context, our identification of the fos-related antigen *fra2* as a component of the biological thumbprint is an intriguing result of our pilot experiment. At the transcription factor level, reports suggest that inhibition of NFKB might involve competition either for coactivators or for one of the NFKB subunits (e.g., [33]), and lipid signaling to the nucleus has highlighted roles in paracrine, endocrine, and autocrine pathways [19,34]. Interestingly, the *wnt* signal pathway, most characterized for its association with cancer progression, is dysregulated after a bout of colitis, as measured by increased levels of β-catenin protein in the nucleus—a defect that can be corrected with a selective synthetic COX-2 inhibitor [16]. While counter-regulatory systems in the nucleus have a substantial impact on the adaptive homeostasis of the GI tract, the molecular mechanisms remain elusive.

Of note, in a study that pooled 13 population-based studies, Archambault et al. (2021) sought to identify non-genetic determinants of risk for the incidence of early-onset colorectal cancer [35]. Factors considered in this large analysis included regular use of NSAIDs, red meat intake, educational attainment, and alcohol intake. This study indicates that adaptation to the environment (adaptive homeostasis) and choice of lifestyle can provide a foundation for target identification of those most at risk. Our hypothesis takes a molecular viewpoint of tapping into the early stages of adaptive homeostasis. The results from our pilot experiment support our hypothesis to offer a basis for further study.

We envision the use of biological thumbprints in the clinic and in future research. Some might consider the use of this biological thumbprint over a lifetime to map the bigger picture of aging over time as the body adapts to the environment (e.g., obesity, non-insulin-dependent diabetes). For example, this might be reflected in key metabolites (surrogate markers) combined with personal, holistic parameters in routine physical examinations (weight, blood work, urine test, blood pressure measurements, etc.). Others might consider the use of biological thumbprints simply as maps of the trajectory of unresolved inflammation to complex diseases, such as cardiovascular disease, COPD, Crohn's disease, and cancer progression. With the rapid advance in computational ability, biological thumbprints could be reflective of autocrine, endocrine, and paracrine systems and measure identities within a tissue, an organ, or an entire body. Simply put, biological thumbprints could be reflective of the dynamics of the health state, measuring progression to disease and also recovery toward the original healthy self.

5. Conclusions

Walter Bradford Cannon [36] coined the classical definition of adaptive homeostasis in 1929. Our hypothesis focuses on this definition with the angle that identity is a fine balance of adaptability with integrity. We defined the scientific concept of biological thumbprints that can measure the success of a defined process of adaptive homeostasis and define health states throughout the dynamic trajectory over a lifetime.

Here, we have used the adaptive response of inflammation as an example. A successful acute inflammation returns the organism back to its original identity $\mathbf{I_0}$. In multifactorial chronic inflammatory diseases, the initial inflammatory response is maladaptive. Incomplete resolution of inflammation returns the organism to a new identity, $\mathbf{I_1}$. The difference in identity, $\mathbf{\Delta I}$, can be measured and provides a thumbprint of disease.

We envision that molecular thumbprints would be particularly useful in complex inflammatory diseases. The progressive approach of our hypothesis helps create personalized maps of health states with potential cost-effective use in the clinic. In a more general sense, these thumbprints would be a measure of wellness. How do these wellness measures synchronize and synergize to maintain health over a lifetime? It's complex. It's personal. It is tangible.

Supplementary Materials: The following supporting information can be downloaded at: https://www.mdpi.com/article/10.3390/biom14101271/s1, Original images of Figures 4 and 5.

Author Contributions: Conceptualization, P.R.D. and J.L.W.; methodology, M.D., J.L.W. and P.R.D. All authors have read and agreed to the published version of the manuscript.

Funding: This research was funded in part by CIHR and the Crohn's Colitis Foundation of Canada (JLW). PRD is funded in part by a K.A.S.H. Scientific Innovation Award. PRD belongs to the PENTACON consortium (funded by the National Institutes of Health HL117798).

Institutional Review Board Statement: All experiments were performed as approved by the Animal Care Committee of the University of Calgary (Protocol #M03140) and in accordance with the guidelines of the Canadian Council on Animal Care, 7 June 2005.

Informed Consent Statement: Not applicable.

Data Availability Statement: All data obtained are presented in the manuscript.

Acknowledgments: Authors are indebted to G. Webb McKnight for his expert contributions. We are fortunate to join this volume in honor of Professor Walter Wahli. As all authors share Walter's spirit of transcending borders in science, it is only fitting that we rein in and contribute a research hypothesis that asks:

"Where's the border?"
-Milan Kundera in *Identity* (1998).

Conflicts of Interest: The authors declare no conflicts of interest.

References

1. Morris, G.P.; Beck, P.L.; Herridge, M.S.; Depew, W.T.; Szewczuk, M.R.; Wallace, J.L. Hapten-induced model of chronic inflammation and ulceration in the rat colon. *Gastroenterology* **1989**, *96*, 795–803. [CrossRef] [PubMed]
2. Wallace, J.L.; MacNaughton, W.K.; Morris, G.P.; Beck, P.L. Inhibition of leukotriene synthesis markedly accelerates healing in a rat model of inflammatory bowel disease. *Gastroenterology* **1989**, *96*, 29–36. [CrossRef] [PubMed]

3. Zamuner, S.R.; Warrier, N.; Buret, A.G.; MacNaughton, W.K.; Wallace, J.L. Cyclooxygenase 2 mediates post-inflammatory colonic secretory and barrier dysfunction. *Gut* **2003**, *52*, 1714–1720. [CrossRef] [PubMed]
4. Devchand, P.R.; Qiu, F.H.; Wada, K.; Serhan, C.N. Aspirin-triggered lipoxin A4 and lipoxin A4 up- regulate transcriptional corepressor NAB1 in human neutrophils. *FASEB J.* **2001**, *15*, 2736–2738.
5. Devchand, P.R.; Hihi, A.K.; Perroud, M.; Schleuning, W.D.; Spiegelman, B.M.; Wahli, W. Chemical probes that differentially modulate peroxisome proliferator-activated receptor alpha and BLTR, nuclear and cell surface receptors for leukotriene B(4). *J. Biol. Chem.* **1999**, *274*, 23341–23348. [CrossRef]
6. Liang, P.; Pardee, A.B. Analysing differential gene expression in cancer. *Nat. Rev. Cancer* **2003**, *3*, 869–876. [CrossRef]
7. Wallace, J.L.; Devchand, P.R. Emerging roles for cyclooxygenase-2 in gastrointestinal mucosal defense. *Br. J. Pharmacol.* **2005**, *145*, 275–282. [CrossRef]
8. Potter, J.D. Colorectal cancer: Molecules and populations. *J. Natl. Cancer Inst.* **1999**, *91*, 916–932. [CrossRef]
9. Glas, J.; Seiderer, J.; Czamara, D.; Pasciuto, G.; Diegelmann, J.; Wetzke, M.; Olszak, T.; Wolf, C.; Muller-Myhso, B.; Balschun, T.; et al. PTGER4 expression-modulating polymorphisms in the 5p13.1 region predispose to Chrohn's disease and affect NFKB and XBP1 binding sites. *PLoS ONE* **2012**, *7*, e52873. [CrossRef]
10. Prager, M.; Buttner, J.; Buning, C. PTGER4 modulating variants in Crohn's disease. *Int. J. Color. Dis.* **2014**, *29*, 909–915. [CrossRef]
11. Drew, D.A.; Kim, A.E.; Lin, Y.; Qu, C.; Morrison, J.; Lewinger, J.P.; Kawaguchi, E.; Wang, J.; Fu, Y.; Zemlianskaia, N.; et al. Two genome-wide interaction loci modify the association of non-steroidal anti-inflammatory drugs with colorectal cancer. *Sci. Adv.* **2024**, *10*, eadk3121. [CrossRef] [PubMed]
12. Na, R.Y.; Jung, D.; Stakenborg, M.; Jang, H.; Gu, G.J.; Jeong, M.R.; Suh, S.Y.; Kim, H.J.; Park, K.J.; Im, J.P.; et al. Prostaglandin E2 receptor PTGER4-expressing macrophages promote intestinal epithelial barrier regeneration upon inflammation. *Gut* **2021**, *70*, 2249–2260. [CrossRef] [PubMed]
13. Argmann, C.; Hou, R.; Ungaro, R.C.; Irizar, H.; Al-Taie, Z.; Huang, R.; Kosoy, R.; Venkat, S.; Song, W.M.; Di, A.F.; et al. Biopsy and blood-based molecular biomarker of inflammation in IBD. *Gut* **2023**, *72*, 1271–1287. [CrossRef] [PubMed]
14. Breyer, R.M.; Bagdassarian, C.K.; Myers, S.A.; Breyer, M.D. Prostanoid receptors: Subtypes and signaling. *Annu. Rev. Pharmacol. Toxicol.* **2001**, *41*, 661–690. [CrossRef]
15. Ricciotti, E.; Haines, P.G.; Chai, W.; FitzGerald, G.A. Prostanoids in cardiac and vascular remodelling. *Ateriosclerosis Thromb. Vasc. Biol.* **2024**, *44*, 558–583. [CrossRef]
16. Zamuner, S.R.; Bak, A.W.; Devchand, P.R.; Wallace, J.L. Predisposition to colorectal cancer in rats with resolved colitis: Role of cyclooxygenase-2-derived prostaglandin D2. *Am. J. Pathol.* **2005**, *167*, 1293–1300. [CrossRef]
17. Forman, J.E.; Sorg, J.M.; McGuinnis, K.S.; Rigas, B.; Williams, J.L.; Clapper, M.L.; Gonzalez, F.J.; Peters, J.M. Regulation of pe-roxisome proliferator-activated receptor b/d by the APC/beta-CATENIN pathway and NSAIDs. *Mol. Carcinog.* **2009**, *48*, 942–952. [CrossRef]
18. Sammuelsson, B.; Dahlen, S.E.; Lindgren, J.A.; Rouzer, C.A.; Serhan, C.N. Leukotrienes and lipoxins: Structures, biosynthesis and biological effects. *Science* **1987**, *237*, 1171–1176. [CrossRef]
19. Devchand, P.R. Lipoxin agonists: Turn right! to path of resolving neutrophil. *Mem. Inst. Oswaldo Cruz.* **2005**, *100* (Suppl. S1), 55–57. [CrossRef]
20. Funk, C.D. Prostaglandins and leukotrienes: Advances in eicosanoid biology. *Science* **2001**, *294*, 1871–1875. [CrossRef]
21. Flower, R.J. The development of COX2 inhibitors. *Nat. Rev. Drug Discov.* **2003**, *2*, 179–191. [CrossRef] [PubMed]
22. FitzGerald, G.A. COX-2 and beyond: Approaches to prostaglandin inhibition in human disease. *Nat. Rev. Drug Discov.* **2003**, *2*, 879–890. [CrossRef] [PubMed]
23. Wallace, J.L.; Miller, M.J. Nitric oxide in mucosal defense: A little goes a long way. *Gastroenterology* **2000**, *119*, 512–520. [CrossRef] [PubMed]
24. Ajuebor, M.N.; Singh, A.; Wallace, J.L. Cyclooxygenase-2-derived prostaglandin D_2 is an early anti- inflammatory signal in experimental colitis. *Am. J. Physiol.* **2000**, *279*, G238–G244. [CrossRef]
25. Krey, G.; Braissant, O.; L'Horset, F.; Kalkhoven, E.; Perroud, M.; Parker, M.G.; Wahli, W. Fatty acids, eicosanoids, and hypolipidemic agents identified as ligands of peroxisome proliferator- activated receptors by coactivator-dependent receptorligand assay. *Mol. Endocrinol.* **1998**, *11*, 779–791. [CrossRef]
26. Kliewer, S.A.; Sundseth, S.S.; Jones, S.A.; Brown, P.J.; Wisely, G.B.; Koble, C.S.; Devchand, P.; Wahli, W.; Willson, T.M.; Lenhard, J.M.; et al. Fatty acids and eicosanoids regulate gene expression through direct interactionswith peroxisome proliferator-acti-vated receptors alpha and gamma. *Proc. Natl. Acad. Sci. USA* **1997**, *94*, 4318–4323. [CrossRef]
27. Forman, B.M.; Chen, J.; Evans, R.M. Hypolipidemic drugs, polyunsaturated fatty acids, and eicosanoids are ligands for peroxisome proliferator-activated receptors alpha and delta. *Proc. Natl. Acad. Sci. USA* **1997**, *94*, 4312–4317. [CrossRef]
28. Devchand, P.R.; Keller, H.; Peters, J.M.; Vasquez, M.; Gonzalez, F.J.; Wahli, W. The PPARa–leukotriene B4 pathway to inflammation control. *Nature* **1996**, *384*, 39–43. [CrossRef]
29. Forman, J.E.; Chang, W.-C.L.; Palkar, P.S.; Zhu, B.; Borland, M.G.; Williams, J.L.; Kramer, L.R.; Clapper, M.L.; Gonzalez, F.J.; Peter, J.M. Functional characterization of peroxisome proliferator-activated receptor b/d expression in colon cancer. *Mol. Carcinog.* **2011**, *50*, 884–900. [CrossRef]
30. Useini, A.; Schwerin, I.K.; Kunze, G.; Strater, N. Structural studies on the binding mode of bisphenols to PPARgamma. *Biomolecules* **2024**, *14*, 640.

31. Ghosh, S.; May, M.J.; Kopp, E.B. NF-kappa B and Rel proteins: Evolutionarily conserved mediators of immune responses. *Annu. Rev. Immunol.* **1998**, *16*, 225–260. [CrossRef] [PubMed]
32. Fenton, M.J.; Clark, B.D.; Collins, K.L.; Webb, A.C.; Rich, A.; Auron, P.E. Transcriptional regulation of the human prointerleukin 1 beta gene. *J. Immunol.* **1988**, *138*, 3972–3979. [CrossRef]
33. Delerive, P.; De Bosscher, K.; Besnard, S.; Vanden Berghe, W.; Peters, J.M.; Gonzalez, F.J.; Fruchart, J.C.; Tedgui, A.; Haegeman, G.; Staels, B. Peroxisome proliferator-activated receptor alpha negatively regulates the vascular inflammatory gene response by negative cross-talk with transcription factors NF-kappaB and AP1. *J. Biol. Chem.* **1999**, *274*, 32048–32054. [CrossRef] [PubMed]
34. Devchand, P.R.; Ziouzenkova, O.; Plutzky, J. Oxidative stress and peroxisome proliferator-activated receptors: Reversing the curse? *Circ. Res.* **2004**, *95*, 1137–1139. [CrossRef] [PubMed]
35. Archambault, A.N.; Lin, Y.; Jeon, J.; Harrison, T.A.; Bishop, D.T.; Brenner, H.; Casey, G.; Chan, A.T.; Chang-Claude, J.; Figueiredo, J.C.; et al. Non-genetic determinants of risk for early onset colorectal cancer. *JNCI Cancer Spectr.* **2021**, *5*, pkab029. [CrossRef]
36. Cannon, W.B. Organisation for physiological homeostasis. *Physiol. Rev.* **1929**, *3*, 399–431. [CrossRef]

biomolecules

MDPI

Article

Structural Studies on the Binding Mode of Bisphenols to PPARγ

Abibe Useini [1], Inken Kaja Schwerin [2], Georg Künze [2,3,4,*] and Norbert Sträter [1,*]

1 Institute of Bioanalytical Chemistry, Centre for Biotechnology and Biomedicine, Leipzig University, Deutscher Platz 5, 04103 Leipzig, Germany; abibe.useini@bbz.uni-leipzig.de
2 Institute for Drug Discovery, Leipzig University, Brüderstraße 34, 04103 Leipzig, Germany; inken.schwerin@medizin.uni-leipzig.de
3 Interdisciplinary Center for Bioinformatics, Leipzig University, 04107 Leipzig, Germany
4 Center for Scalable Data Analytics and Artificial Intelligence (ScaDS.AI), Leipzig University, 04105 Leipzig, Germany
* Correspondence: georg.kuenze@uni-leipzig.de (G.K.); strater@bbz.uni-leipzig.de (N.S.)

Abstract: Bisphenol A (BPA) and bisphenol B (BPB) are widely used in the production of plastics, and their potential adverse health effects, particularly on endocrine disruption and metabolic health, have raised concern. Peroxisome proliferator-activated receptor gamma (PPARγ) plays a pivotal role in metabolic regulation and adipogenesis, making it a target of interest in understanding the development of obesity and associated health impacts. In this study, we employ X-ray crystallography and molecular dynamics (MD) simulations to study the interaction of PPARγ with BPA and BPB. Crystallographic structures reveal the binding of BPA and BPB to the ligand binding domain of PPARγ, next to C285, where binding of partial agonists as well as antagonists and inverse agonists of PPARγ signaling has been previously observed. However, no interaction of BPA and BPB with Y437 in the activation function 2 site is observed, showing that these ligands cannot stabilize the active conformation of helix 12 directly. Furthermore, free energy analyses of the MD simulations revealed that I341 has a large energetic contribution to the BPA and BPB binding modes characterized in this study.

Keywords: peroxisome proliferator-activated receptor gamma; bisphenols; obesity; adverse health; structural analysis

1. Introduction

Obesity is a widespread epidemic impacting both adults and children. It poses a significant risk for the development of physical and mental disorders and is a threat to public health. The number of overweight and obese people has escalated to alarming levels, with 40% of the global adult population being affected. According to the World Health Organization, the incidence of obesity has nearly tripled since 1975. As of 2016, more than 1.9 billion adults were overweight, and over 650 million adults were classified as obese [1]. Emerging research suggests that endocrine disruptors contribute to the rise in obesity rates. Exposure to these chemicals, which are found in many everyday products, can alter metabolic processes and increase the risk of obesity [2,3]. Both in vivo and in vitro investigations have demonstrated the promotion of adipogenesis, lipid accumulation and acceleration of pre-adipocyte differentiation as a result of exposure to these chemicals used as plasticizers [4–8]. They interfere with the body's hormonal system and with metabolic pathways that play a key role in regulating hormonal activity, lipid homeostasis and adipogenesis. Among other pathways, adipogenesis is regulated by peroxisome proliferator-activated receptor gamma (PPARγ) signaling [9,10].

PPARγ, an intensively studied nuclear receptor (NR), is intricately involved in many essential pathways, including adipocyte differentiation and the regulation of glucose and lipid metabolism [11]. Upon forming heterodimers with retinoic acid receptor (RXR),

PPARγ orchestrates the transcriptional regulation of various genes. The signaling activity of PPARγ can be fine-tuned by ligands, both endogenous and synthetic, which interact with the ligand binding domain (LBD) of PPARγ [12–17]. Besides the LBD, PPARγ contains a DNA-binding domain and a hinge region. Given its significance in drug design and pharmaceutical treatments, the structure and ligand interactions of PPARγ LBD have been extensively studied. It consists of 13 α-helices and a four-stranded β-sheet. In contrast to other NRs, PPARγ features an additional α-helix, named H2′, positioned between the initial β-strand and H3. Notably, the orientation of H2 in PPARγ differs from other NR structures, thereby facilitating enhanced ligand accessibility. H3, H7 and H10 form a large cavity for ligand binding [13]. The flexible Ω loop between H3 and H2′ may serve as a gate influencing access of ligands to the binding pocket [18].

Effective ligand-dependent gene transcription relies on a highly conserved motif, termed activating function-2 (AF-2). Positioned at the C-terminus of the ligand binding domain, this motif resides within a large Y-shaped ligand binding pocket (LBP), encompassing the AF-2 sub-pocket adjacent to helix 12 and the Ω sub-pocket [13,19]. The binding of ligands to the LBP of PPARγ promotes complex formation with co-activator proteins binding to the AF-2 surface (outside of the LBP). This in turn favors the formation of the heterodimer with RXR, DNA binding and transcription of the target genes [20]. The activation mechanism of PPARγ by endogenous lipids, including fatty acids or their metabolites such as 15-deoxy-delta-12,14-prostaglandin J2 (15d-PGJ2) and synthetic compounds such as thiazolidinediones, has been characterized by X-ray crystallography [12–15,19,21].

Bisphenols have been extensively utilized in the plastic industry as plasticizers for over half a century. Bisphenol A (BPA) and related bisphenols are mostly used in the production of polycarbonate plastics, epoxy resins and thermal papers [22]. BPA was first synthesized in 1891 and has been used since the 1950s–1960s in plastic fabrication, although its estrogenic properties have been known since the 1930s [23]. BPA became one of the chemicals produced in the highest volume worldwide, reaching a global consumption of 5.6 million metric tons in 2022 [24]. The increasing use of BPA led to a closer investigation and biomonitoring of human exposure. BPA is non-covalently dispersed in plastic polymers and diffuses into food and drinks. Therefore, exposure to BPA is believed to primarily occur through diet. Food warming in plastic containers accelerates BPA leaching. Ingestion is not the only root for BPA uptake. It can occur also through skin penetration and inhalation.

The amount of BPA in urine and serum has been assessed by several studies for different groups of people [25]. Studies report average BPA levels in urine in a range from 0.2–5.7 ng/mL. Based on the 2013–14 National Health and Nutrition Examination Survey (NHANES), urinary BPA is detectable in over 90% of the United States population, with children having significantly higher BPA levels than adults [26]. In the latest release of the BPA re-evaluation by the European Food Safety Authority (EFSA) in 2023, the tolerable daily intake of BPA for humans was lowered from 4 μg to 0.2 ng per kilogram of body weight per day (kg/bw/day). People with average or high exposure exceed this threshold, indicating health concerns [27]. The use of BPA is now limited not just in items designed for infants (EU Commission Regulation 2018/213) [28] but also in other food contact materials and other products, such as thermal paper, where it should not exceed 0.02% by weight [22]. BPA is substituted by other bisphenols like bisphenol B, C, AF, F, S.

With the restriction of BPA, the use of BPB as a substitute is gradually increasing. BPB is a very close structural analog of BPA and is more resistant to biodegradation [29]. This stresses the need to assess its potential endocrine properties and to evaluate if it is safer than BPA. BPB has been detected in human urine, blood and serum [30–32], food and the environment [29,33–35]. BPB was found at a serum level of up to 5.15 ng/mL mean concentration, which is higher than BPA (concentration 2.91 ± 1.74 ng/mL) [36].

BPB has been investigated for health risks by many studies, and BPB effects on reproductive function have been reviewed [37]. Exposure of zebrafish to BPB induced developmental toxic effects and oxidative damage [36]. Another study showed that BPB exposure adversely affects the uterus morphology in mice [38]. Furthermore, BPB is known to exert higher estrogenic effects than BPA, which is mediated through the G protein-coupled estrogen receptor pathway [39]. In the cited study, the binding affinity of BPB to GPER was 9-fold higher compared to that of BPA.

BPA induces a significant increase in pre-adipocyte proliferation, lipid accumulation and increased expression of PPARγ [8,16,40]. A study on murine pre-adipocytes concluded that BPA and BPS induce adipogenesis through direct activation of PPARγ [41]. In contrast to this, a decrease in adipogenesis of human adipocytes and antagonistic effects of bisphenol A and B on human PPARγ signaling were reported [16]. The presence of BPA did not promote adipogenesis in mesenchymal stromal stem cells (MSCs), in contrast to 3T3-L1 cells, where induced adipogenesis was observed already at 10 nM BPA concentration [42]. An antagonistic effect on rosiglitazone activation of PPARγ was observed at 10 μM of BPA. Activation of PPARγ depends on the degree of halogenation of BPA analogs [43]. Brominated BPA derivatives exhibited increasing potency and agonistic activity on PPARγ signaling with an increasing degree of bromination. For tetrachlorobisphenol A (TCBPA) and tetrabromobisphenol A (TBBPA), half-maximal inhibition (IC_{50}) of rosiglitazone binding (at 3 nM) to cells expressing PPARγ was detected at 0.7 μM and 6 μM, respectively [43]. In a surface plasmon resonance assay, an increasing signal was observed in the range of 25 to 400 μM BPA or BPB [16].

Tetrachlorobisphenol A (TCBPA) and tetrabromobisphenol A (TBBPA) bind in very similar binding modes to the LBP of PPARγ [43]. Furthermore, co-crystal structures of ERRγ, which is another member of the group of nuclear receptors, with BPA (PDB 2P7G) and BPB (PDB 6I61) are available, too. ERRγ and PPARγ have similar folds and a sequence similarity of 26% (Table S1). In addition to bisphenols, other plasticizers or their metabolites such as mono(ethylhexyl) phthalate (MEHP) and 1,2-cyclohexanedicarboxylic acid mono-4-methyloctyl ester (MINCH), have been identified as agonists of PPARγ. We previously determined the binding modes of MEHP and MINCH to PPARγ [17].

In this work, we aimed to study the binding modes of BPA and BPB on PPARγ and compare them with those of TCBPA, TCBPB, MEHP, MINCH and other PPARγ ligands (Figure S1). We co-crystallized PPARγ with BPA or BPB and determined the complex structures by X-ray crystallography. Molecular dynamics (MD) simulation was used to study the stability of the binding modes and the ligand interactions with the coordinating protein side chains.

2. Materials and Methods

2.1. Chemicals

For PPARγ LBD expression, the gene was obtained by gene synthesis from Invitrogen. Restriction enzymes and buffers used for subcloning were purchased from New England Biolabs (Frankfurt am Main, Germany). The pre-packed columns for protein isolation were ordered from Cytiva (Freiburg, Germany). Chemicals for buffers were from either Roth (Karlsruhe, Germany) or Sigma Aldrich (Taufkirchen, Germany). 4,4′-isopropylidendiphenol (Bisphenol A, 97% purity, Cas-No.: 80057, Product Number: 133027) and 2,2-Bis-(4-hydroxyphenyl)-butan, 4,4′-*sec*-butylidendiphenol (Bisphenol B, Cas-No.: 77407, Product Number: 50877, 98% purity) were purchased from Sigma Aldrich (Taufkirchen, Germany).

2.2. Protein Expression, Purification and Characterization

Human PPARγ LBD was prepared as described before [17]. Briefly, PPARγ was expressed for 4 hours, in 3 L *E. coli* culture. After harvesting and cell lysis, PPARγ was isolated via the His_6-tag, using a 5 mL HiTrap-chelating-Ni^{2+} (Cytiva) column. Thereafter, PPARγ was purified to homogeneity using size exclusion chromatography. The pure protein was concentrated to 10 mg/mL and stored flash-frozen at −80 degrees. Sample purity and quality were assessed via SDS-PAGE and dynamic light scattering.

2.3. Crystallization and Structure Determination

For co-crystallization of PPARγ with BPA and BPB, roughly 800 different crystallization conditions were screened in 96-well format plates. These experiments were performed by sitting drop vapor diffusion with a drop volume of 200 nL and a reservoir volume of 100 μL using a Mosquito Xtal3 pipetting robot (SPT Labtech, Melbourn, England). The conditions giving the best crystals were manually reproduced at a microliter scale, in a hanging drop vapor diffusion setup, mixing equal volumes of protein sample and reservoir solution (1 μL each). Prior to crystallization, the protein sample was concentrated at 10 mg/mL in PBS (phosphate-buffered saline) buffer pH 7.4 and mixed with 1 mM peptide PPARγ co-activator 1α (PGC-1α, QEAEEPSLLKKLLLAPANT) and BPA or BPB to a final concentration of 15 mM. Both ligands were prepared as 1 M stock solutions in DMSO. PPARγ × BPA crystals grew at 19 °C in 20% *w/v* PEG 3350 and 0.2 M magnesium acetate as crystallization reservoir, whereas PPARγ × BPB crystals grew in a reservoir solution composed of 20% *w/v* PEG 3350 and 0.2 M NaCl. Both PPARγ × BPA and PPARγ × BPB crystals appeared after about 5 days.

Single crystals of 0.5–1 mm length were flash-frozen in liquid nitrogen without cryoprotectant. Data were collected at beamline P14 at PETRA III, at the DESY synchrotron (Hamburg, Germany). The diffraction data were indexed, integrated and scaled with XDS (version 10 January 2022) [44] and STARANISO (version 2.3.74) [45] as implemented in ISPyB [46] at DESY. Statistics of data collection and refinement are listed in Table S2. The structure of unliganded PPARγ of PDB id 8BF1 [17] was used as a starting model for rigid-body refinement (REFMAC version 5.8.0425, [47]) of PPARγ × BPA. For the PPARγ × BPB structure, the protein chains were placed by molecular replacement using Phaser (version 2.8.3) [48] as implemented in CCP4i2 (of CCP4 version 8.0) [49] using PDB id 8BF1 as the search model. Both crystals belong to a $P2_1$ crystal form that has been observed in several PPARγ crystal structures before and which contains one PPARγ molecule in the asymmetric unit. Helix H12 is in the active ("up") conformation. The structures were further refined with jelly-body refinement. Coot (version 0.9.8.93) [50] was used for model building, and phenix (version 1.20.1_4487) [51] for the final refinement steps. The stereochemical restraints for BPA and BPB were generated with the Grade Web Server (https://grade.globalphasing.org, accessed on 16 January 2023). Similar to previous PPARγ co-crystal structures with natural ligands such as 15d-PGJ2 or nonanoic acid [21], the electron density of the ligands is weak, most likely due to partial occupancy and flexibility. The ligand occupancy was refined to 0.58 (BPA1), 0.61 (BPA2) and 0.89 (BPB).

2.4. Molecular Dynamic Simulations

The procedure used for the MD simulations is summarized in Scheme S1. The PPARγ × BPA co-crystal structure was used as the initial protein model to generate all protein complex models for the MD simulations. Three complex models were generated: (1) a model (denoted as PPARγ_BPA) with the two BPA binding modes observed in the PPARγ × BPA crystal structure, (2) a model (denoted PPARγ_BPA-TCBPA) with BPA bound in the binding mode observed for PPARγ × TCBPA crystal structure (PDB id 3osi) [43] and (3) a model (denoted PPARγ_BPB) of the BPB binding mode as observed in the PPARγ×BPB crystal structure. Starting from the PPARγ × BPA co-crystal structure, alternate side-chain conformations with the lower occupancy were removed, and crystallographic water molecules that had either no contact with other atoms isolated or too short distances to non-hydrogen atoms ($d \leq 2.4$ Å) were removed. The structure was further prepared via the structure preparation and protonate3D modules in MOE (version 2022.02) [52] to place polar hydrogen atoms. This yielded the PPARγ_BPA model with two bound BPA molecules (BPA1 in the LBP and BPA2 outside of the LBP). To create the input model for the PPARγ_BPA-TCBPA system, BPA was superimposed onto the TCBPA binding mode via real space refinement in the TCBPA electron density using the program COOT [50]. This complex was then superimposed onto the PPARγ_BPA model based on the Cα coordinates of the protein chains. The BPA-TCBPA ligand was inserted into the model, and clashing water molecules and the two BPA ligands of the PPARγ_BPA model were removed. For modeling PPARγ_BPB, the PPARγ × BPB crystal structure was superimposed onto the PPARγ_BPA model, and the BPB ligand was inserted into the protein model. Again, the two BPA molecules and clashing water molecules were deleted. Thus, for all three simulations, the same protein model was used as the starting structure.

MD simulations of the PPARγ structure bound to BPA1 and BPA2, BPA-TCBPA and BPB were performed using Amber20 [53]. The ff19SB force field was used for the protein [54], and the general Amber force field (GAFF) for the ligand [55]. The assignment of Amber atom types for the BPA and BPB ligands was performed using Antechamber (version 21.0) [56]. RESP atomic charges of the ligands were calculated using Gaussian 16 [57] and Antechamber [56]. The PPARγ–ligand complex structure was surrounded by a cubic TIP3P water box with a thickness of at least 13 Å between any protein or ligand atom and the edge of the box. Crystallographic waters that did not clash with other water molecules or protein atoms were kept in the model. The charge of the system was neutralized by adding Na^+ or Cl^- ions. SHAKE bond length constraints [58] were applied to all bonds involving hydrogen atoms. Nonbonded interactions were evaluated with a 10 Å cutoff, and electrostatic interactions were calculated by the particle-mesh Ewald method [59].

The energy of the MD system was first minimized using a two-step minimization procedure: 20,000-step minimization of water molecules and ions and 20,000-step minimization of the whole system. With protein and ligand atoms restrained to their minimized coordinates, the system was then heated from 0 K to 300 K over 75 ps in the NVT ensemble with a step size of 1 fs. After changing to the NPT ensemble, the system was equilibrated at 300 K with a reference pressure of 1 bar for 10 ns using weak positional restraints (with a force constant of 10 kcal mol^{-1} $Å^{-2}$) applied to protein backbone and ligand heteroatoms. Langevin dynamics with a collision frequency of 1 ps^{-1} and an integration time step size of 2 fs were used in these steps. Positional restraints on protein and ligand atoms were then removed stepwise in a total of 15 ns and the system was equilibrated for another 5 ns without Cartesian restraints. Production MD was conducted for 100 ns using constant pressure periodic boundary conditions and Langevin dynamics. Three independent replicas were carried out for each PPARγ × ligand complex.

The binding free energy ($\Delta G_{binding}$) of the BPA and BPB ligands as well as the per-residue contributions to $\Delta G_{binding}$ were computed using the MM/GBSA procedure of the MMPBSA.py program (version 14.0) [60]. To this end, molecular conformations were sampled at 50 ps intervals from the last 95 ns of each MD simulation, corresponding to 5700 conformations per system, and used to compute the molecular mechanics energy and solvation free energies. The single trajectory mode was applied; i.e., snapshots of the protein, ligand, and protein–ligand complex were taken from the same trajectory. The ionic strength of water was set to 150 mM.

Analysis of the MD trajectories was carried out using CPPTRAJ (version 6.4.4) [61]. This included calculation of the root-mean-square deviations (RMSDs) and root-mean-square fluctuations (RMSFs) for protein and ligand and enumeration of protein–ligand hydrogen bonds. Trajectories were visualized and molecular graphics generated using ChimeraX (version 1.6.1) [62].

2.5. Conservation Analysis

Previously, we analyzed the LBP of PPARγ for the evolutionary conservation of amino acids by comparing vertebrate ortholog sequences [17] using the ConSurf server [63,64]. For analysis of the residues involved in the binding of BPA and BPB, we used the same analysis for all residues within a 5 Å distance from the ligand.

3. Results

3.1. Binding of BPA to PPARγ

To characterize the binding mode of BPA, the PPARγ LBD was co-crystallized with BPA and PGC-1α co-activator peptide, which contains the motif LXXLL, important for the interaction with the protein. PGC-1α binds near helix H12 and keeps the structure in the active conformation. Crystals of the $P2_1$ space group with one protein in the asymmetric unit were obtained, which diffracted to 1.4 Å (Figure 1A, Table S2). The overall fold of the structure is almost identical to the apo structure of PPARγ (Figure 1B), with some minor differences in flexible loops. Residual difference electron density indicated the presence of two BPA molecules bound to the LBP (Figure 1A,C). One BPA molecule (BPA1) binds in the Ω sub-pocket of PPARγ. The binding of BPA to this sub-pocket is supported by the Ω loop (residues I262-V277). This loop is flexible and usually not resolved in PPARγ crystal structures. In the PPARγ × BPA structure, the loop could be modeled completely (Figures 1A and S2A,B). Hydrophobic interaction of BPA in the Ω sub-pocket with I262 and K263 stabilizes the loop conformation in addition to crystal packing interactions of H266 and P269. In comparison to the ligand-free structure, I262 has moved out of the ligand binding pocket, ensuring significant space for the bound BPA. Differences are also observed for residues E259 and Q283 (Figure 1F).

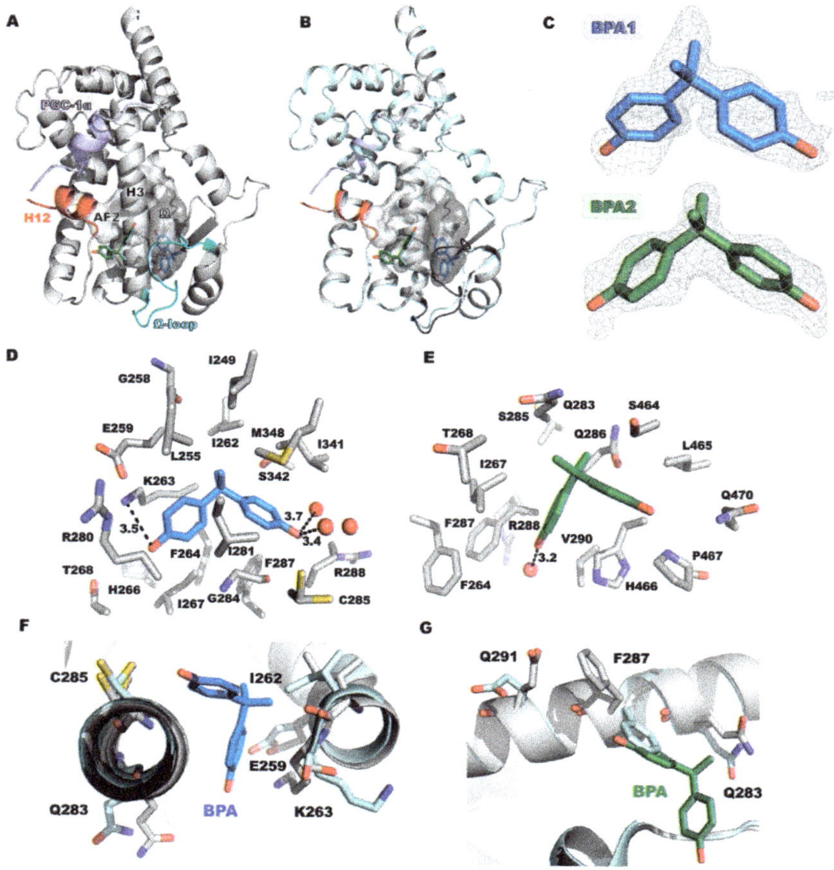

Figure 1. Binding mode of BPA to PPARγ. (**A**) PPARγ × BPA crystal structure with two bound BPA molecules. BPA1 (blue) binds in the Ω sub-pocket (grey, labeled "Ω"), and BPA2 (green) binds outside

the LBP. The peptide PGC1α is shown in blue, H12 in red and the Ω loop in cyan. (**B**) Alignment of PPARγ $\times$ BPA (grey) with the unliganded PPARγ structure (cyan, PDB: 8BF1). (**C**) Residual difference electron density (polder omit as calculated with phenix) of BPA molecules contoured at $3.2\sigma_{rmsd}$ for BPA in Ω sub-pocket (blue) and $3.0\sigma_{rmsd}$ for BPA near H12. (**D**) Interactions of BPA in the Ω pocket and near H12 (**E**). (**F**,**G**) Superpositions of the BPA-bound (white) and unliganded (light blue) PPARγ structures at the BPA binding sites in the Ω pocket (**F**) and near helix H12 (**G**). Side chains that are positioned differently in the two structures are shown.

BPA in the Ω pocket forms a hydrogen bond with K263 of the Ω loop and with two water molecules. However, the C_δ, C_ε and N_ζ atoms of this side chain have weak density indicating flexibility (Figure S2C). The remaining contacts involve essentially hydrophobic interactions (Figure 1D). The second BPA molecule (BPA2) binds at a crystal contact (Figure S3). It binds near helices H3 and H12, interacting with Q286 and F287 of helix H3. The BPA is hydrogen-bonded to a water molecule (Figure 1E). Residues Q283, F287 and E291 are oriented differently than in the apo structure to accommodate the bound BPA (Figure 1G). Both BPA molecules have weaker density compared to the surrounding protein residues (Figure S2C,D). The ligand occupancy was refined to 0.58 (BPA1) and 0.61 (BPA2) with average B factors of 43.1 Å^2 (BPA1) and 36.3 Å^2 (BPA2). The average B-value of the protein atoms within a 5 Å distance from BPA1 and BPA2 is 42.7 Å^2 and 42.3 Å^2.

3.2. Binding Mode of BPB to PPARγ

A co-crystal structure of the PPARγ LBD in complex with the co-activator peptide PGC-1 α and BPB was analyzed at a resolution of 1.7 Å. The crystals contain one molecule in the asymmetric unit, and helix H12 is in the active conformation (Figure 2A). A superposition with the unliganded structure of PPARγ shows only differences in loop regions (Figure 2B). The electron density showed the binding of one BPB molecule near C285 (Figure 2A). Weak electron density indicated partial occupancy (refined to 95%) and/or flexibility of the BPB ligand (Figures 2C and S2E). The average B factor of BPB is 38.5 Å^2, and that of the protein atoms within a 5 Å distance from BPB is 23.3 Å^2. One of the BPB phenol groups interacts with both helix 3 and helix 7 via S289 and K367, while the other phenol moiety forms weak polar interactions with a water molecule and the backbone of I281. Hydrophobic interactions further support the positioning of BPB in the hydrophobic ligand binding pocket (Figure 2D). The binding of this phenol is probably mostly supported by the π-π-stacking interaction with the peptide bond between G284 and C285. In a comparison to the apo structure of PPARγ (Figure 2E), slight rearrangements of residues surrounding BPB are observed. This includes mostly R288 and C285 (and nearby residues Q286, F282 and F363).

BPB binds similarly to the halogenated BPA derivatives tetrachlorobisphenol A and tetrabromobisphenol A (Figure 2F) [43]. The phenol group that is coordinated to K367 shares the same binding site with the halogenated analogs, whereas the other phenol group is more oriented for an interaction with the side chain of I341.

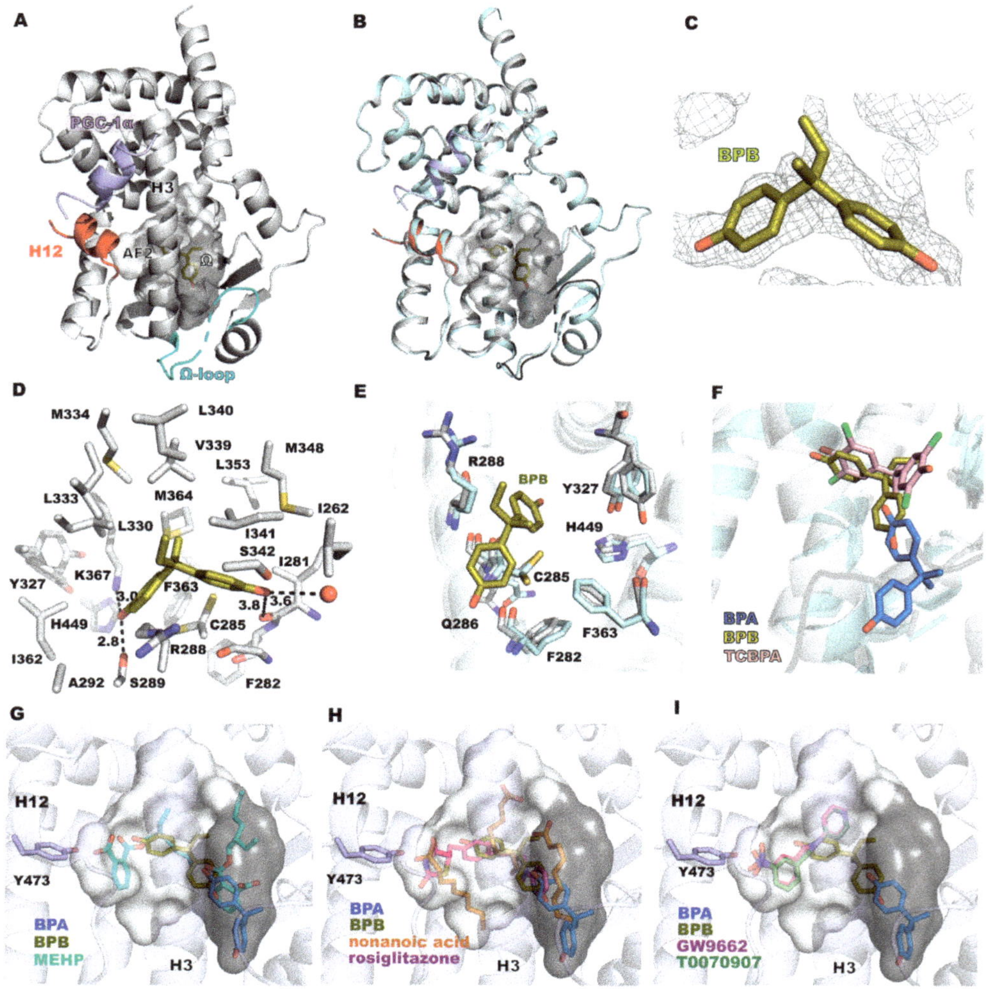

Figure 2. Crystal structure of PPARγ in complex with BPB. (**A**) Overall structure of PPARγ × BPB. Helix 12 is displayed in red, PGC-1α in blue, BPB in yellow and the Ω loop in cyan. (**B**) Superposition of PPARγ × BPB and the apo structure of PPARγ (PDB 8BF1). (**C**) (2Fo-Fc)-type electron density of BPB contoured at $0.71\sigma_{rmsd}$ level. (**D**) Interactions of BPB. Distances of polar interactions are specified in Å. (**E**) shows the residues that undergo slight changes upon BPB binding (grey), compared to the apo structure of PPARγ (cyan). (**F**) Superposition of binding modes of BPA, BPB and tetrachlorobisphenol A (TCBPA, PDB 3OSI). (**G**) Superposition of the PPARγ × BPA, PPARγ × BPB and PPARγ × MEHP (PDB 8BF2) structures. (**H**) Superposition of binding modes of BPA, BPB, nonanoic acid (PDB 4EM9) and rosiglitazone (PDB 4EMA). (**I**) Superposition of binding modes of BPA, BPB, GW9662 (PDB 6MD1) and T0070907 (PDB 6ICI). The LBP is displayed as a transparent surface with the AF2 sub-pocket in white and the Ω sub-pocket in grey.

3.3. Molecular Dynamics (MD) Simulations of the BPA and BPB Binding Modes

Since the weak density of the BPA and BPB binding modes indicates partial occupancy and alternate conformations and since one of the BPA binding sites (BPA2) is supported by a crystal contact, we decided to study the binding modes of the bisphenol molecules using MD simulations. In addition to the crystallographically observed BPA and BPB binding modes, we also modeled BPA in the binding mode observed for TCBPA [30]. TCBPA binds similarly to BPB near C285 (Figure 1F). The MD simulation of the BPA1 binding mode resulted in a stable binding mode within the 100 ns time period (Figure 3B). The phenol ring hydrogen that is bonded to K263 in the crystal structure (labeled O_2 in Figure 3) exhibited more flexibility resulting from a switch between three different hydrogen bonding interactions with the side chains of E272, Q283 and E259 (Figure 3C,D). This phenol ring is oriented towards the flexible Ω loop, which might contribute to its flexibility. Hydrophobic interactions with I341, I281 and G284 contribute mostly to the low energy of this binding mode (Figure 3A). The binding mode of BPA2, in contrast, displays much higher RMSF values (Figure 3B). The BPA2 molecule even escaped its binding site in two out of three MD replicates, in one run after ca. 60 ns and in another after ca. 70 ns. The BPA2 molecule only remained at its crystallographically observed binding site in one MD run, making intermittent hydrogen bonds with H266, S464 and I267. For the BPA1 binding mode, the total MM/GBSA energy was −31.0 ± 3.1 kcal/mol, whereas this value amounted to −20.2 ± 3.9 kcal/mol for the BPA2 binding mode.

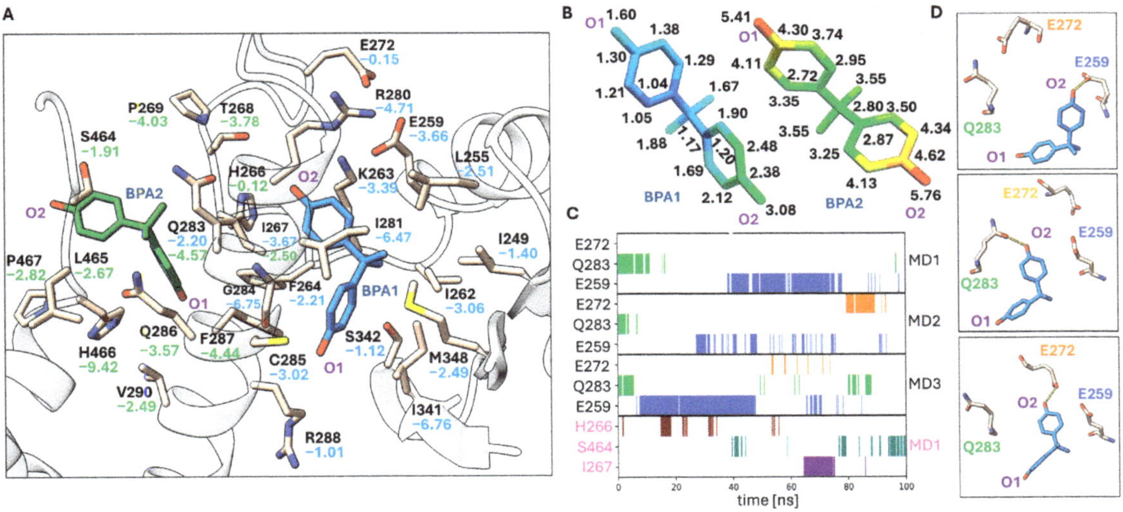

Figure 3. MD analysis of the BPA binding mode. (**A**) Depiction of the BPA binding pocket including the most important residues that take part in ligand binding based on the computed per-residue MM/GBSA energy. Residues are labeled with their respective MM/GBSA energies in kJ/mol (BPA1 blue, BPA2 green). (**B**) RMSF values mapped onto the BPA structures (color coding from blue (low values) over green to red (high values)). The RMSF values in Å are indicated next to the atoms. (**C**) Occurrence of the three most frequent hydrogen bonds of BPA observed over the time course of the three MD replicates for BPA1 and one replicate for BPA2 (pink). (**D**) Representative conformations of BPA1 in the ligand binding pocket. Hydrogen bond interactions are indicated in yellow.

MD simulations of the BPB binding mode also resulted in relatively low RMSF values indicating a stable binding mode in the 100 ns time period (Figure 4B). The hydrogen bonding interaction of the O_2 phenol ring with the side chain of S289 is observed in most frames of the MD simulations, in particular in replicas 1 and 2 (Figure 4C). Other hydrogen bonding contacts are only occasionally observed. Hydrophobic interactions with C285, R288 and I341 contribute most strongly to the stability of this binding mode (Figure 4A). A total MM/GBSA energy of −30.0 ± 2.9 kcal/mol was obtained for the BPB binding mode.

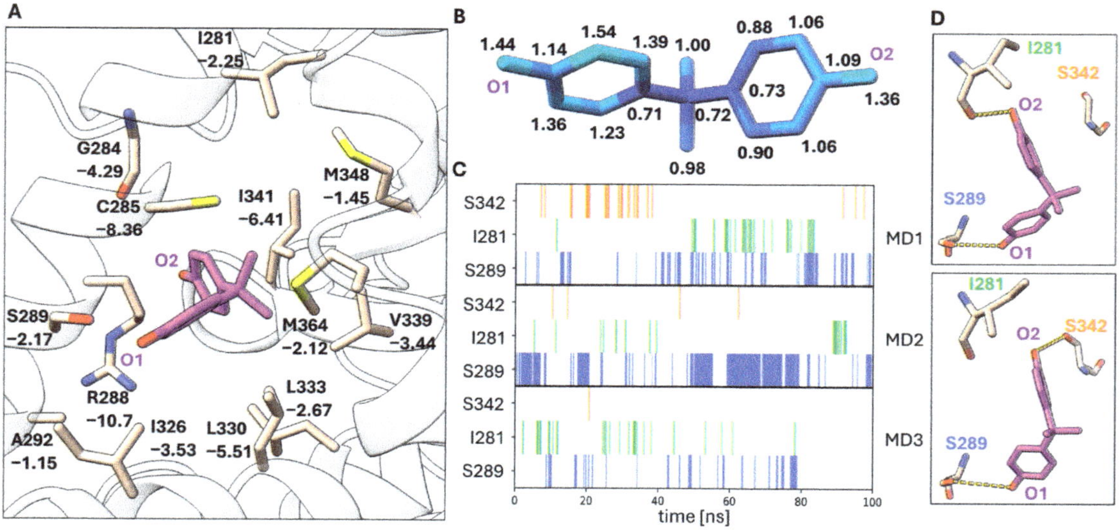

Figure 4. MD analysis of the BPB binding mode. (**A**) Depiction of the BPB binding pocket including the residues that are most important for ligand binding based on the computed per-residue MM/GBSA energies. Residues are labeled with their respective MM/GBSA energies in kJ/mol. (**B**) RMSF values are mapped onto the BPB structure (blue to green to red). The RMSF values in Å are indicated next to the atoms. (**C**) The occurrence of the three most abundant hydrogen bonds of BPB mapped over the time course of three MD replicates. (**D**) Representative conformations of BPB in the ligand binding pocket. Hydrogen bond interactions are indicated in yellow.

The binding mode of BPA observed in the PPARγ × BPA crystal structure is stabilized by interactions with the Ω loop, a flexible loop that is fixed in one conformation by crystal packing interactions in this crystal structure. Since these crystals were obtained by co-crystallization of PPARγ with BPA, it might be that ligand binding stabilized the observed conformation of the Ω loop and induced the formation of the obtained crystals. Nevertheless, the BPA binding mode is in turn also indirectly stabilized by crystal packing interactions and it might be that other low-energy binding modes exist in the LBP. For the BPA derivative TCBPA, a binding mode near C285 has been previously observed [30]. Therefore, we modeled BPA in a binding mode resembling that of TCBPA and studied its stability and interactions via MD simulations (Figure 5).

An analysis of the RMSF values shows that this binding mode is the least flexible in the MD simulations (Figure 5B). The TCBPA binding mode is similar to the BPB binding mode (Figure 2F) in particular concerning the phenol group that interacts by hydrogen bonding with S289. This hydrogen bonding interaction is also the most frequently observed hydrogen bonding interaction in the MD simulations of BPA in the TCBPA binding mode (Figure 5C,D). As for BPB, the interactions with C285, R288 and I341 contribute most strongly to this binding mode. A total MM/GBSA energy of −27.9 + −2.3 kcal/mol was obtained for the BPA-TCBPA binding mode.

3.4. Conservation of the BPA and BPB Binding Sites in Other Species

Exposure to plasticizers is not only a concern for human health, the leaking of these compounds into the environment might also affect animal wildlife. We therefore studied the conservation of the BPA and BPB binding sites in different species (Table S3) [17]. We checked residues that are within 5 Å distance from each ligand. Following the overall evolutionary distance, these residues are highly conserved in the investigated vertebrates, but differences occur in fish (Figure S4).

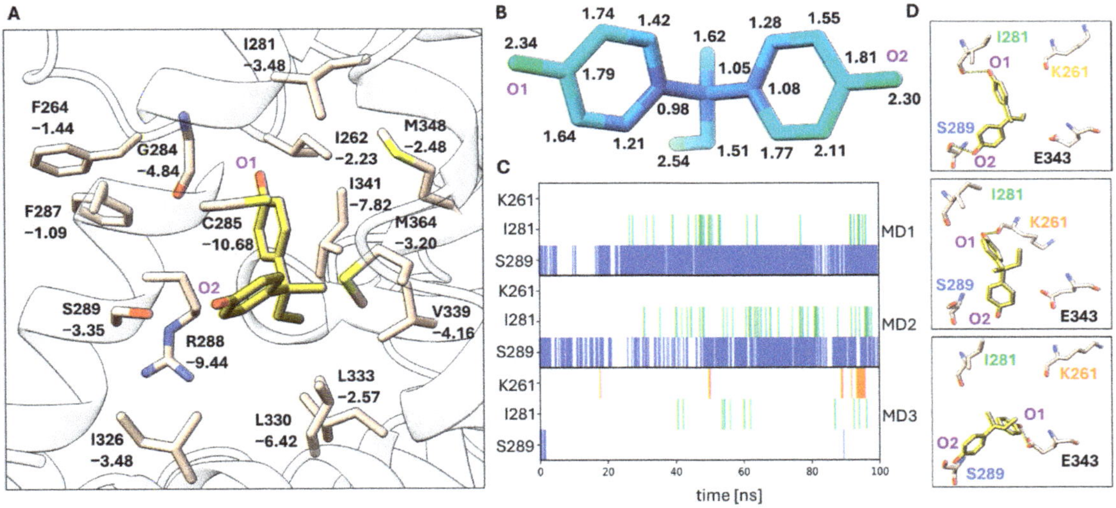

Figure 5. MD analysis of the BPA bound in the TCBPA binding pocket. (**A**) Depiction of BPA in the TCBPA binding pocket including the most important residues for ligand binding based on the computed per-residue MM/GBSA energies. Residues are labeled with their respective MM/GBSA energies in kJ/mol. (**B**) RMSF values are mapped onto the BPA structure (from blue over green to red). The RMSF values in Å are indicated next to the atoms. (**C**) The occurrence of the three most abundant hydrogen bonds of BPA mapped over the time course of three MD replicates. (**D**) Representative conformations of BPA in the TCBPA binding pocket. Hydrogen bond interactions are indicated in yellow.

4. Discussion

Crystal structure analysis of the binding modes of BPA and BPB on PPARγ showed that these ligands bind to the LBP, but fail to interact with Y473 of helix 12. PPARγ ligands which make polar interactions with Y473 can stabilize helix 12 in an active conformation, increasing the affinity for interaction of co-activator proteins binding to the AF2 surface. This is considered a key interaction that determines the activity of full agonists such as fatty acids as endogenous ligands or synthetic agonists such as rosiglitazone (Figure 2H) [13,65,66]. The plasticizer metabolites MINCH and MEHP also interact with Y473 (Figure 2G), and their agonistic activity on PPARγ signaling probably depends on this interaction [17]. Binding of BPA was observed at two sites: (1) in the Ω pocket and (2) outside the LBP (Figure 1). The latter binding mode is supported by direct crystal contacts (Figure S3), and MD simulations indicated a low stability of BPA bound at this site in the absence of crystal contacts (Figure 3B). The magnitude of the calculated total MM/GBSA energy is -20.2 ± 3.9 kcal/mol, significantly smaller in magnitude than that of the other binding modes. These findings indicate that the BPA2 binding mode is not physiologically relevant.

The BPA binding mode in the LBP is supported by the flexible Ω loop, which is stabilized in the PPARγ × BPA crystal structure by crystal packing contacts and interaction with the ligand. In the MD simulations, this binding mode was quite stable, in particular the interaction with I341 and G284, whereas the other phenol group that interacts with the Ω loop was more flexible. We also showed that BPA modeled in the binding mode observed for TCBPA [43] has high stability in the MD simulations, as indicated by its low RMSF values. While the binding energies of the BPA1 and BPA-TCBPA binding modes are overall similar, the former is slightly preferred energetically. This is in agreement with the observation of only the BPA1 binding mode in the PPARγ × BPA crystal structure. The two binding modes are too closely spaced to be occupied simultaneously. It was previously shown that the halogenated BPA derivatives TCBPA and TBBPA activate PPARγ at micromolar concentrations, whereas no significant effect was observed for BPA up to 10 μM concentration in this assay [43]. BPB bound to PPARγ in a similar binding mode to that of TCBPA (Figure 2F), and the MD simulations indicated this binding mode has good stability (Figure 1B).

Although the binding modes and interactions of agonists, antagonists and inverse agonists with PPARγ have been characterized in many studies, it is difficult to predict the influence of the BPA and BPB binding modes observed in this study on PPARγ transactivation activity. This is due to the fact that the conformational changes involved in PPARγ activation of full and partial agonists as well as antagonists or inverse agonists are not well understood. It was previously shown that partial agonists, which fail to stabilize the active helix 12 conformation because of a lack of Y473 interaction, instead stabilize the β-sheet region and helix 3, which was indicated by the kinetics of amide H/D exchange [65]. I341, S342 and M348 are part of the β-sheet region of PPARγ, and the partial agonism of MRL-24 was attributed to these interactions [65]. Interaction with I341 has a large contribution to all three BPA and BPB binding modes, as shown by our per-residue MM/GBSA energy analysis (Figures 3A, 4A and 5A). These interactions and the similarity of the BPB binding mode to that of TCBPA and TBBPA suggest that BPA and BPB may exert partial agonistic activity through their binding modes. On the other hand, it has also been demonstrated that small structural changes in ligand structure such as a methine-to-nitrogen substitution can change an antagonist (GW9662) into an inverse agonist (T0070907) [67]. These compounds covalently attach to C285 and bind near the BPA or BPB binding sites (Figure 2I). Differential interactions of these compounds with R288 have been implicated in the observed differences in PPARγ activity. Interactions with C285 and R288 also contribute strongly to the binding modes of BPB (Figure 4A) and of BPA in the TCBPA site (Figure 5A) in the MD simulations. Differential stabilization of PPARγ conformational states for interaction with co-repressors and co-activators is the basis for the differences between GW9662 and T0070907 in PPARγ transactivation activity [67]. In the present work, we could characterize the interactions of BPA and BPB with the active state of PPARγ stabilized by co-crystallization with a co-activator peptide. This was important in experimental work to obtain well-diffracting crystals. Elucidation of the binding modes of bisphenols paves the way for a deeper understanding of how these compounds influence PPARγ signaling. Sensitive binding assays combined with biophysical studies in solution may clarify these open questions.

Supplementary Materials: The following supporting information can be downloaded at https://www.mdpi.com/article/10.3390/biom14060640/s1: Table S1: Alignment of human PPARγ and human ERRγ; Table S2: Diffraction data and refinement statistics; Table S3: Sequence IDs used for the alignment analyses shown in Figure S4; Scheme S1: Overview of the methodology used for the MD simulations. Figure S1: Schematic representation of BPA, BPB, TCBPA, MEHP, MINCH, endogenous (nonanoic acid) and synthetic (rosiglitazone, GW9662, T007090) PPARγ ligands; Figure S2: Electron density of the Ω loop and the ligand binding sites; Figure S3: BPA2 binds at a crystal contact in the PPARγ × BPA structure; Figure S4: Amino acid sequence alignment of selected species for mammalian orthologs (*human, dog, pig, mouse, rat,* and *whale*), for birds (*chick*), for fish (*zebrafish* and *kryma*), and for reptiles (*lizard*).

Author Contributions: Conceptualization, A.U., G.K. and N.S.; methodology, A.U., I.K.S. and G.K.; validation, A.U., I.K.S., G.K. and N.S.; formal analysis, A.U. and I.K.S.; investigation, A.U. and I.K.S.; resources, G.K. and N.S.; writing—original draft preparation, A.U. and I.K.S.; writing—review and editing, N.S. and G.K.; visualization, A.U. and I.K.S.; supervision, N.S. and G.K.; project administration, N.S. and G.K.; funding acquisition, G.K. and N.S. The crystal structure analysis was mostly carried out by A.U. and N.S., and the molecular dynamics analysis by I.K.S. and G.K. All authors have read and agreed to the published version of the manuscript.

Funding: This research was funded by the Deutsche Forschungsgemeinschaft (CRC1052, project Z6 to N.S. and G.K., project number 209933838). I.K.S. and G.K. further acknowledge DFG funding through project 421152132, CRC1423, subproject C07. Furthermore, G.K. and I.K.S. acknowledge financial support by the Federal Ministry of Education and Research of Germany and by Sächsische Staatsministerium für Wissenschaft, Kultur und Tourismus in the programme Center of Excellence for AI-research "Center for Scalable Data Analytics and Artificial Intelligence Dresden/Leipzig", project identification number: ScaDS.AI. Publication was supported by the Open Access Publishing Fund of Leipzig University and CRC1052.

Institutional Review Board Statement: Not applicable.

Informed Consent Statement: Not applicable.

Data Availability Statement: The coordinates and structure factors of the crystal structure analysis have been submitted to the protein data bank with the identifiers 9F7W (PPARγ × BPA) and 9F7X (PPARγ × BPB).

Acknowledgments: We thank the EMBL beamlines of the DESY synchrotron in Hamburg and the MX Laboratory at the Helmholtz Zentrum Berlin (BESSY II) for synchrotron beamtime. We gratefully acknowledge the Leipzig University Computing Center and the NHR Center of TU Dresden (projects *p_sdslnmrdata* and *p_peptide*) for providing the computational resources for this work. We acknowledge the work of Christian Zocher as part of a student practical.

Conflicts of Interest: The authors declare no conflicts of interest.

References

1. World Health Organization (WHO). Obesity and Overweight. 2024. Available online: https://www.who.int/news-room/fact-sheets/detail/obesity-and-overweight (accessed on 3 May 2024).
2. Ribeiro, C.M.; Beserra, B.T.S.; Silva, N.G.; Lima, C.L.; Rocha, P.R.S.; Coelho, M.S.; Neves, F.d.A.R.; Amato, A.A. Exposure to endocrine-disrupting chemicals and anthropometric measures of obesity: A systematic review and meta-analysis. *BMJ Open* **2020**, *10*, e033509. [CrossRef] [PubMed]
3. Darbre, P.D. Endocrine Disruptors and Obesity. *Curr. Obes. Rep.* **2017**, *6*, 18–27. [CrossRef]
4. Miyawaki, J.; Sakayama, K.; Kato, H.; Yamamoto, H.; Masuno, H. Perinatal and postnatal exposure to bisphenol a increases adipose tissue mass and serum cholesterol level in mice. *J. Atheroscler. Thromb.* **2007**, *14*, 245–252. [CrossRef]
5. Masuno, H.; Iwanami, J.; Kidani, T.; Sakayama, K.; Honda, K. Bisphenol a accelerates terminal differentiation of 3T3-L1 cells into adipocytes through the phosphatidylinositol 3-kinase pathway. *Toxicol. Sci.* **2005**, *84*, 319–327. [CrossRef]
6. Hugo, E.R.; Brandebourg, T.D.; Woo, J.G.; Loftus, J.; Alexander, J.W.; Ben-Jonathan, N. Bisphenol A at environmentally relevant doses inhibits adiponectin release from human adipose tissue explants and adipocytes. *Environ. Health Perspect.* **2008**, *116*, 1642–1647. [CrossRef] [PubMed]
7. Ohlstein, J.F.; Strong, A.L.; McLachlan, J.A.; Gimble, J.M.; Burow, M.E.; Bunnell, B.A. Bisphenol A enhances adipogenic differentiation of human adipose stromal/stem cells. *J. Mol. Endocrinol.* **2014**, *53*, 345–353. [CrossRef] [PubMed]
8. Ariemma, F.; D'Esposito, V.; Liguoro, D.; Oriente, F.; Cabaro, S.; Liotti, A.; Cimmino, I.; Longo, M.; Beguinot, F.; Formisano, P.; et al. Low-Dose Bisphenol-A Impairs Adipogenesis and Generates Dysfunctional 3T3-L1 Adipocytes. *PLoS ONE* **2016**, *11*, e0150762. [CrossRef]
9. Chawla, A. Peroxisome proliferator-activated receptor (PPAR) gamma: Adipose- predominant expression and induction early in adipocyte differentiation. *Endocrinology* **1994**, *135*, 798–800. [CrossRef]
10. Tontonoz, P.; Hu, E.; Spiegelman, B.M. Stimulation of adipogenesis in fibroblasts by PPAR gamma 2, a lipid-activated transcription factor. *Cell* **1994**, *79*, 1147–1156. [CrossRef]
11. Ballav, S.; Biswas, B.; Sahu, V.K.; Ranjan, A.; Basu, S. PPAR-γ Partial Agonists in Disease-Fate Decision with Special Reference to Cancer. *Cells* **2022**, *11*, 3215. [CrossRef]
12. Kliewer, S.A.; Lenhard, J.M.; Willson, T.M.; Patel, I.; Morris, D.C.; Lehmann, J.M. A prostaglandin J2 metabolite binds peroxisome proliferator-activated receptor gamma and promotes adipocyte differentiation. *Cell* **1995**, *83*, 813–819. [CrossRef]
13. Nolte, R.T.; Wisely, G.B.; Westin, S.; Cobb, J.E.; Lambert, M.H.; Kurokawa, R.; Rosenfeld, M.G.; Willson, T.M.; Glass, C.K.; Milburn, M.V. Ligand binding and co-activator assembly of the peroxisome proliferator-activated receptor-gamma. *Nature* **1998**, *395*, 137–143. [CrossRef]
14. Waku, T.; Shiraki, T.; Oyama, T.; Morikawa, K. Atomic structure of mutant PPARgamma LBD complexed with 15d-PGJ2: Novel modulation mechanism of PPARgamma/RXRalpha function by covalently bound ligands. *FEBS Lett.* **2009**, *583*, 320–324. [CrossRef]
15. Shang, J.; Brust, R.; Griffin, P.R.; Kamenecka, T.M.; Kojetin, D.J. Quantitative structural assessment of graded receptor agonism. *Proc. Natl. Acad. Sci. USA* **2019**, *116*, 22179–22188. [CrossRef]
16. Schaffert, A.; Krieg, L.; Weiner, J.; Schlichting, R.; Ueberham, E.; Karkossa, I.; Bauer, M.; Landgraf, K.; Junge, K.M.; Wabitsch, M.; et al. Alternatives for the worse: Molecular insights into adverse effects of bisphenol a and substitutes during human adipocyte differentiation. *Environ. Int.* **2021**, *156*, 106730. [CrossRef]
17. Useini, A.; Engelberger, F.; Künze, G.; Sträter, N. Structural basis of the activation of PPARγ by the plasticizer metabolites MEHP and MINCH. *Environ. Int.* **2023**, *173*, 107822. [CrossRef]

18. Genest, D.; Garnier, N.; Arrault, A.; Marot, C.; Morin-Allory, L.; Genest, M. Ligand-escape pathways from the ligand-binding domain of PPARgamma receptor as probed by molecular dynamics simulations. *Eur. Biophys. J. EBJ* **2008**, *37*, 369–379. [CrossRef]
19. Lee, M.A.; Tan, L.; Yang, H.; Im, Y.-G.; Im, Y.J. Structures of PPARγ complexed with lobeglitazone and pioglitazone reveal key determinants for the recognition of antidiabetic drugs. *Sci. Rep.* **2017**, *7*, 16837. [CrossRef]
20. Jeninga, E.H.; Gurnell, M.; Kalkhoven, E. Functional implications of genetic variation in human PPARgamma. *Trends Endocrinol. Metab. TEM* **2009**, *20*, 380–387. [CrossRef]
21. Liberato, M.V.; Nascimento, A.S.; Ayers, S.D.; Lin, J.Z.; Cvoro, A.; Silveira, R.L.; Martínez, L.; Souza, P.C.T.; Saidemberg, D.; Deng, T.; et al. Medium chain fatty acids are selective peroxisome proliferator activated receptor (PPAR) γ activators and pan-PPAR partial agonists. *PLoS ONE* **2012**, *7*, e36297. [CrossRef]
22. Bousoumah, R.; Leso, V.; Iavicoli, I.; Huuskonen, P.; Viegas, S.; Porras, S.P.; Santonen, T.; Frery, N.; Robert, A.; Ndaw, S. Biomonitoring of occupational exposure to bisphenol A, bisphenol S and bisphenol F: A systematic review. *Sci. Total Environ.* **2021**, *783*, 146905. [CrossRef]
23. Motarjemi, Y. *Encyclopedia of Food Safety*; Elsevier Science: Burlington, MA, USA, 2014.
24. ChemAnalyst. Bisphenol A Market Analysis. 2023. Available online: https://www.chemanalyst.com/industry-report/bisphenol-a-market-57 (accessed on 3 May 2024).
25. Rosenfeld, C.S. Neuroendocrine disruption in animal models due to exposure to bisphenol A analogues. *Front. Neuroendocrinol.* **2017**, *47*, 123–133. [CrossRef]
26. Wells, E.M.; Bisphenol, A. *Encyclopedia of Environmental Health*; Elsevier: Amsterdam, The Netherlands, 2019; pp. 424–428.
27. Lambré, C.; Barat Baviera, J.M.; Bolognesi, C.; Chesson, A.; Cocconcelli, P.S.; Crebelli, R.; Gott, D.M.; Grob, K.; Lampi, E.; Mengelers, M.; et al. Re-evaluation of the risks to public health related to the presence of bisphenol A (BPA) in foodstuffs. *EFSA J. Eur. Food Saf. Auth.* **2023**, *21*, e06857.
28. Gu, J.; Guo, M.; Yin, X.; Huang, C.; Qian, L.; Zhou, L.; Wang, Z.; Wang, L.; Shi, L.; Ji, G. A systematic comparison of neurotoxicity of bisphenol A and its derivatives in zebrafish. *Sci. Total Environ.* **2022**, *805*, 150210. [CrossRef]
29. Chang, B.-V.; Liu, J.-H.; Liao, C.-S. Aerobic degradation of bisphenol-A and its derivatives in river sediment. *Environ. Technol.* **2014**, *35*, 416–424. [CrossRef]
30. Liu, Y.; Zhang, S.; Song, N.; Guo, R.; Chen, M.; Mai, D.; Yan, Z.; Han, Z.; Chen, J. Occurrence, distribution and sources of bisphenol analogues in a shallow Chinese freshwater lake (Taihu Lake): Implications for ecological and human health risk. *Sci. Total Environ.* **2017**, *599–600*, 1090–1098. [CrossRef]
31. Cunha, S.C.; Fernandes, J.O. Quantification of free and total bisphenol A and bisphenol B in human urine by dispersive liquid-liquid microextraction (DLLME) and heart-cutting multidimensional gas chromatography-mass spectrometry (MD-GC/MS). *Talanta* **2010**, *83*, 117–125. [CrossRef]
32. Cobellis, L.; Colacurci, N.; Trabucco, E.; Carpentiero, C.; Grumetto, L. Measurement of bisphenol A and bisphenol B levels in human blood sera from healthy and endometriotic women. *Biomed. Chromatogr. BMC* **2009**, *23*, 1186–1190. [CrossRef]
33. Yan, Z.; Liu, Y.; Yan, K.; Wu, S.; Han, Z.; Guo, R.; Chen, M.; Yang, Q.; Zhang, S.; Chen, J. Bisphenol analogues in surface water and sediment from the shallow Chinese freshwater lakes: Occurrence, distribution, source apportionment, and ecological and human health risk. *Chemosphere* **2017**, *184*, 318–328. [CrossRef]
34. Liao, C.; Kannan, K. Concentrations and profiles of bisphenol A and other bisphenol analogues in foodstuffs from the United States and their implications for human exposure. *J. Agric. Food Chem.* **2013**, *61*, 4655–4662. [CrossRef]
35. Cunha, S.C.; Almeida, C.; Mendes, E.; Fernandes, J.O. Simultaneous determination of bisphenol A and bisphenol B in beverages and powdered infant formula by dispersive liquid-liquid micro-extraction and heart-cutting multidimensional gas chromatography-mass spectrometry. *Food Addit. Contam. Part A Chem. Anal. Control Expo. Risk Assess.* **2011**, *28*, 513–526. [CrossRef]
36. Wang, F.; Ma, X.; Sun, Q.; Zhang, Y.; Liu, Y.; Gu, J.; Wang, L. Bisphenol B induces developmental toxicity in zebrafish via oxidative stress. *Environ. Sci. Pollut. Res. Int.* **2023**. [CrossRef]
37. Serra, H.; Beausoleil, C.; Habert, R.; Minier, C.; Picard-Hagen, N.; Michel, C. Evidence for Bisphenol B Endocrine Properties: Scientific and Regulatory Perspectives. *Environ. Health Perspect.* **2019**, *127*, 106001. [CrossRef]
38. Wu, X.; Yang, X.; Tian, Y.; Xu, P.; Yue, H.; Sang, N. Bisphenol B and bisphenol AF exposure enhances uterine diseases risks in mouse. *Environ. Int.* **2023**, *173*, 107858. [CrossRef]
39. Cao, L.-Y.; Ren, X.-M.; Li, C.-H.; Zhang, J.; Qin, W.-P.; Yang, Y.; Wan, B.; Guo, L.-H. Bisphenol AF and Bisphenol B Exert Higher Estrogenic Effects than Bisphenol A via G Protein-Coupled Estrogen Receptor Pathway. *Environ. Sci. Technol.* **2017**, *51*, 11423–11430. [CrossRef]
40. Alharbi, H.F.; Algonaiman, R.; Alduwayghiri, R.; Aljutaily, T.; Algheshairy, R.M.; Almutairi, A.S.; Alharbi, R.M.; Alfurayh, L.A.; Alshahwan, A.A.; Alsadun, A.F.; et al. Exposure to Bisphenol A Substitutes, Bisphenol S and Bisphenol F, and Its Association with Developing Obesity and Diabetes Mellitus: A Narrative Review. *Int. J. Environ. Res. Public Health* **2022**, *19*, 15918. [CrossRef]
41. Ahmed, S.; Atlas, E. Bisphenol S- and bisphenol A-induced adipogenesis of murine preadipocytes occurs through direct peroxisome proliferator-activated receptor gamma activation. *Int. J. Obes.* **2016**, *40*, 1566–1573. [CrossRef]
42. Chamorro-García, R.; Kirchner, S.; Li, X.; Janesick, A.; Casey, S.C.; Chow, C.; Blumberg, B. Bisphenol A diglycidyl ether induces adipogenic differentiation of multipotent stromal stem cells through a peroxisome proliferator-activated receptor gamma-independent mechanism. *Environ. Health Perspect.* **2012**, *120*, 984–989. [CrossRef]

43. Riu, A.; Grimaldi, M.; Le Maire, A.; Bey, G.; Phillips, K.; Boulahtouf, A.; Perdu, E.; Zalko, D.; Bourguet, W.; Balaguer, P. Peroxisome proliferator-activated receptor γ is a target for halogenated analogs of bisphenol A. *Environ. Health Perspect.* **2011**, *119*, 1227–1232. [CrossRef]
44. Kabsch, W. XDS. *Acta Crystallogr. Sect. D Biol. Crystallogr.* **2010**, *66*, 125–132. [CrossRef]
45. Vonrhein, C.; Tickle, I.J.; Flensburg, C.; Keller, P.; Paciorek, W.; Sharff, A.; Bricogne, G. Advances in automated data analysis and processing within autoPROC, combined with improved characterisation, mitigation and visualisation of the anisotropy of diffraction limits using STARANISO. *Acta Crystallogr. Sect. A* **2018**, *74*, a360. [CrossRef]
46. Delagenière, S.; Brenchereau, P.; Launer, L.; Ashton, A.W.; Leal, R.; Veyrier, S.; Gabadinho, J.; Gordon, E.J.; Jones, S.D.; Levik, K.E.; et al. ISPyB: An information management system for synchrotron macromolecular crystallography. *Bioinformatics* **2011**, *27*, 3186–3192. [CrossRef] [PubMed]
47. Vagin, A.A.; Steiner, R.A.; Lebedev, A.A.; Potterton, L.; McNicholas, S.; Long, F.; Murshudov, G.N. REFMAC5 dictionary: Organization of prior chemical knowledge and guidelines for its use. *Acta Crystallogr. Sect. D Biol. Crystallogr.* **2004**, *60*, 2184–2195. [CrossRef] [PubMed]
48. McCoy, A.J.; Grosse-Kunstleve, R.W.; Adams, P.D.; Winn, M.D.; Storoni, L.C.; Read, R.J. Phaser crystallographic software. *J. Appl. Crystallogr.* **2007**, *40*, 658–674. [CrossRef]
49. The CCP4 suite: Programs for protein crystallography. *Acta Crystallogr. Sect. D Biol. Crystallogr.* **1994**, *50*, 760–763. [CrossRef] [PubMed]
50. Emsley, P.; Lohkamp, B.; Scott, W.G.; Cowtan, K. Features and development of Coot. *Acta Crystallogr. Sect. D Biol. Crystallogr.* **2010**, *66*, 486–501. [CrossRef]
51. Liebschner, D.; Afonine, P.V.; Baker, M.L.; Bunkóczi, G.; Chen, V.B.; Croll, T.I.; Hintze, B.; Hung, L.W.; Jain, S.; McCoy, A.J.; et al. Macromolecular structure determination using X-rays, neutrons and electrons: Recent developments in Phenix. *Acta Crystallogr. Sect. D Struct. Biol.* **2019**, *75*, 861–877. [CrossRef]
52. *Molecular Operating Environment (MOE)*; Chemical Computing Group ULC: Montreal, QC, Canada, 2024.
53. Case, D.A.; Aktulga, H.M.; Belfon, K.; Cerutti, D.S.; Cisneros, G.A.; Cruzeiro, V.W.D.; Forouzesh, N.; Giese, T.J.; Götz, A.W.; Gohlke, H.; et al. AmberTools. *J. Chem. Inf. Model.* **2023**, *63*, 6183–6191. [CrossRef]
54. Tian, C.; Kasavajhala, K.; Belfon, K.A.A.; Raguette, L.; Huang, H.; Migues, A.N.; Bickel, J.; Wang, Y.; Pincay, J.; Wu, Q.; et al. ff19SB: Amino-Acid-Specific Protein Backbone Parameters Trained against Quantum Mechanics Energy Surfaces in Solution. *J. Chem. Theory Comput.* **2020**, *16*, 528–552. [CrossRef]
55. Wang, J.; Wolf, R.M.; Caldwell, J.W.; Kollman, P.A.; Case, D.A. Development and testing of a general amber force field. *J. Comput. Chem.* **2004**, *25*, 1157–1174. [CrossRef]
56. Wang, J.; Wang, W.; Kollman, P.A.; Case, D.A. Automatic atom type and bond type perception in molecular mechanical calculations. *J. Mol. Graph. Model.* **2006**, *25*, 247–260. [CrossRef] [PubMed]
57. Frisch, M.J.; Trucks, G.W.; Schlegel, H.B.; Scuseria, G.E.; Robb, M.A.; Cheeseman, J.R.; Scalmani, G.; Barone, V.; Petersson, G.A.; Nakatsuji, H.; et al. *Gaussian 16*; Gaussian Inc.: Wallingford, CT, USA, 2016.
58. Ryckaert, J.-P.; Ciccotti, G.; Berendsen, H.J. Numerical integration of the cartesian equations of motion of a system with constraints: Molecular dynamics of n-alkanes. *J. Comput. Phys.* **1977**, *23*, 327–341. [CrossRef]
59. Darden, T.; York, D.; Pedersen, L. Particle mesh Ewald: An N ·log(N) method for Ewald sums in large systems. *J. Chem. Phys.* **1993**, *98*, 10089–10092. [CrossRef]
60. Miller, B.R.; McGee, T.D.; Swails, J.M.; Homeyer, N.; Gohlke, H.; Roitberg, A.E. MMPBSA.py: An Efficient Program for End-State Free Energy Calculations. *J. Chem. Theory Comput.* **2012**, *8*, 3314–3321. [CrossRef]
61. Roe, D.R.; Cheatham, T.E. PTRAJ and CPPTRAJ: Software for Processing and Analysis of Molecular Dynamics Trajectory Data. *J. Chem. Theory Comput.* **2013**, *9*, 3084–3095. [CrossRef]
62. Pettersen, E.F.; Goddard, T.D.; Huang, C.C.; Meng, E.C.; Couch, G.S.; Croll, T.I.; Morris, J.H.; Ferrin, T.E. UCSF ChimeraX: Structure visualization for researchers, educators, and developers. *Protein Sci. A Publ. Protein Soc.* **2021**, *30*, 70–82. [CrossRef] [PubMed]
63. Glaser, F.; Pupko, T.; Paz, I.; Bell, R.E.; Bechor-Shental, D.; Martz, E.; Ben-Tal, N. ConSurf: Identification of functional regions in proteins by surface-mapping of phylogenetic information. *Bioinformatics* **2003**, *19*, 163–164. [CrossRef] [PubMed]
64. Landau, M.; Mayrose, I.; Rosenberg, Y.; Glaser, F.; Martz, E.; Pupko, T.; Ben-Tal, N. ConSurf 2005: The projection of evolutionary conservation scores of residues on protein structures. *Nucleic Acids Res.* **2005**, *33*, W299–W302. [CrossRef] [PubMed]
65. Bruning, J.B.; Chalmers, M.J.; Prasad, S.; Busby, S.A.; Kamenecka, T.M.; He, Y.; Nettles, K.W.; Griffin, P.R. Partial agonists activate PPARgamma using a helix 12 independent mechanism. *Structure* **2007**, *15*, 1258–1271. [CrossRef]
66. Hughes, T.S.; Chalmers, M.J.; Novick, S.; Kuruvilla, D.S.; Chang, M.R.; Kamenecka, T.M.; Rance, M.; Johnson, B.A.; Burris, T.P.; Griffin, P.R.; et al. Ligand and receptor dynamics contribute to the mechanism of graded PPARγ agonism. *Structure* **2012**, *20*, 139–150. [CrossRef]
67. Brust, R.; Shang, J.; Fuhrmann, J.; Mosure, S.A.; Bass, J.; Cano, A.; Heidari, Z.; Chrisman, I.M.; Nemetchek, M.D.; Blayo, A.-L.; et al. A structural mechanism for directing corepressor-selective inverse agonism of PPARγ. *Nat. Commun.* **2018**, *9*, 4687. [CrossRef] [PubMed]

biomolecules

MDPI

Review

Cladosporols and PPARγ: Same Gun, Same Bullet, More Targets

Roberta Rapuano [1], Antonella Mercuri [1], Sabrina Dallavalle [2], Salvatore Moricca [3], Antonio Lavecchia [4,*] and Angelo Lupo [1,*]

1 Dipartimento di Scienze e Tecnologie, Università del Sannio, Via dei Mulini, 82100 Benevento, Italy; rrapuano@unisannio.it (R.R.); a.mercuri1@studenti.unisannio.it (A.M.)

2 Dipartimento di Scienze per gli Alimenti, la Nutrizione e l'Ambiente, Università degli Studi di Milano, Via Celoria 2, 20133 Milano, Italy; sabrina.dallavalle@unimi.it

3 Dipartimento di Scienze e Tecnologie Agrarie, Alimentari, Ambientali e Forestali (DAGRI), Università degli Studi di Firenze, Piazzale delle Cascine 28, 50144 Firenze, Italy; salvatore.moricca@unifi.it

4 Dipartimento di Farmacia "Drug Discovery Laboratory", Università di Napoli "Federico II", Via D. Montesano 49, 80131 Napoli, Italy

* Correspondence: antonio.lavecchia@unina.it (A.L.); lupo@unisannio.it (A.L.); Tel.: +39-0816-78613 (A.L.); +39-0824-305103 (A.L.)

Abstract: Several natural compounds have been found to act as PPARγ agonists, thus regulating numerous biological processes, including the metabolism of carbohydrates and lipids, cell proliferation and differentiation, angiogenesis, and inflammation. Recently, Cladosporols, secondary metabolites purified from the fungus *Cladosporium tenuissimum*, have been demonstrated to display an efficient ability to control cell proliferation in human colorectal and prostate cancer cells through a PPARγ-mediated modulation of gene expression. In addition, Cladosporols exhibited a strong anti-adipogenetic activity in 3T3-L1 murine preadipocytes, preventing their in vitro differentiation into mature adipocytes. These data interestingly point out that the interaction between Cladosporols and PPARγ, in the milieu of different cells or tissues, might generate a wide range of beneficial effects for the entire organism affected by diabetes, obesity, inflammation, and cancer. This review explores the molecular mechanisms by which the Cladosporol/PPARγ complex may simultaneously interfere with a dysregulated lipid metabolism and cancer promotion and progression, highlighting the potential therapeutic benefits of Cladosporols for human health.

Keywords: Cladosporols; PPARγ agonists; proliferation; migration; apoptosis; lipogenesis

1. PPARγ: Gene Structure, mRNAs, Proteins, and Related Functions

Peroxisome Proliferator-Activated Receptors (PPARs) belong to nuclear receptor class 2 and usually remain inactive in the nucleus until they get in touch with the ligand to which they bind. Three members of its own family, i.e., PPARα (NR1C1), PPARβ/δ NR1C2), and PPARγ (NR1C3), are included [1,2]. Similar to other nuclear receptors, the protein sequence of PPARs includes three different structural and functional domains: the N-terminal transactivation domain (AF1), the DNA-binding domain (DBD) and the C-terminal ligand binding domain, which contains the ligand-dependent transactivation function (AF2). The interaction of the ligand with the LBD (ligand binding domain) of the PPARs generates, in its structure, a conformational change that, in turn, is responsible for heterodimerization with the Retinoid X receptor (RXR) and, sequentially, the migration of the activated receptor in the nucleus and the binding to the PPAR response element (PPRE) on the DNA. This binding stimulates the transcription of various gene targets and, consequently, the regulation of various biological processes [3]. Although PPARs show structural analogies, they display distinctive functions due to different tissue distributions, a differential response to specific ligands, and, finally, diverse biochemical features. PPARα, which is present in the liver, heart, and skeletal muscle, stimulates the fatty acid transport and oxidation and, in such a way, regulates lipid metabolism [4]. PPARβ/δ, which is expressed in the skeletal muscle, adipocytes, macrophages, skin, and gastrointestinal tissue, regulates lipid metabolism, but it is also involved in keratinocyte differentiation and macrophage-derived inflammation [5]. Finally, PPARγ, which is mainly expressed in the adipose tissue, immune cells, and colon, is the master gene of adipocyte differentiation and plays a significant role in controlling insulin resistance and energy homeostasis [6]. The complexity of the PPAR family is, then, further increased as a result of the demonstration that the four distinctive mRNAs originated in humans from the same PPARγ gene through the use of different promoters and alternative splicing. Among these, PPARγ1, PPARγ3, and PPARγ4 mRNAs produce the same protein product, namely, PPARγ1 (477 aa), while PPARγ2 mRNA generates the PPARγ2 isoform containing 28 additional amino acids at the N-terminus (505 aa) (Figure 1) [6]. While the PPARγ1 isoform is expressed in various tissues including the skeletal muscle; pancreas; heart; colon; white (WAT) and brown adipose tissue (BAT); immune cells, such as macrophages; and kidney, PPARγ2 is strongly restricted to the adipose tissue, although a high-fat diet may also induce it in other tissues [7,8]. Due to the characteristic tissue distribution, PPARγ is considered to be the master regulatory gene of adipocyte differentiation, playing a fundamental role in fatty acid synthesis and the transport and formation of triglyceride depots [9]. Indeed, the lack of adipose tissue in a PPARγ-null mouse definitively confirmed the crucial function of PPARγ in this tissue [10]. Additionally, PPARγ expression was also verified within the WAT perivascular niche in an adipocyte ancestor population, and this discovery indicated that it may display a crucial function in adipocyte self renewal [11]. Finally, maintaining a mature adipocyte phenotype has been shown to require the presence of PPARγ, since its selective removal in this cell type led to survival for only a few days [12]. Beyond its function in lipid metabolism and adipocyte differentiation, PPARγ regulates the expression of genes controlling glucose homeostasis and uptake, including Glut1, Glut4, and the c-Cbl–associated protein (CAP) [13,14]. Moreover, PPARγ modulates the synthesis and secretion of various adipokines, such as adiponectin, resistin, leptin, and tumor necrosis factor-α (TNF-α, which controls insulin sensitivity [15–18]. Recent studies have shown that PPARγ can induce the expression of adipose-related factors FGF1 and FGF21, members of the fibroblast growth factor family, which are able to maintain adipose homeostasis and counteract insulin resistance [19,20]. Intriguingly, these factors can act in both autocrine and paracrine manners, locally transmitting PPARγ signals to stimulate adipogenesis and sustain the insulin response [20]. It is very interesting that the distinct PPAR isoforms, despite their structural and functional similarities, can display peculiar properties due to their specific tissue distribution and selective responses to different ligands, thus generating such a wide range of actions. In accordance with this reasoning, PPARγ is, therefore, able to activate the expression of different genes controlling glucose homeostasis including Glut1, Glut4, and CAP, but at the same time PPARγ also modulates the synthesis and secretion of different adipokines (adiponectin, leptin, and TNF-α) from the adipose tissue, thus enhancing the insulin response. Therefore, it is reasonable to argue that, through PPARγ activation, different protein factors might stimulate distinctive pathways that, in turn, lead to improved glucose homeostasis. A PPARγ-driven targeted therapeutical approach requires selective ligands that activate the receptor, resulting in beneficial effects in fighting diabetes and obesity.

Furthermore, PPARγ is involved in the differentiation of immune cells (antigen-presenting myeloid dendritic cells and macrophages) and in the stimulation of both the immunological response and metabolic processes. Indeed, in dendritic cells, the activated PPARγ modulates lipid transport and metabolism, but also stimulates antigen uptake, dendritic differentiation, cytokine production, and antigen presentation [21]. In macrophages, PPARγ is involved in lipid metabolism and the differentiation of monocytes into macrophages [22]. Additionally, PPARγ is also involved in the polarization of M1 to M2 macrophages, thus blocking the secretion of proinflammatory cytokines, such as TNF-α, interleukin-1β (IL-1β), and interleukin-6 (IL-6), and promoting anti-inflammatory effects [23]. This is achieved through a transrepression mechanism, which activates inflammatory genes such as NF-kB and AP-1 in the M1 macrophage.

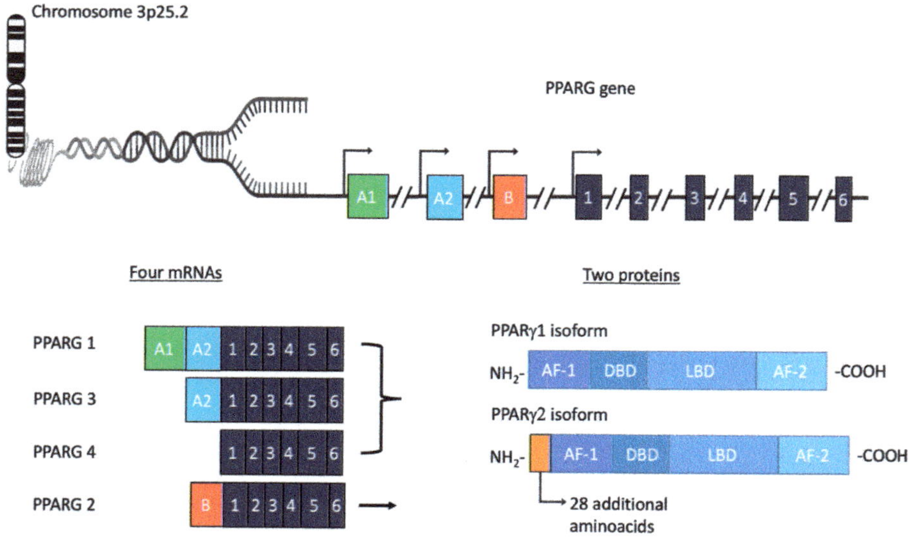

Figure 1. Schematic representation of PPARγ gene, mRNA species, and protein isoforms.

Several studies demonstrated that PPARγ can inhibit angiogenesis during the tumor formation. For example, thiazolidinediones-activated PPARγ is able to arrest tumor angiogenesis and in vivo cell growth in ovarian carcinoma and pancreatic cancer [24,25]. In gliomas, a sustained PPARγ expression has been shown to positively correlate with thrombospondin-1 (TSP-1) expression [26]. TSP-1, a glycoprotein present in the extracellular matrix and involved in the growth and adhesion of cancer and epithelial cells, counteracts angiogenesis also inhibiting tumor growth and metastasis formation [26].

It is well known that the dysfunction of nuclear receptors (NRs), due to their over-expression or under-expression, mutations, truncations, post-transductional modifications, and other events modifying the structure and the function of NRs, can induce the activation of different pathways and provoke the initiation and promotion of several cancers. Understanding of these complicated processes usually provides unexpected solutions in the therapeutic treatment of tumor patients. The expression of PPARγ has been reported in various tumor tissues and cell lines. Targeting PPARγ, indeed, through its activation mediated by specific agonists, displayed an efficient cell growth inhibition [27,28]. As mentioned above, PPARγ is particularly expressed in the colon, where it plays a protective role in in vitro and in vivo colorectal cancer (CRC) models [29–31]. In the gastrointestinal tract, in fact, PPARγ inhibits cellular proliferation, promotes differentiation, and stimulates apoptosis [32]. Several studies in mice and humans show that PPARγ prevents tumor initiation and progression, acting as a tumor suppressor gene [33]. In CRC-derived cells and transplanted tumors in nude mice, PPARγ activation, after the administration of selective ligands such as thiazolinediones (TZDs), provoked cell growth inhibition, G0/G1 arrest, and caspase-activated apoptosis [34,35].

It is reasonable to think that the multiple PPARγ functions depend on the distinctive binding mode of the selective ligands to the receptor ligand binding domain (LBD), which causes the recruitment of distinct protein cofactors capable, in turn, of activating different pathways involved in the control of metabolism, proliferation, or differentiation. The choice to activate (or not) these pathways is also likely to be influenced by the amount of the available receptors in a given cell, the chemical features of the ligands, and their affinity for the receptor. Ligands can induce conformational changes in the PPARγ LBD, thus affecting the recruitment of functional effectors (coactivators, corepressors, molecular adapters, chromatin modifying enzyme activities, etc.), which, in turn, facilitates chromatin opening and the activation of numerous target genes.

2. PPARγ Activation by Natural and Synthetic Ligands

Although the first pioneering studies about PPARs date back to the 1950s, it was necessary to wait until the 1990s to finally confirm the existence of PPARs through cloning experiments carried out in the laboratory of Issemann and Green [36]. Afterwards, the hunt started for finding natural molecules that bind specifically to PPARs, considered as biological sensors capable of detecting environmental changes and developing an adaptive response via modulated gene expression. Among the endogenous agonists of PPARs, the first natural molecules were fatty acids and their derivatives, which are usually produced by de novo lipogenesis, derived from lipolysis, or provided by the diet [37]. Dietary lipids, including unsaturated fatty acids produced from plants, are able to activate all three types of PPARs. Phospholipids usually activate PPARα, whereas 15d-PGJ2 prefers PPARγ and prostacyclin I2 binds to PPARβ/δ. Many studies have been carried out to identify other endogenous lipids, such as PPARγ ligands, but it has been difficult to define the mechanisms by which these molecules can activate the receptor and generate biological effects on lipid metabolism [38,39]. On the contrary, searching synthetic ligands for PPARγ was a more productive work, leading to the discovery of thiazolinediones (TZDs), potent activators of PPARγ with an elevated ability to sensitize cells to insulin stimulation [40]. In 1983, Ciglitazone was discovered as the first molecule displaying a certain ability of insulin sensitizing, but its clinical use has never been approved because of its weak therapeutic effect on diabetics [41]. Among TZDs, Rosiglitazone (BRL49653) was the first shown to induce the differentiation of C3H10T1/2 stem cells into adipocytes. This was, indeed, the first experimental evidence that PPARγ is a TZD target [42]. Later, Troglitazone was synthesized and used in therapeutical protocols to sensitize diabetic patients to insulin, stimulating glucose transport and metabolism. Afterwards, Pioglitazone followed and, together with Rosiglitazone, was definitively approved for the clinical treatment of Type 2 Diabetes Mellitus (T2DM) patients in 1999. Unfortunately, the oral administration of these molecules, which are able to strongly activate PPARγ, provoked unwanted side effects at the same time (weight gain, fluid retention, osteoporosis, cardiovascular diseases, and bladder cancer) [43–46]. For these reasons, both Rosiglitazone and Pioglitazone were removed from the market in the United States and Europe [47,48]. Despite these drawbacks, TZDs have significantly advanced our understanding of diabetes management, particularly T2DM. New powerful and sophisticated technologies promise to clarify every single aspect of PPARγ activation and signaling in tissues to regulate different and vital processes. Many efforts will be led to develop novel molecules that will reduce or eliminate the adverse effects associated with TZDs and improve insulin sensitization through new PPARγ-mediated molecular mechanisms.

PPARγ requires ligand binding to express its properties, inducing conformational changes that facilitate the activated receptor entry into the nucleus. Hence, PPARγ heterodimerizes with RXR, which, in turn, is also bound by its specific ligand and, together as the PPARγ/RXR complex, they recognize the PPRE motif present in the promoter regions of several gene targets, activating their transcription. In the absence of a ligand, the PPARγ/RXR complex associates with co-repressors like NCoR (nuclear receptor corepressor 1) and SMRT (silencing mediator of retinoic acid and thyroid hormone receptor), inhibiting PPARγ-mediated transcription. When an agonist binds to PPARγ, co-repressors are removed from the PPARγ/RXR complex and co-activators are recruited. Co-activators, such as CBP (CREB-binding protein), MED1 (Mediator 1), SRC1 (steroid receptor coactivator1), SRC2, SRC3, and PGC1α (peroxisome proliferator-activated receptor gamma coactivator 1α), closely interact with the ternary PPRE/PPARγ/RXR complex-promoting transcription through modifying the chromatin structure and recruiting transcription factors to the promoters of gene targets (Figure 2) [49]. Understanding how PPARγ ligands enhance the signal transduction of the activated receptor and sustain the target gene transcription requires focusing the binding mode of these molecules to the PPARγ LBD.

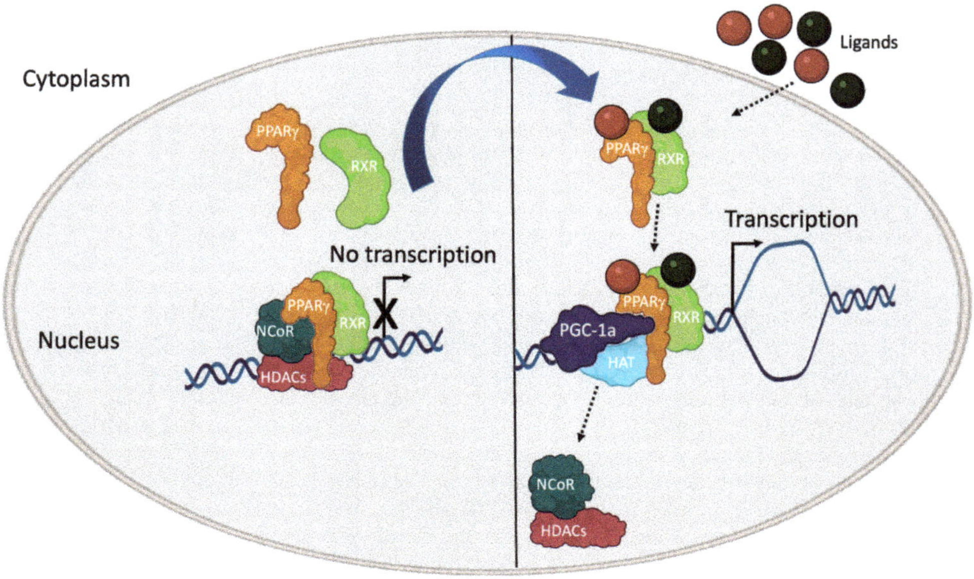

Figure 2. Molecular mechanisms of the transactivation of PPARγ-mediated gene transcription.

3. PPARγ Full Agonists, Partial Agonists, and Antagonists: Distinctive Modes of Binding

Similar to most nuclear receptors, beyond the other structural and functional domains, PPARγ shows a highly complex and conserved LBD at the C-terminus. Within this large domain (formed by about 270 amino acids) are included four subdomains displaying different functions. Firstly, there is an additional dimerization module, which is different from the one in the DNA-binding domain and works in a DNA-dependent manner. Afterwards, within the LBD is located the ligand binding pocket of PPARγ, which stabilizes the structure of the domain [50]. Thirdly, a surface necessary to recognize and bind to transcriptional co-regulators has been revealed. Finally, a fourth subdomain was found in the LBD, bearing the activation function 2 (AF2), which is responsible of the stabilization of the ligand-activated PPARγ, permitting the interaction with transcriptional co-regulators. Among these functions, the one showing the most evident differences, also compared to other nuclear receptors, is properly the ligand binding pocket. Both the primary structure and volume of the cavity of the PPARγ LBD pocket are really different from the classic nuclear receptors and even other PPAR family members, suggesting that in the tridimensional structure of the PPARγ ligand binding pocket resides a specific sensitivity towards certain ligands. Indeed, the PPARγ ligand binding pocket is the largest among the other nuclear receptors. The fundamental work of Nolte's group in 1998 confirmed these previous details and proposed additional new evidences about the PPARγ LBD pocket, which proved to be truly important to understand its function [51]. On the basis of this study, the first X-ray crystal structure of PPARγ was obtained, and, hence, we learned about many features of the PPARγ ligand binding pocket. For example, now we know that this subdomain occupies the center of the PPARγ LBD and can be described as a large Y- or T-shaped cavity (its size is approximately 1200 Å^3) formed by 13 α-helices and four β-sheet strands. It is divided into three branches with different abilities and specific sensitivities towards various ligands (Figure 3). The first branch, surrounded by the H3, H5, H11, and H12 helices, displays a hydrophilic feature that interacts with the acidic head group of ligands like Rosiglitazone, a well-known member of the thiazolidinedione TZD family. Branch I is located very close to the H12 helix, which is part of the activation function-2 (AF-2) co-regulator binding surface and is stabilized through ligand-dependent conformational changes. Branch II, formed by the H2′, H3, H6, and H7 helices and a β-sheet, is, instead, strongly hydrophobic. Branch III, finally, delimited by a β-sheet and the H2, H3, and H5 helices, has both hydrophilic and hydrophobic features. Understanding the differential structures within the PPARγ ligand binding pocket has been crucial in identifying how various ligands interact with it and elicit specific biological responses. Researchers have extensively studied which ligands enter each branch and how they bind to specific side chains of amino acids within the pocket. These interactions ultimately determine the ligand's biological effects and have guided the development of new PPARγ ligands with peculiar desired actions.

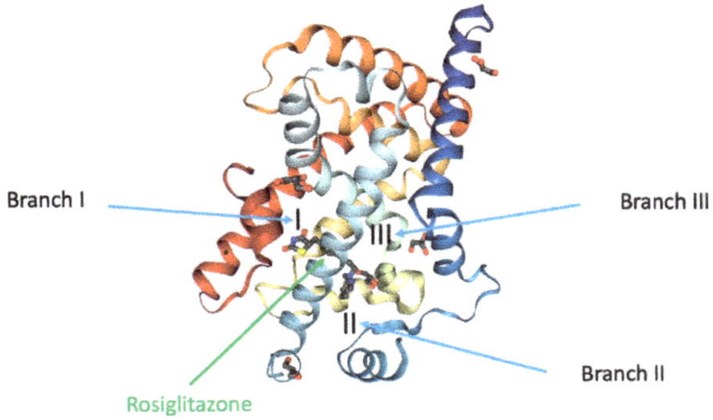

Figure 3. Localization of the different branches in the PPARγ LBD. The branches of the ligand binding pocket have been labelled with Roman numerals. The interaction of full agonist Rosiglitazone to PPARγ is also reported. Rosiglitazone is located around helix 3 (PDB-4XLD modified by Lupo et al.).

Starting from the early 2000s, the clinical use of TZDs as activators of PPARγ has significantly impacted the management of dyslipidemia and diabetes [52]. Rosiglitazone, the first TZD used in patients affected by T2DM, was demonstrated to show high efficiency in enhancing insulin sensitivity and controlling the peripheral uptake of glucose [53]. In addition, Rosiglitazone treatment caused a strong reduction in circulating fatty acids and other proteins involved in cardiovascular disease [54]. Other members of the TZD family have been, successively, used as insulin sensitizers in the treatment of T2DM patients, raising great hopes for a targeted diabetes therapy. The thiazolidinedione head group is a crucial feature of TZDs for entry and binding to the hydrophilic part of branch I, limited by the H3, H5, H6, and H12 helices. Rosiglitazone, fitting this model perfectly, enters the PPARγ cavity and binds to the side chains of certain amino acids within the LBD pocket, stabilizing the fluctuating and dynamic H12 helix through hydrogen bonds residues H323, H449, and Y473 on the inner surface of helix 12 (Figure 4). This stabilization promotes the engagement of coactivators and PPARγ transactivation [55]. Rosiglitazone contacts other areas of the PPARγ binding pocket, turning all around the H3 helix and placing itself perpendicular to it. In such a way, Rosiglitazone also increases binding affinity through hydrophobic interactions and van der Walls contacts with amino acid residues of H3, H5, H6, and H7 and the β-sheet lying on branches II and III of the PPARγ binding pocket [55]. Ligands like Rosiglitazone, which bind to the same subdomain of the PPARγ pocket with the same mechanism, have been defined as PPARγ full agonists. In transient transfection assays carried out in HEK293 cells (which do not express any type of PPARγ), only in the presence of ectopic receptors and chimeric plasmids bearing PPRE motifs upstream of the reporter luciferase gene, PPARγ full agonists display maximal PPARγ–induced transcriptional activity. This suggests that they are able to elicit and organize a stable surface onto the AF2 domain for the coactivator recruitment in the proximity of most PPARγ target genes. Additional TZD members, such as Troglidazone, Pioglitazone, Rivoglitazone, and variants of the same pharmacocore were used in the treatment of T2DM patients, and all behave as PPARγ full agonists. Although they showed a strong potential as insulin sensitizers in T2DM patients, due to their high affinity for binding to PPARγ, they were withdrawn from market, because different adverse side effects in TZD-treated patients affected by T2DM have been evidenced. Among these, renal fluid retention, heart failure, bone fractures, hepatotoxicity, and cancer raised particular interest and, at the same time, strong concern. Indeed, observational trials focusing on the use of Pioglitazone and Rosiglitazone indicated that some of these side effects are common (fluid retention, congestive heart failure, and bone fractures), thus suggesting that these effects were mediated by PPARγ [56,57]. Conversely, a meta-analysis, carried out in other clinical studies, demonstrated that Rosiglitazone had negligible effects on cardiovascular events like infarction or heart failure, whereas Pioglitazone reduced cardiovascular disease risk [44,58,59]. Moreover, a relevant risk of causing bladder cancer was evidenced in Pioglitazone-treated patients, while no carcinogenic effect was detected with Rosiglitazone [60,61]. Troglitazone, in contrast, was associated with hepatotoxicity, an effect not revealed with Rosiglitazone or Pioglitazone [62,63]. Altogether these findings strongly suggest that part of the effects of TZDs were attributable to the primary target of the drugs (PPARγ), the adverse effects were probably generated from the secondary targets of TZDs. For example, it has been reported that PPARγ full agonists, like TZDs, can interact with other nuclear receptors or functional proteins causing additional side effects [64–66]. It is conceivable that a differential distribution in the tissues and organs of the targets of PPARγ full agonists can lead to different effects, including adverse effects. It is also noteworthy to consider that many experiments to identify PPARγ full agonists have been carried out in in vitro systems (for example, cellular models or cell-free systems), but only a comparison with in vivo models could confirm or refute these results. The studies of PPARγ full agonist effects in animal models, including in mouse or, even more significantly, human models, should not neglect a deeper understanding of the distribution and concentration of PPARγ in the various tissues, as well as its own activation in the metabolism and in the regulation of vital processes.

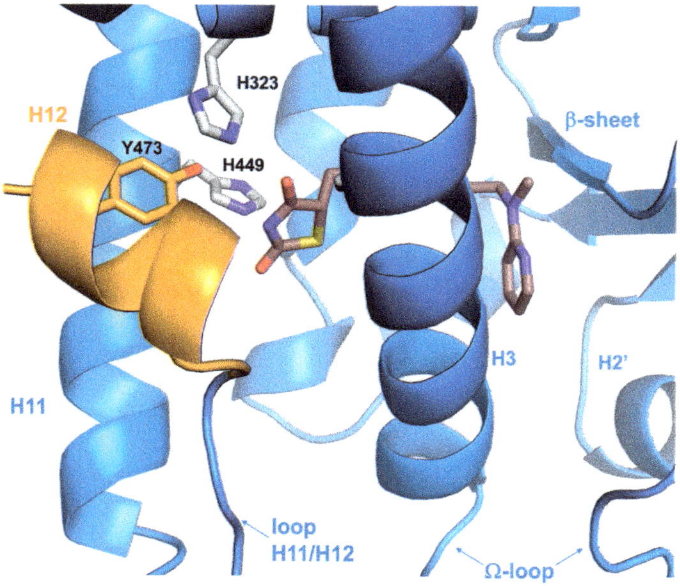

Figure 4. Crystal structure of Rosiglitazone bound to the PPARγ LBD (PDB: 2PRG). H12 is shown in orange. The Ω–loop, a flexible loop region between H2′ and H3, and the β-sheet region of the LBD are displayed.

Bearing in mind the difficulty of fully managing these contradictory results, it was necessary to implement a strategy to distinguish between the beneficial and adverse effects caused by PPARγ full agonists. With this aim, many efforts have been conducted to investigate new PPARγ ligands, permitting the isolation and identification of the primary PPARγ target genes that cause beneficial consequences in controlling glucose resistance from those responsible for adverse side effects. Several molecules capable of partially activating PPARγ-mediated gene transcription were identified and named PPARγ partial agonists to distinguish them from full agonists [67]. Partial agonists do not usually occupy branch I of the PPARγ LBD pocket, which is surrounded by the H3, H5, H11, and H12 helices. Instead, they often enter branch II (delimited by H2, H3, H6, and H7 and a β sheet) and may also occupy part of branch III (enclosed by H2, H3, and H5 and a β sheet). The common feature of these different partial agonists consists in the exclusion of the fluctuating H12 helix, resulting in no contact with the H12 residues unlike full agonists. Most of the partial agonists prefer to occupy the space between the H3 helix and β-sheet, interacting with specific amino acid residues belonging to these two structures and, consequently, preferentially stabilizing the β-sheet region rather than the H12 helix [68]. Data obtained by nuclear magnetic resonance (NMR) spectroscopy and hydrogen–deuterium exchange (HDX) mass spectrometry indeed suggest that other PPARγ protein domains may play a role in the partial agonist-mediated receptor activation. Specific co-regulators, recruited by partial agonist binding, recognize the new interacting surfaces on PPARγ, inducing, in such a way, a partial or, more precisely, graded transcription, which is directed at a restricted set of PPARγ target genes. A seminal study carried out by Oberfield's group in 1999 identified one of the first PPARγ partial agonisst, reported as GW0072 from the thiazolidine family (Figure 5A) [69]. The crystal structure of the PPARγ LBD/GW0072 complex clarified that this partial agonist occupied the region close to the H3 and H7 helices and the β-sheet, without interacting with residues H323, H449, and Y473 present on the inner surface of H12. This lack of interaction prevented the conformational changes in the H12 helix necessary for its stabilization. Similar binding patterns to the PPARγ LBD have been observed for other classes of partial agonists. nTZDpa (Figure 5B) and MRL24 (Figure 5C), belonging to the indole family compounds, interact with H3 through hydrogen bonds with Cys285 and Arg 288 and with the β-sheet region through hydrogen bonds with Ser242 and Ile341. These interactions allow them to occupy branches II and III of the PPARγ LBD pocket rather than branch I, stabilizing helix H3 and the β-sheet, resulting in a partial activation of the PPARγ pathway [70]. Telmisartan, an angiotensin II type I receptor inhibitor used to control blood pressure and treat cardiovascular diseases, is another example of a PPARγ partial agonist [71]. This molecule from the benzimidazole family binds to PPARγ in a manner similar to Rosiglitazone, wrapping around the H3 helix and making contact with the H3 helix through one of its benzimidazole group (Figure 5D). Unlike nTZDpa and MRL24, Telmisartan occupies branch I of the PPARγ LBD pocket rather than II and III, making a weak contact with H12. Although Telmisartan forms a hydrogen bond to the H12 residue Tyr473, this interaction is weaker than that brought about by Rosiglitazone due to the different distances between the groups. As a result, Telmisartan is unable to stabilize the dynamic H12 helix and to stimulate a full transcriptional activity, categorizing it as a partial agonist of PPARγ. INT131, a novel partial agonist of PPARγ from the sulfonamide family, fills the pocket space by inserting itself at the boundary of branches I, II, and III, partially wrapping around the H3 helix (Figure 6A) [71]. INT131 weakly binds to Tyr327 and makes hydrophobic contacts with Met364 of H3, Leu330 of H5, and Ile341 of the β-sheet. Altogether, these investigations demonstrated that full PPARγ agonists, such as TZDs, utilize an H12-dependent mechanism to recruit co-regulators and modulate the transcriptional response. In contrast, PPARγ partial agonists, such as GW0072, MRL24, Telmisartan, INT131, and others that have not been mentioned here, use an H12-independent mechanism to stimulate the transcription of target genes through the involvement of specific transcriptional co-modulators [72,73].

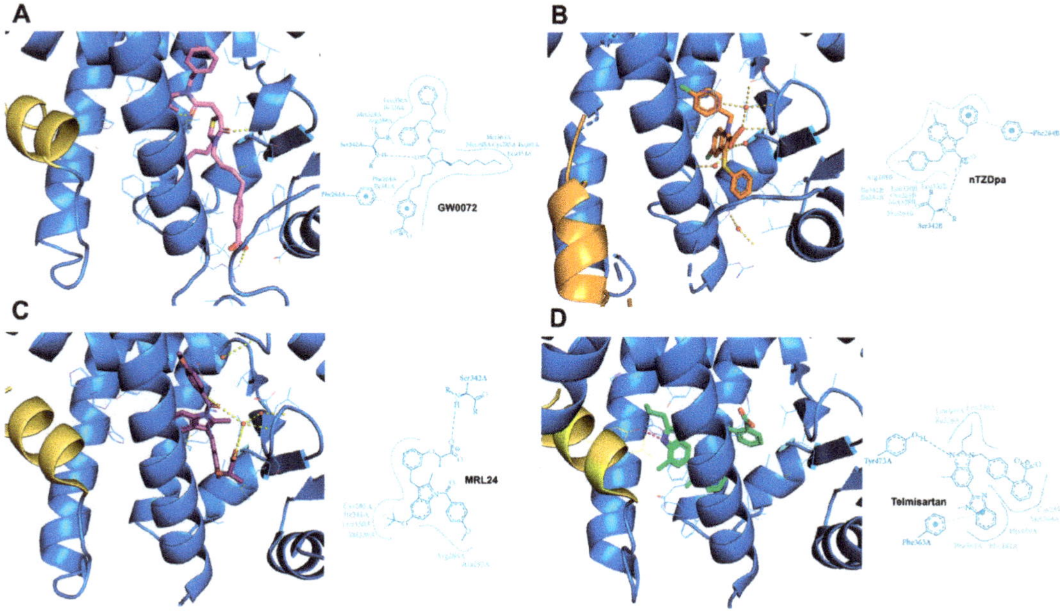

Figure 5. Crystal structures and PoseView maps highlighting key interactions of various ligands to the PPARγ LBD: (**A**) GWOO72 (PDB: 4PRG); (**B**) nTZDpa (PDB: 2Q5S); (**C**) MRL24 (PDB: 2Q5P); and (**D**) Telmisartan (PDB: 3VN2). H12 is shown in orange.

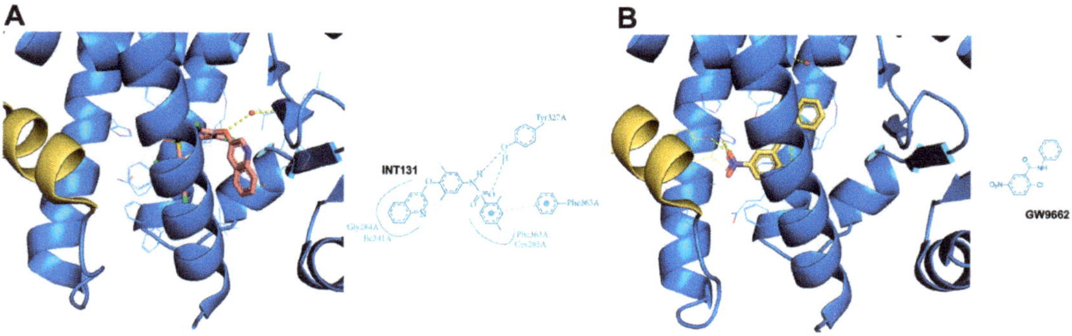

Figure 6. Crystal structures and PoseView maps highlighting key interactions of various ligands bound to the PPARγ LBD: (**A**) INT131 (PDB: 3FUR); and (**B**) GW9662 (PDB: 3B0R). H12 is shown in orange.

The studies on natural or synthetic PPARγ ligands, both full and partial agonists, carried out through crystal structures analysis, strongly support the idea that these molecules differentially occupy the three branches of the PPARγ LBD pocket. In addition, it has been largely demonstrated that these ligands usually share common spaces in the PPARγ LBD pocket, being in competition for the occupancy of it [74–79]. Biological competition assays in the presence of GW9662, an irreversible PPARγ antagonist, have been instrumental in elucidating the behavior of PPARγ ligands and the related biological consequences. GW9662 covalently binds to Ser285 of the H3 helix, a key residue in branch I of the PPARγ LBD pocket, thereby causing the exclusion of any other ligand from occupying this site (Figure 6B). Numerous natural and synthetic PPARγ agonists have been, indeed, identified and characterized for their ability to occupy the same canonical site in branch I of the Tor Y-shaped PPARγ LBD pocket [75–79]. The use of GW9662 or other PPARγ antagonists like T0070907 in biological competition experiments has provided significant insights into the interactions between PPARγ and new ligand molecules. These studies have elucidated the functions of full or partial agonists in the activation of PPARγ-mediated gene transcription and their beneficial effects on the metabolism, cell proliferation, and other vital processes. Recent work by Hughes's group demonstrated that PPARγ ligands, in addition, might interact not only with the canonical site but also with an alternate or allosteric site within the PPARγ LBD (Figure 7) [80,81]. This allosteric site partially overlaps with the canonical cavity near the β-sheet belonging to branch II but is more specifically located in the solvent-exposed space near the H3 helix, limited by the Ω-loop. Three interaction modes with the alternate site by the PPARγ ligands have been proposed. Firstly, a ligand that binds to the canonical (orthosteric) site may also bind to the alternate site, but at lower concentrations. Studies on the synthetic PPARγ ligands MRL20 and MRL24 revealed a differential modulation of these two sites in the LBD pocket, with a higher affinity (nM) for the canonical site versus a lower affinity (μM) for the alternate site. This differential binding may generate a more regulated biological response by these synthetic PPARγ ligands. In the second binding mode, the alternate or allosteric site becomes relevant when the canonical or orthosteric site is occupied by a PPARγ antagonist (GW9662 or T0070907) [76]. Indeed, daidzein, a PPARγ agonist, displayed a stronger neuroprotective effect through ligand binding independent activation when the canonical site in the PPARγ LBD pocket was blocked by a PPARγ antagonist [82]. The third model involves endogenous ligands, such as oxidized fatty acids, which covalently bind to the canonical site. In this scenario, the alternate site is also targeted by the same ligands, leading to allosteric control by these PPARγ ligands, thus increasing the beneficial effects in terms of anticancer properties [83]. This regulatory mechanism is often observed in cancer cells, which increase lipid biosynthesis to sustain their proliferation. Binding to the alternate site of PPARγ by fatty acids, produced by de novo biosynthesis or massively derived from diet, may support anti-cancer effects. The existence of two different sites within the PPARγ LBD pocket is plausible, given that the ligand binding domain of PPAR is larger than that of other nuclear receptors and can accommodate more than just one molecule. The presence of a second functional binding site in the large cavity of the PPAR receptor binding domain could reasonably explain the side effects, especially those from synthetic ligands as partial PPAR agonists competing with full agonists such as Rosiglitazone. Although it is intriguing that there is a regulatory site, i.e., an alternate site, that can influence the structure and function of PPARγ through the binding of new ligands, this hypothesis, however, still remains under investigation and not fully accepted. Few works have demonstrated the concomitant occupancy of both canonical and alternate sites by PPARγ ligands [74,84]. Further studies will be necessary to better explain how multiple molecules can bind to PPARγ, the mechanisms inducing conformational changes in the receptor structure, and the specific and differential biological effects these interactions provoke.

Recently, we have focused on determining the specific biological features of Cladosporols, a family of compounds acting as PPARγ ligands in regulating cell proliferation and migration in cancer. The study of these natural molecules, extracted from *Fungi*, furnished the clear demonstration that much can be learned from them and from other potential PPARγ ligands. This research promises to reveal new modes of binding and generate new hypotheses to explain the differential ability of promoting the activation or repression of the PPARγ-dependent gene expression program.

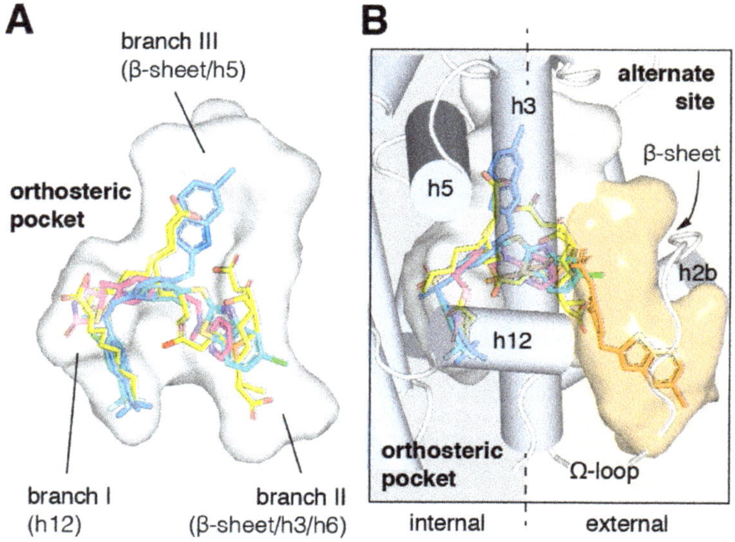

Figure 7. Structural locations of PPARγ orthosteric pocket and alternate site. (**A**) The T- or Y-shaped orthosteric pocket can accommodate one or more natural ligands such as nonanoic acid (C9; yellow sticks; PDB: 4EM9) and synthetic ligands such as Rosiglitazone (pink sticks; PDB: 4EMA) or T2384 (light and dark blue sticks representing different crystallized binding modes; PDB: 3K8S, chains A and B, respectively). (**B**) The orthosteric pocket (the white pocket surface) is completely enclosed within the alpha helical sandwich fold of the ligand binding domain (LBD). Ligands such as T2384 (orange sticks; PDB: 3K8S, chain B) can also bind to a solvent-accessible alternate site (the orange pocket surface) distinct from the orthosteric pocket, structurally defined as the region between helix 3 (H3) and the flexible Ω-loop (the dotted line separating the region internal to the LBD with a gray background and external to the LBD with a light orange background). From "Cooperative cobinding of synthetic and natural ligands to the nuclear receptor PPARγ" by Shang et al. (2018) eLife [81]. This article is distributed under the terms of the Creative Commons Attribution License, which permits unrestricted use and redistribution provided that the original author and source are credited "CC BY 4.0".

4. Natural PPARγ Ligands: Cladosporols as the Model of a PPARγ Ligand

Cladosporols A and B, part of a small family of compounds, have been identified and isolated as secondary metabolites of *Cladosporium tenuissimum* (Figure 8) [85,86]. The selected fungal strain ITT21, derived from the destructive mycoparasite *Cladosporium tenuissimum* Cooke, was recovered from *Cronartium flaccidum* aeciospore samples collected in 1996 on *Pinus pinaster* in Tuscany (Italy) [85]. This fungus effectively controls the development of diseases in the field by parasitizing spores of rusts and other harmful plant fungal pathogens [86]. Cladosporols almost totally inhibit spore germination in various rust fungi, such as *Cronartium flaccidum*, *Melampsora pinitorqua*, and *Uromyces appendiculatus*, and decrease the radial growth of non-rust fungi, as in colonies of a phytopathogenetic fungi, like *Cercospora beticola*, *Botrytis cinerea*, and *Alternaria alternata*, and among *Oomycota*, *Phytophtora erytroseptica*, *Phytophtora cinnamomi*, and *Phytophtora capsici*. They also influence the growth of human pathogenetic strains of *Candida* sp. The antifungal properties of Cladosporols is likely to be associated with the intrinsic toxicity of the 2-tetralone chromophore and their highly reactive functionalities, such as the epoxy–alcohol moiety, which shares structural features with (+)-isoepoxidon, a well-known β-1,3-glucan biosynthesis inhibitor [87].

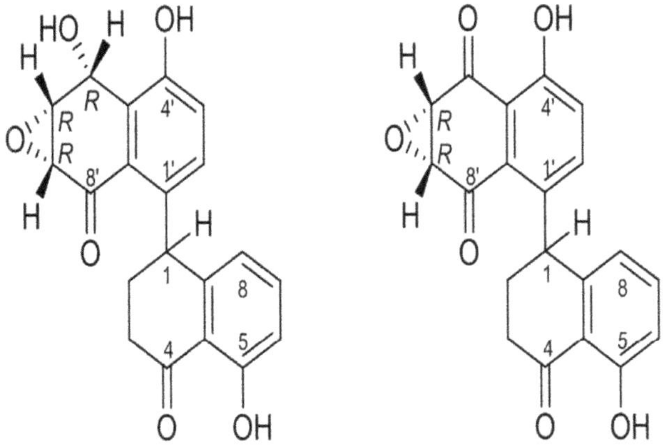

Cladosporol A **Cladosporol B**

Figure 8. Chemical structures of Cladosporols A and B.

In our initial studies, we showed that Cladosporol A inhibited colorectal cancer (CRC) cell proliferation, particularly in HT-29 cells, by inducing a G1 phase arrest. This was accompanied by the simultaneous elevation in p21waf1/cip1 protein levels and the decrease in some cell cycle regulators (cyclin D1, CDK2, and CDK4) [88]. In addition, we identified Cladosporol A as a new PPARγ ligand and characterized its antiproliferative properties, demonstrating that it activated p21waf1/cip1 gene expression in an Sp1-dependent manner [89]. We also showed that Cladosporol A directed β-catenin to proteasomal degradation, reducing the nuclear protein level and, in turn, the transcription of oncogenetic target genes (c-myc and cyclin D1). β-catenin typically forms a complex with E-cadherin, generating cell–cell interactions and preserving the epithelial features of cells. We revealed that E-cadherin was a PPARγ target in HT-29 cells, further supporting the role of Cladosporol A as a new compound with antiproliferative and antimetastatic potential (Figure 9) [89]. Subsequently, we investigated the biological properties of Cladosporol B, the oxidized form of Cladosporol A. Cladosporol B also appeared to display antiproliferative and pro-apoptotic characteristics. Our studies demonstrated that Cladosporol B showed anticancer properties in HT-29 cells, which were not only dependent on the arrest of the cell cycle in the G0/G1 phase but also on a strong activation of programmed cell death [90]. The effects of Cladosporol B were more pronounced than those of Cladosporol A and appeared to be related to a different affinity for the PPARγ LBD, which is associated with a reduced potential for PPRE-mediated transactivation. The structural analysis of the binding of Cladosporols A and B to the PPARγ LBD domain revealed that Cladosporol B formed interactions with different amino acids compared to Cladosporol A and Rosiglitazone [90]. In particular, it emerged that, while Rosiglitazone was arranged perpendicular to the H3 helix, Cladosporol A assumed a north–south orientation parallel to the H3 helix, forming hydrogen bonds with the amino acid residues Y473, H449, and H327 with consequent stabilization of H12. On the contrary, Cladosporol B did not directly contact the H12 helix but stabilized the H3 helix through hydrophobic interactions with residues belonging to this helix (S289, F282, Q283, and Q286) (Figure 10) [90]. This different binding mode altered the LBD region structure, exposing new receptor surfaces and favoring their stabilization, which is necessary for the recruitment of the different transcriptional complexes. In other words, the different way of binding to the receptor with the recruitment of different co-regulators could lead to a distinctive activation of the genes eliciting specific biological functions [90]. Other compounds have been shown to display the same binding mode of Cladosporol B (when compared to the full agonist rosiglitazone) demonstrating reduced transactivation ability and stronger antiproliferative activity, typical features of partial agonists [91–94]. In detail, the weak interactions of these PPARγ partial agonists occur within the LBD pocket, with residues residing at a considerable distance from the H12 helix. The distance between Cladosporol B or similar partial agonists and the H12 helix is remarkable, thus preventing the binding to the three residues H323, H449, and Y473, which are characteristic of Rosiglitazone's full agonism, and the maximal transcriptional activation of the target genes. Ligand docking assays suggested that the binding mode of these partial agonists rather implies the involvement of the side chains of the residues exposed from H3 and H11 and the 11/12 loop, preferably stabilizing H3 through hydrophobic bonds to the specific amino acids belonging to this helix (S289, F282, Q283 and Q286). The authors of these studies proposed designating this hydrophobic region in the branch I cavity of the PPARγ LBD pocket as the "diphenyl pocket", suggesting that the stabilization of H3 is a specific feature and a crucial requirement of the partial agonists binding to the PPARγ LBD pocket. Beyond the classifications proposed by the different research groups, based on the type of binding and the specific branch of the pocket occupied by the different ligands (agonists, antagonists, or partial agonists), it is reasonable to conclude that most of the partial agonists do not occupy branch I of the pocket but prefer branches II and III between the H3 helix and the β-sheet, primarily stabilizing the H3 helix. Full agonists, instead, are closer to H12, making strong contacts with His 323, Tyr 473, and His 449. Some partial agonists may also occupy branch I of the pocket like rosiglitazone, but they no longer contact the His 323, Tyr 473, and His 449 triad. Although some contacts have been demonstrated for some partial agonists with H12, these interactions do not involve the triad, because the distance does not allow any bonding. In summary, we may conclude that partial agonists, like Cladosporol B, ureidofibrate-like derivatives, or chiral phenoxyacetic acid analogues, by moving away from the H12 helix and closer to the H3 helix, stabilize the latter instead of the former, thereby exposing novel interfaces with other hypothetical co-regulators of transcription different from those of full agonists [91–94].

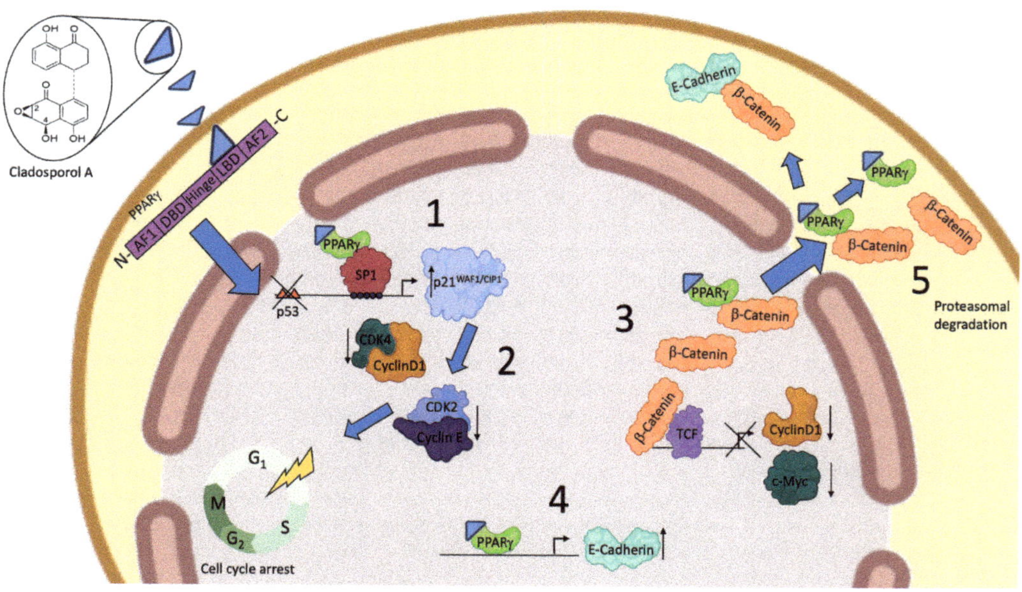

Figure 9. Cladosporol A actions regulating cell proliferation in colon cancer-derived cells. (1) Increase in p21waf1/cip1 transcription mediated by PPARγ and Sp1. (2) Down-regulation of cyclin D1, cyclin E, CDK2, and CDK4 expression and their phosphorylation activities. (3) Down-regulation of the β-catenin-dependent target gene transcription, i.e., cyclin D1 and c-myc. (4) PPARγ-dependent increase in E-cadherin mRNA transcription. (5) Proteasomal degradation of PPARγ and β-catenin.

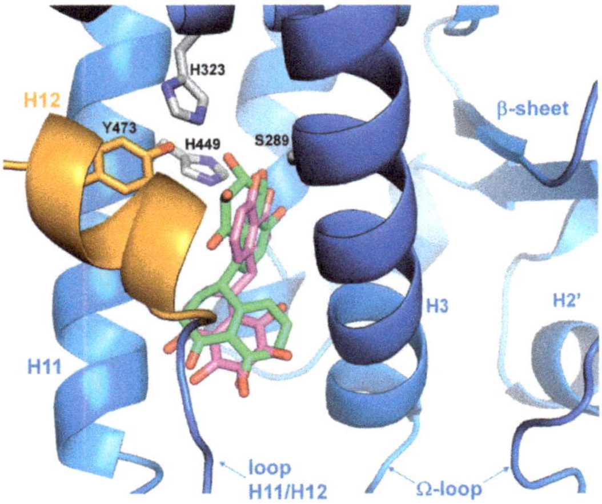

Figure 10. Docking of Cladosporol A (green stick) and Cladosporol B (violet stick) into the PPARγ LBD. The Ω–loop, a flexible loop region between H2 and H3, and β-sheet region of the LBD are displayed.

5. Different Tissues, Different Biological Processes, but a Unique Target: PPARγ

Considering the aforementioned results, which elucidate the differential interaction of Cladosporols with PPARγ and the regulation of crucial biological processes, we further investigated the properties of Cladosporols A and B in 3T3-L1 cells through an in vitro murine model of preadipocytes committed to differentiation. Our study demonstrated that Cladosporols A and B inhibit adipogenesis in 3T3-L1 preadipocytes, which have the ability to differentiate into mature adipocytes. Specifically, both Cladosporols prevented the mitotic clonal expansion (MCE) of preadipocytes by arresting the cell cycle in the G0/G1 phase [95]. Using real-time polymerase chain reaction (PCR) and western blotting analyses, we also verified that the mRNA and protein levels of early and late adipogenetic markers (PPARγ and C/EBPα; aP2; and FASN) were significantly reduced by treatment with Cladosporol A or Cladosporol B (Figure 11).

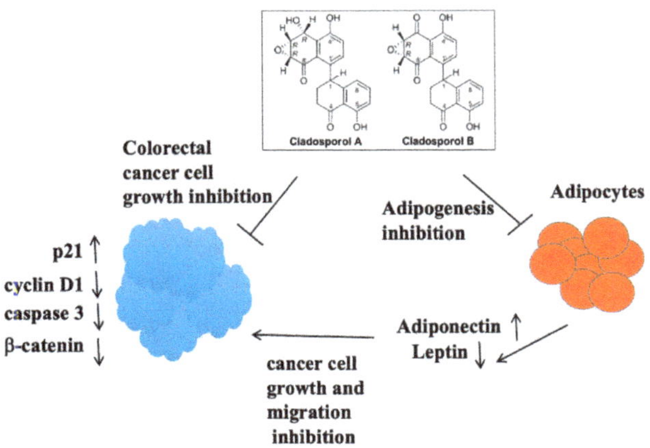

Figure 11. Cladosporol actions on 3T3-L1 cell differentiation and HT-29 CRC cell proliferation and migration.

Given the role of adipose tissue and its fatty acid depots in the tumor microenvironment, we investigated whether secreted factors (adipokines) from 3T3-L1 cells could influence HT-29 cell growth, invasion, and migration. We demonstrated that the proliferation and migration of HT-29 cells were inhibited by a medium derived from 3T3-L1 cells cultured in the presence of Cladosporol A or Cladosporol B. Specifically, we found an evident reduction in cyclin D1 and β-catenin and a simultaneous increase in p21waf1/cip1 expression in the HT-29 cells, indicating an evident cell cycle blockage in the HT-29 cells grown in the 3T3-L1 conditioned medium from cells exposed to either Cladosporols. Moreover, a significant stimulation of apoptosis was revealed and, in fact, a reduction in caspase-3 precursor levels was found in the HT-29 cells cultured in the conditioned medium from 3T3-L1 cells previously treated with either Cladosporols, suggesting an enhanced programmed cellular death. These findings prompted a molecular mechanism through which a dysregulated lipid metabolism in mature 3T3-L1 adipocytes and cancer promotion and progression in HT-29 cells could simultaneously be influenced by the use of the same therapeutic tool, i.e., Cladosporols [95]. Leptin and adiponectin are two adipokines produced by mature adipose tissue that integrate peripheral and central signals, but they also represent part of the milieu that stimulates the tumor microenvironment and thus regulates the signaling pathways related to cancer progression. We analyzed the presence of these important adipokines in 3T3-L1 cells and in the culture medium of undifferentiated and differentiated 3T3-L1 cells treated with Cladosporol A or Cladosporol B, respectively. Both Cladosporols reduced the leptin expression level in the cells and in the medium, while being responsible for a decreased adiponectin expression in the cells, but promoted adiponectin secretion in the medium [95]. These results strongly suggest that Cladosporols can regulate the production and availability of crucial hormones that differentiated adipocytes secrete in order to influence the biological properties of neighboring cells. Our findings suggested a Cladosporol treatment-dependent break in the crosstalk between 3T3-L1 mature adipocytes and HT-29 cells. The increased adiponectin synthesis and secretion in the medium from differentiated 3T3-L1 cells treated with Cladosporol A or Cladosporol B is responsible for the inhibition of HT-29 cell growth and migration (Figure 11) [95].

More recent data also show that PC-3 cells, derived from metastatic prostate cancer, were highly sensitive to the same regulation mediated by Cladosporol-treated 3T3-L1 adipocytes [96]. We firstly demonstrated that the inhibition of the growth of PC-3 cells, caused by both Cladosporols A and B, is mediated by PPARγ. This aligns with the results from other studies indicating that targeting the PPARγ2 isoform induces a strong apoptosis in the most aggressive prostate tumor cells [97,98]. The inhibition of PC-3 cell proliferation was, indeed, obtained through an increase in apoptosis as proved by the diminished expression of caspase-3, Bcl-2, and caspase-9, well-known apoptotic markers [96]. In addition, Cladosporols A and B, by regulating the expression of the proteins involved in cell proliferation and migration (β-catenin, E-cadherin, and MMP-9), were responsible for a reduced capability of the PC-3 cells in proliferating and migrating [96]. Cladosporols A and B modulated in the opposite manner the synthesis and secretion of leptin and adiponectin in the 3T3-L1 cells. Therefore, when PC-3 prostate cancer cells were cultured in the medium from Cladosporol-treated 3T3-L1 mature adipocytes, through an indirect mechanism both Cladosporols also influenced the proliferation and migration of the PC-3 cells [96]. Notably, the higher concentration of adiponectin protein levels in the medium culture of the Cladosporol-treated 3T3-L1 cells correlated well with the increase in the adiponectin receptor 1 (Adipo-R1) expression in the PC-3 cells cultured in this Cladosporol-conditioned medium from the 3T3-L1 cells [96]. These encouraging data interestingly indicated that adiponectin, poured off in the medium from the 3T3-L1 cells treated with Cladosporols, interrupts the dialogue between mature 3T3-L1 adipocytes and PC-3 cells in the tumor microenvironment, promoting a strong inhibition of cell growth and migration [96].

Lipid metabolism and cancer are strictly related, and, thereby, a dysregulated metabolism of lipids constitutes one of the most relevant hallmarks of cancer. To sustain the excessive proliferation rate, cancer cells continuously need fatty acids for membrane synthesis [99–101]. Prostate cancer cells, in particular, rely on increased de novo lipogenesis to meet the demands for membrane synthesis, redox balance, and the energy storage and activation of the intracellular pathways. We investigated whether Cladosporol A and Cladosporol B played a regulatory role in lipid metabolism. The exposure of PC-3 prostate cancer cells to Cladosporols dramatically inhibited lipid accumulation through a negative regulation of the enzymes directly involved in lipid metabolism, including acetyl-CoenzymeA carboxylase (ACC), fatty acid synthase (FAS), fatty acid binding protein 4 (FABP4), hydroxi-methyl-glutaril-CoenzymeA reductase (HMG-CoAr), and sterol responsive element binding protein 1 (SREBP-1) [96]. Our experiments showed that the Cladosporol treatments efficiently diminished the protein amount of these enzymes. Pre-exposure to GW9662 led to an additional reduction, indicating that the negative regulatory effect on lipid synthesis was due to Cladosporol binding to the PPARγ LBD pocket [96]. Interestingly, the reduction in the synthesis of the enzymes involved in SREBP-1-mediated cholesterol biosynthesis in PC-3 prostate cancer cells suggests that the inhibition of this pathway, due to the Cladosporol–PPARγ complex action, also impedes the de novo intratumoral biosynthesis of androgens. This is particularly relevant for prostate cancer re-growth, when patients are usually kept in a low androgen concentration condition, because they are under hormonal ablation therapy [102,103]. We may argue that the Cladosporol–PPARγ complex-mediated blockage of the metabolic pathway stimulating cholesterol synthesis also inhibited the intratumoral de novo androgen biosynthesis in PC-3 metastatic prostate cancer-derived cells.

Altogether, the results for cells deriving from different tissues allow us to state that, in distinctive cell contexts, Cladosporols can modulate different biological processes through the binding to the same molecular mediator, i.e., PPARγ [95,96]. However, the mode of binding to the PPARγ LBD pocket, which determines the choice of the transduction pathway that is primarily affected by the Cladosporol treatment, is quite different from that described for PPARγ full agonists. In accordance with this way of reasoning, the non-canonical binding of Cladosporols into the PPARγ LBD pocket could be the reason for their ability to inhibit adipogenesis in 3T3-L1 cells and also to interfere in the lipid metabolism dysregulation in metastatic PC-3 prostate cancer cells but, at the same time, to limit cell proliferation and migration in HT-29 CRC as well as in PC-3 cells through the inhibition of diverse pathways. At this point, a question is obligatory. Is Cladosporol A (or Cladosporol B) alone responsible for its actions directed towards multiple targets? The answer is likely no, because, after the interaction of the Cladosporols with the PPARγ receptor, a series of events follow, one after another, to realize the action of these natural PPARγ ligands. The stabilization of an interaction surface on the PPARγ structure, indeed, firstly needs a few, but more efficient, contacts between the functional groups of the ligands and amino acid determinants exposed in the PPARγ LBD pocket. It is reasonable to think that a perfect binding of Cladosporols, for example, to the PPARγ pocket is obtained through an adaptation that takes place across time. In a very interesting recent work, Shang and Kojetin proposed that the binding to the PPARγ pocket by ligands happens through a two-phase fit mechanism: an initial encounter complex is followed by a conformational change into the final bound state [104]. This mechanism could also be followed by Cladosporols in their searching for a more stable complex with PPARγ. Cladosporol A, indeed, enters branch I, sits parallel to the H3 helix, assuming a north–south orientation, and interacts with the amino acid residues Y473, H449, and H327 of the H12 helix with its consequent stabilization. On the contrary, Cladosporol B does not directly contact the H12 helix (which is located a little further away) but stabilizes the H3 helix through hydrophobic interactions with the residues belonging to this helix (S289, F282, Q283, and Q286) [91]. The diverse binding mode of the two Cladosporols, which arises as a result of searching for the best and most stable interaction, can be the cause of the LBD structure conformational change with the exposure of new receptor surfaces necessary for the recruitment of the new and specific transcriptional co-factors.

These co-factors are capable of directing the action of the Cladosporol/PPARγ complex towards the signal transduction pathways, which can discriminate and choose different molecular targets in different tissue types, thus generating multiple but specific effects.

However, the identity of these co-regulators still remains unknown. In this scenario, covalent modifications of the PPARγ receptor, which may occur before or after the co-regulators' recruitment, could also play a role in enabling new differential transduction pathways. Finally, the ability to direct the function of PPARγ towards new pathways could also be dependent on the concentration of the PPARγ isoforms in the target tissue and on the quality of the receptor itself. For instance, understanding which isoform is most prevalent in a specific tissue and which of these PPARγ isoforms preferably binds Cladosporols is an essential topic. Our experiments in 3T3-L1 mature adipocytes and HT-29 CRC and prostate PC-3 cells have clarified various aspects of Cladosporol's role in modulating complex phenomena occurring in different tissues in the tumor microenvironment. At the same time, our studies have raised further questions that we must address to understand how these natural molecules achieve distinct actions in different tissues but with the common goal of inhibiting cell growth and migration in colon and prostate tumors.

6. Conclusions and Perspectives

We have not fully identified and finely elucidated the differential pathways through which Cladosporols regulate functions such as cell proliferation and migration and lipid metabolism in the cells derived from distinctive tissues. However, it is evident that this behavior of Cladosporols is partially due to their mode of binding to PPARγ. While our studies, alongside those of other investigators, proved a relationship between PPARγ and partial agonists in simple 2D cellular models and in cell-based assays, further experiments are needed to confirm these findings in in vivo or ex vivo models. The adoption of three-dimensional (3D) organotypic cell models could bridge the existing gap between the results coming out from the in vitro and in vivo results. Future research should focus on combining novel 3D phenotypic cell-based assays, using confocal or multiphoton microscopy with advanced analytical techniques such as mass spectrometry and a microarray for proteomic, transcriptomic, and metabolomics analyses. This integrative approach will greatly enhance our comprehension of the role of Cladosporols in tumor biology and aid in the development of new therapies targeting novel molecular players.

Author Contributions: Conceptualization, A.L. (Angelo Lupo) and A.L. (Antonio Lavecchia); drafting of the manuscript, A.L. (Angelo Lupo), A.L. (Antonio Lavecchia), A.M., R.R., S.D. and S.M.; review and editing, A.L. (Angelo Lupo), A.L. (Antonio Lavecchia) and R.R.; and funding acquisition, A.L. (Angelo Lupo) and A.L. (Antonio Lavecchia). All authors have read and agreed to the published version of the manuscript.

Funding: This work was supported by funds from the DST—University of Sannio (FRA 2022–2023). A.Lav. acknowledges funding from the Italian Ministry of Education, University and Research (MIUR), Progetti di Rilevante Interesse Nazionale (PRIN), grant no. 2022P5LPHS.

Institutional Review Board Statement: Not applicable.

Informed Consent Statement: Not applicable.

Data Availability Statement: Details regarding the results reported in the present review are available in the data bank (PubMed). These data refer to the papers on Cladosporols whose references are present at the end of the discussion (Refs. [88–90,95,96]).

Conflicts of Interest: The authors declare no conflicts of interest.

References

1. Tontonoz, P.; Spiegelman, B.M. Fat and beyond: The diverse biology of PPARγ. *Annu. Rev. Biochem.* **2008**, *77*, 289–312. [CrossRef] [PubMed]
2. Michalik, L.; Auwerx, J.; Berger, J.P.; Chatterjee, V.K.; Glass, C.K.; Gonzalez, F.J.; Grimaldi, P.A.; Kadowaki, T.; Lazar, M.A.; O'Rahilly, S.; et al. International Union of Pharmacology. LXI. Peroxisome proliferator-activated receptors. *Pharmacol. Rev.* **2006**, *58*, 726–741. [CrossRef] [PubMed]
3. Tyagi, S.; Gupta, P.; Saini, A.S.; Kaushal, C.; Sharma, S. The peroxisome proliferator-activated receptor: A family of nuclear receptors role in various diseases. *J. Adv. Pharm. Technol. Res.* **2011**, *2*, 236–240. [CrossRef] [PubMed]

4. Mandard, S.; Müller, M.; Kersten, S. Peroxisome proliferator-activated receptor alpha target genes. *Cell Mol. Life Sci.* **2004**, *61*, 393–416. [CrossRef] [PubMed]
5. Barish, G.D.; Narkar, V.A.; Evans, R.M. PPAR delta: A dagger in the heart of the metabolic syndrome. *J. Clin. Investig.* **2006**, *116*, 590–597. [CrossRef]
6. Janani, C.; Ranjitha Kumari, B.D. PPAR gamma gene—A review. *Diabetes Metab. Syndr.* **2015**, *9*, 46–50. [CrossRef]
7. Medina-Gomez, G.; Gray, S.L.; Yetukuri, L.; Shimomura, K.; Virtue, S.; Campbell, M.; Curtis, R.K.; Jimenez-Linan, M.; Blount, M.; Yeo, G.S.; et al. PPAR gamma 2 prevents lipotoxicity by controlling adipose tissue expandability and peripheral lipid metabolism. *PLoS Genet.* **2007**, *3*, e64. [CrossRef]
8. Saraf, N.; Sharma, P.K.; Mondal, S.C.; Garg, V.K.; Singh, A.K. Role of PPARg2 transcription factor in thiazolidinedione-induced insulin sensitization. *J. Pharm. Pharmacol.* **2012**, *64*, 161–171. [CrossRef]
9. Tontonoz, P.; Hu, E.; Spiegelman, B.M. Stimulation of adipogenesis in fibroblasts by PPARγ2, a lipidactivated transcription factor. *Cell* **1994**, *79*, 1147–1156. [CrossRef]
10. Barak, Y.; Nelson, M.C.; Ong, E.S.; Jones, Y.Z.; Ruiz-Lozano, P.; Chien, K.R.; Koder, A.; Evans, R.M. PPAR gamma is required for placental, cardiac, and adipose tissue development. *Mol. Cell* **1999**, *4*, 585–595. [CrossRef]
11. Tang, W.; Zeve, D.; Suh, J.M.; Bosnakovski, D.; Kyba, M.; Hammer, R.E.; Tallquist, M.D.; Graff, J.M. White fat progenitor cells reside in the adipose vasculature. *Science* **2008**, *322*, 583–586. [CrossRef]
12. Imai, T.; Takakuwa, R.; Marchand, S.; Dentz, E.; Bornert, J.M.; Messaddeq, N.; Wendling, O.; Mark, M.; Desvergne, B.; Wahli, W.; et al. Peroxisome proliferator-activated receptor gamma is required in mature white and brown adipocytes for their survival in the mouse. *Proc. Natl. Acad. Sci. USA* **2004**, *101*, 4543–4547. [CrossRef]
13. Bauman, C.A.; Chokshi, N.; Saltiel, A.R.; Ribon, V. Cloning and characterization of a functional peroxisome proliferator-activated receptor gamma-responsive element in the promoter of CAP gene. *J. Biol. Chem.* **2000**, *275*, 9131–9135. [CrossRef] [PubMed]
14. Nugent, C.; Prins, J.B.; Whitehead, J.P.; Wentworth, J.M.; Chatterjee, V.K.; O'Rahilly, S. Arachidonic Acid Stimulates Glucose Uptake in 3T3-L1 Adipocytes by Increasing GLUT1 and GLUT4 Levels at the Plasma Membrane Evidence for involvement of lipoxygenase metabolites and peroxisome proliferator-activated receptor. *J. Biol. Chem.* **2000**, *276*, 9149–9157. [CrossRef]
15. Hollenberg, A.N.; Susulic, V.S.; Madura, J.P.; Zhang, B.; Moller, D.E.; Tontonoz, P.; Sarraf, P.; Spiegelman, B.M.; Lowell, B.B. Functional antagonism between CCAAT/enhancer binding protein-α and peroxisome proliferator-activated receptor-γ on the leptin promoter. *J. Biol. Chem.* **1997**, *272*, 5283–5290. [CrossRef] [PubMed]
16. Iwaki, M.; Matsuda, M.; Maeda, N.; Funahashi, T.; Matsuzawa, Y.; Makishima, M.; Shimomura, I. Induction of adiponectin, a fat-derived antidiabetic and antiatherogenic factor, by nuclear receptors. *Diabetes* **2003**, *52*, 1655–1663. [CrossRef] [PubMed]
17. Hofmann, C.; Lorenz, K.; Braithwaite, S.S.; Colca, J.R.; Palazuk, B.B.; Hotamisligil, G.S.; Spiegelman, B.M. Altered gene expression for tumor necrosis factor-α and its receptors during drug and dietary modulation of insulin resistance. *Endocrinology* **1994**, *134*, 264–270. [CrossRef] [PubMed]
18. Tomaru, T.; Steger, D.J.; Lefterova, M.I.; Schupp, M.; Lazar, M.A. Adipocyte-specific expression of murine resistin is mediated by synergism between peroxisome proliferator-activated receptor γ and and CCAAT/enhancer-binding proteins. *J. Biol. Chem.* **2009**, *284*, 6116–6125. [CrossRef]
19. Jonker, J.W.; Suh, J.M.; Atkins, A.R.; Ahmadian, M.; Li, P.; Whyte, J.; He, M.; Juguilon, H.; Yin, Y.; Phillips, C.T.; et al. A PPARγ-FGF1 axis is required for adaptive adipose remodelling and metabolic homeostasis. *Nature* **2012**, *485*, 391–394. [CrossRef]
20. Dutchak, P.A.; Katafuchi, T.; Bookout, A.L.; Choi, J.H.; Yu, R.T.; Mangelsdorf, D.J.; Kliever, S.A. Fibroblast growth factor-21 regulates PPARγ activity and the antidiabetic actions of thiazolidinediones. *Cell* **2012**, *148*, 556–567. [CrossRef]
21. Szatmari, I.; Töröcsik, D.; Agostini, M.; Nagy, T.; Gurnell, M.; Barta, E.; Chatterjee, K.; Nagy, L. PPARgamma regulates the function of human dendritic cells primarily by altering lipid metabolism. *Blood* **2007**, *110*, 3271–3280. [CrossRef] [PubMed]
22. Chawla, A.; Barak, Y.; Nagy, L.; Liao, D.; Tontonoz, P.; Evans, R.M. PPAR-gamma dependent and independent effects on macrophage-gene expression in lipid metabolism and inflammation. *Nat. Med.* **2001**, *7*, 48–52. [CrossRef] [PubMed]
23. Bouhley, M.A.; Derudas, B.; Rigamonti, E.; Dievart, R.; Brozek, J.; Haulon, S.; Zawadzki, C.; Jude, B.; Torpier, G.; Marx, N.; et al. PPArgamma activation primes human monocytes into alternative M2 macrophages with anti-inflammatory properties. *Cell Metab.* **2007**, *6*, 137–143. [CrossRef] [PubMed]
24. Yokoyama, Y.; Xin, B.; Shigeto, T.; Mizunuma, H. Combination of ciglitazone, a peroxisome proliferator-activated receptor gamma ligand, and cisplatin enhances the inhibition of growth of human ovarian cancers. *J. Cancer Res. Clin. Oncol.* **2011**, *137*, 1219–1228. [CrossRef] [PubMed]
25. Dong, Y.W.; Wang, X.P.; Wu, K. Suppression of pancreatic carcinoma growth by activating peroxisome proliferator-activated receptor gamma involves angiogenesis inhibition. *World J. Gastroenterol.* **2009**, *15*, 441–448. [CrossRef] [PubMed]
26. Zhang, J.; Yang, W.; Zhao, D.; Han, Y.; Liu, B.; Zhao, H.; Wang, H.; Zhang, Q.; Xu, G. Correlation between TSP-1, TGF-β and PPAR-γ expression levels and glioma microvascular density. *Oncol. Lett.* **2011**, *7*, 95–100. [CrossRef] [PubMed]
27. Dhiman, V.K.; Bolt, M.J.; White, K.P. Nuclear receptors in cancer—Uncovering new and evolving roles through genomic analysis. *Nat. Rev.Genet.* **2018**, *19*, 160–174. [CrossRef]
28. Wagner, N.; Wagner, K. Peroxisome Proliferator-Activated Receptors and the Hallmarks of Cancer. *Cells* **2022**, *11*, 2432. [CrossRef] [PubMed]
29. Chen, G.G.; Lee, J.F.; Wang, S.H.; Chan, U.P.; Ip, P.C.; Lau, W.Y. Apoptosis induced by activation of peroxisome-proliferator activated receptor-gamma is associated with Bcl-2 and NF-kappaB in human colon cancer. *Life Sci.* **2002**, *70*, 2631–2646. [CrossRef]

30. Panigrahy, D.; Singer, S.; Shen, L.Q.; Butterfield, C.E.; Freedman, D.A.; Chen, E.J.; Moses, M.A.; Kilroy, S.; Duensing, S.; Fletcher, C.; et al. PPARgamma ligands inhibit primary tumor growth and metastasis by inhibiting angiogenesis. *J. Clin. Investig.* **2002**, *110*, 923–932. [CrossRef]
31. Pascual, G.; Fong, A.L.; Ogawa, S.; Gamliel, A.; Li, A.C.; Perissi, V.; Rose, D.W.; Wilson, T.M.; Rosenfeld, M.G.; Glass, C.K. A SUMOylation-dependent pathway mediates transrepression of inflammatory response genes by PPAR-gamma. *Nature* **2005**, *437*, 759–763. [CrossRef] [PubMed]
32. Shimada, T.; Kojima, K.; Yoshiura, K.; Hiraishi, H.; Terano, A. Characteristics of the peroxisome proliferator activated receptor gamma (PPARgamma) ligand induced apoptosis in colon cancer cells. *Gut* **2002**, *50*, 658–664. [CrossRef] [PubMed]
33. Thompson, E.A. PPARgamma physiology and pathology in gastrointestinal epithelial cells. *Mol. Cells* **2007**, *24*, 167–176. [CrossRef] [PubMed]
34. Sarraf, P.; Mueller, E.; Jones, D.; King, F.J.; DeAngelo, D.J.; Partridge, J.B.; Holden, S.A.; Chen, L.B.; Singer, S.; Fletcher, C.; et al. Differentiation and reversal of malignant changes in colon cancer through PPARgamma. *Nat. Med.* **1998**, *4*, 1046–1052. [CrossRef] [PubMed]
35. Voutsadakis, I.A. Peroxisome proliferator-activated receptor gamma (PPARgamma) and colorectal carcinogenesis. *J. Cancer Res. Clin. Oncol.* **2007**, *133*, 917–928. [CrossRef] [PubMed]
36. Issemann, I.; Green, S. Activation of a member of the steroid hormone receptor superfamily by peroxisome proliferators. *Nature* **1990**, *347*, 645–650. [CrossRef] [PubMed]
37. Gross, B.; Pawlak, M.; Lefebvre, P.; Staels, B. PPARs in obesity-induced t2dm, dyslipidaemia and nafld. *Nat. Rev. Endocrinol.* **2017**, *13*, 36–49. [CrossRef]
38. Forman, B.M.; Tontonoz, P.; Chen, J.; Brun, R.P.; Spiegelman, B.M.; Evans, R.M. 15-Deoxy-δ 12, 14-prostaglandin J2 is a ligand for the adipocyte determination factor PPARγ. *Cell* **1995**, *83*, 803–812. [CrossRef]
39. Forman, B.M.; Chen, J.; Evans, R.M. Hypolipidemic drugs, polyunsaturated fatty acids, and eicosanoids are ligands for peroxisome proliferator-activated receptors α and δ. *Proc. Natl. Acad. Sci. USA* **1997**, *94*, 4312–4317. [CrossRef]
40. Giglio, R.V.; Papanas, N.; Rozvi, A.A.; Ciaccio, M.; Patti, A.M.; Ilias, I.; Stoian, A.P.; Sahebkar, A.; Janez, A.; Rizzo, M. An update on the current and emerging use of thiazolidinediones for Type 2 Diabetes. *Medicina* **2022**, *58*, 1475. [CrossRef]
41. Soccio, R.E.; Chen, E.R.; Lazar, M.A. Thiazolidinediones and the promise of insulin sensitization in type 2 diabetes. *Cell Metab.* **2014**, *20*, 573–591. [CrossRef] [PubMed]
42. Lehmann, J.M.; Moore, L.B.; Smith-Oliver, T.A.; Wilkison, W.O.; Wilson, T.M.; Kliewer, S.A. An antidiabetic thiazolidinedione is a high affinity ligand for peroxisome proliferator-activated receptor gamma (PPAR gamma). *J. Biol. Chem.* **1995**, *270*, 12953–12956. [CrossRef] [PubMed]
43. Kung, J.; Henry, R.R. Thiazolidinedione safety. *Expert Opin. Drug Saf.* **2012**, *11*, 565–579. [CrossRef] [PubMed]
44. Nissen, S.E.; Wolski, K. Effect of rosiglitazone on the risk of myocardial infarction and death from cardiovascular causes. *N. Engl. J. Med.* **2007**, *356*, 2457–2471. [CrossRef]
45. Graham, D.J.; Ouellet-Hellstrom, R.; MaCurdy, T.E.; Ali, F.; Sholley, C.; Worral, C.; Kelman, J.A. Risk of acute myocardial infarction, stroke, heart failure, and death in elderly Medicare patients treated with rosiglitazone or pioglitazone. *J. Am. Med. Assoc.* **2010**, *304*, 411–418. [CrossRef]
46. Neumann, A.; Weil, A.; Ricordeau, P.; Fagot, J.P.; Alla, F.; Allemand, H. Pioglitazone and risk of bladder cancer among diabetic patients in France: A population-based cohort study. *Diabetologia* **2012**, *55*, 1953–1962. [CrossRef]
47. Wilcox, R.; Kupfer, S.; Erdmann, E. Effects of pioglitazone on major adverse cardiovascular events in high-risk patients with type 2 diabetes: Results from PROspective pioglitAzone Clinical Trial In macro Vascular Events (PROactive 10). *Am. Heart J.* **2008**, *155*, 712–717. [CrossRef]
48. European Medicines Agency. European Medicines Agency Recommends Suspension of Avandia, Avandamet and Avaglim. 2010. Available online: https://www.ema.europa.eu/en/news/european-medicines-agency-recommends-suspension-avandia-avandamet-avaglim (accessed on 25 July 2024).
49. Mouchiroud, L.; Eichner, L.J.; Shaw, R.J.; Auwerx, J. Transcriptional coregulators: Fine-tuning metabolism. *Cell Metab.* **2014**, *20*, 26–40. [CrossRef]
50. Helsen, C.; Claessens, F. Looking at nuclear receptors from anewangle. *Mol. Cell. Endocrinol.* **2014**, *382*, 97–106. [CrossRef]
51. Nolte, R.T.; Wisely, G.B. Ligand binding and co-activator assembly of the peroxisome proliferator-activated receptorγ. *Nature* **1998**, *395*, 137–143. [CrossRef]
52. Yki-Jarvinen, H. Thiazolidinediones. *N. Engl. J. Med.* **2004**, *351*, 1106–1118. [CrossRef] [PubMed]
53. Pearson, S.L.; Cawthorne, M.A.; Claphametal, J.C. The thiazolidinedione insulin sensitiser, BRL 49653, increases the expression of PPAR-γ and aP2 in adipose tissue of high-fat-fed rats. *Biochem. Biophys. Res. Commun.* **1996**, *229*, 752–757. [CrossRef] [PubMed]
54. Day, C. Thiazolidinediones: A new class of antidiabetic drugs. *Diabet. Med.* **1999**, *16*, 179–192. [CrossRef] [PubMed]
55. Farce, A.; Renault, N.; Chavatte, P. Structural insight into PPARgamma ligands binding. *Curr. Med. Chem.* **2009**, *16*, 1768–1789. [CrossRef]
56. Nesto, R.W.; Bell, D.; Bonow, R.O.; Fonseca, V.; Grundy, S.M.; Horton, E.S.; Le Winter, M.; Porte, D.; Semenkovich, C.F.; Smith, S.; et al. Thiazolidinedione use, fluid retention, and congestive heart failure: A consensus statement from the American Heart Association and American Diabetes Association. *Circulation* **2003**, *108*, 2941–2948. [CrossRef] [PubMed]

57. Betteridge, D.J. Thiazolidinediones and fracture risk in patients with type 2 diabetes. *Diabet. Med.* **2011**, *28*, 759–771. [CrossRef] [PubMed]
58. Lincoff, A.M.; Wolski, K.; Nicholls, S.J.; Nissen, S.E. Pioglitazone and risk of cardiovascular events in patients with type 2 diabetes mellitus: A meta-analysis of randomized trials. *JAMA* **2007**, *298*, 1180–1188. [CrossRef]
59. Mahaffey, K.W.; Hafley, G.; Dickerson, S.; Burns, S.; Tourt-Uhlig, S.; Whyte, J.; Newby, L.K.; Komajda, M.; McMurray, J.; Bigelow, R.; et al. Results of a reevaluation of cardiovascular outcomes in the RECORD trial. *Am. Heart J.* **2013**, *166*, 240–249. [CrossRef]
60. Ryder, R.E.J. Pioglitazone has a dubious bladder cancer risk but an undoubted cardiovascular benefit. *Diabet. Med.* **2015**, *32*, 305–313. [CrossRef]
61. Friedland, S.N.; Leong, A.; Filion, K.B.; Genest, J.; Lega, I.C.; Mottillo, S.; Poirier, P.; Reoch, J.; Eisenberg, M.J. The cardiovascular effects of peroxisome proliferator-activated receptor agonists. *Am. J. Med.* **2012**, *125*, 126–133. [CrossRef] [PubMed]
62. Ikeda, T. Drug-induced idiosyncratic hepatotoxicity: Prevention strategy developed after the troglitazone case. *Drug Metab. Pharmacokinet.* **2011**, *26*, 60–70. [CrossRef] [PubMed]
63. Scheen, A.J. Thiazolidinediones and liver toxicity. *Diabetes Metab.* **2001**, *27*, 305–313. [PubMed]
64. Matthews, L.; Berry, A.; Tersigni, M.; D'Acquisto, F.; Ianaro, A.; Ray, D. Thiazolidinediones are partial agonists for the glucocorticoid receptor. *Endocrinology* **2009**, *150*, 75–86. [CrossRef] [PubMed]
65. Mieczkowska, A.; Basle, M.F.; Chappard, D.; Mabilleau, G. Thiazolidinedione induced osteocyte apoptosis by a GPR40-dependent mechanism. *J. Biol. Chem.* **2012**, *287*, 23517–23526. [CrossRef]
66. Morrison, A.; Yan, X.; Tong, C.; Li, J. Acute rosiglitazone treatment is cardioprotective against ischemia- reperfusion injury by modulating AMPK, Akt and JNK signalling in nondiabetic mice. *Am. J. Physiol.* **2011**, *301*, H895–H902.
67. Krocker, A.J.; Bruning, J.B. Review of structural and dynamic mechanisms of PPARγ partial agonism. *PPAR Res.* **2015**, *2015*, 816856. [CrossRef]
68. Wright, M.B.; Bortolini, M.; Tadayyon, M.; Bopst, M. Minireview: Challenges and Opportunities in Development of PPAR Agonists. *Mol. Endocrinol.* **2014**, *28*, 1756–1768. [CrossRef]
69. Oberfield, J.L.; Collins, J.L.; Holmes, C.P.; Goreham, D.M.; Cooper, J.P.; Cobb, J.E.; Lenhard, J.M.; Hull-Ryde, E.A.; Mohr, C.P.; Blanchard, S.G.; et al. A peroxisome proliferator-activated receptor gamma ligand inhibits adipocyte differentiation. *Proc. Natl. Acad. Sci. USA* **1999**, *96*, 6102–6106. [CrossRef]
70. Bruning, J.B.; Chalmers, M.J.; Prasad, S.; Busby, S.A.; Kamenecka, T.M.; He, Y.; Nettles, K.W.; Griffin, P.R. Partial agonists activate PPARgamma using a helix 12 independent mechanism. *Structure* **2007**, *15*, 1258–1271. [CrossRef]
71. Amano, Y.; Yamaguchi, T.; Ohno, K.; Niimi, T.; Orita, M.; Sakashita, H.; Takeushi, M. Structural basis for telmisartan-mediated partial activation of PPAR gamma. *Hypertens. Res.* **2012**, *35*, 715–719. [CrossRef]
72. Pochetti, G.; Godio, C.; Mitro, N.; Caruso, D.; Galmozzi, A.; Scurati, S.; Loiodice, F.; Fracchiolla, G.; Tortorella, P.; Laghezza, A.; et al. Insights into the mechanism of partial agonism: Crystal structures of the peroxisome proliferatoractivated receptor gamma ligand-binding domain in the complex with two enantiomeric ligands. *J. Biol. Chem.* **2007**, *282*, 17314–17324. [CrossRef] [PubMed]
73. Waku, T.; Shiraki, T.; Oyama, T.; Fujimoto, Y.; Maebara, K.; Kamiya, N.; Jingami, H.; Morikawa, K. Structural insight into PPARgamma activation through covalent modification with endogenous fatty acids. *J. Mol. Biol.* **2008**, *385*, 188–199. [CrossRef]
74. Itoh, T.; Fairall, L.; Amin, K.; Inaba, Y.; Szanto, A.; Balint, B.L.; Nagy, L.; Yamamoto, K.; Schwabe, J.W. Structural basis for the activation of PPARgamma by oxidized fatty acids. *Nat. Struct. Mol. Biol.* **2008**, *15*, 924–931. [CrossRef] [PubMed]
75. Malapaka, R.R.; Khoo, S.; Zhang, J.; Choi, J.H.; Zhou, X.E.; Xu, Y.; Gong, Y.; Li, J.; Yong, E.L.; Chalmers, M.J.; et al. Identification and mechanism of 10-carbon fatty acid as modulating ligand of peroxisome proliferator-activated receptors. *J. Biol Chem.* **2012**, *287*, 183–195. [CrossRef]
76. Leesnitzer, L.M.; Parks, D.J.; Bledsoe, R.K.; Cobb, J.E.; Collins, J.L.; Consler, T.G.; Davis, R.G.; Hull-Ride, E.A.; Lenhard, J.M.; Patel, L.; et al. Functional consequences of cysteine modification in the ligand binding sites of peroxisome proliferator activated receptors by GW9662. *Biochemistry* **2002**, *41*, 6640–6650. [CrossRef]
77. Nagy, L.; Tontonoz, P.; Alvarez, J.G.A.; Chen, H.; Evans, R.M. Oxidized LDL regulates macrophage gene expression through ligand activation of PPARγ. *Cell* **1998**, *93*, 229–240. [CrossRef]
78. Xu, H.E.; Lambert, M.H.; Montana, V.G.; Parks, D.J.; Blanchard, S.G.; Brown, P.J.; Sternbach, D.D.; Lehmann, J.M.; Wisely, G.B.; Willson, T.M.; et al. Molecular recognition of fatty acids by peroxisome proliferator activated receptors. *Mol. Cell* **1999**, *3*, 397403. [CrossRef] [PubMed]
79. Ziouzenkova, O.; Perrey, S.; Asatryan, L.; Hwang, J.; MacNaul, K.L.; Moller, D.E.; Rader, D.J.; Sevanian, A.; Zechner, R.; Hoefler, G.; et al. Lipolysis of triglyceride-rich lipoproteins generates PPAR ligands: Evidence for an antiinflammatory role for lipoprotein lipase. *Proc. Natl. Acad. Sci. USA* **2003**, *100*, 2730–2735. [CrossRef] [PubMed]
80. Hughes, T.S.; Giri, P.K.; de Vera, I.M.S.; Marciano, D.P.; Kuruvilla, D.S.; Shin, Y.; Blayo, A.L.; Kamenecka, T.M.; Burris, T.P.; Griffin, P.R.; et al. An alternate binding site for PPARγ ligands. *Nat. Commun.* **2014**, *5*, 3571. [CrossRef] [PubMed]
81. Shang, J.; Brust, R.; Mosure, S.A.; Bass, J.; Tello, P.M.; Lin, H.; Hughes, T.S. Cooperative cobinding of synthetic and natural ligands to the nuclear receptor PPARγ. *eLife* **2018**, *7*, e43320. [CrossRef]
82. Hurtado, O.; Ballesteros, I.; Cuartero, M.I.; Moraga, A.; Pradillo, J.M.; Ramirez-Franco, J.; Bartolomè-Martin, D.; Pascual, D.; Torres, M.; Sanchez-Prieto, J.; et al. Daidzein has neuroprotective effects through ligand-binding independent PPARgamma activation. *Neurochem. Int.* **2012**, *61*, 119–127. [CrossRef]

83. Robbins, G.T.; Nie, D. PPAR gamma, bioactive lipids, and cancer progression. *Front. Biosci.* **2012**, *17*, 1816–1834. [CrossRef] [PubMed]
84. Waku, T.; Shiraki, T.; Oyama, T.; Maebara, K.; Nakamori, R.; Morikawa, K. The nuclear receptor PPARgamma individually responds to serotonin- and fatty acid-metabolites. *EMBO J.* **2010**, *29*, 3395–3407. [CrossRef] [PubMed]
85. Nasini, G.; Arnone, A.; Assante, G.; Bava, A.; Moricca, S.; Ragazzi, A. Secondary metabolites of *Cladosporium tenuissimum*, a hyperparasite of rust fungi. *Phytochemistry* **2004**, *65*, 2104–2111. [CrossRef]
86. Moricca, S.; Ragazzi, A.; Assante, G. Biocontrol of rust fungi by *Cladosporium tenuissimum*. In *Rust Diseases of Willow and Poplar*; Hao Pei, M., McCracken, A.R., Eds.; CABI Publishing: Wallingford, UK, 2005; pp. 213–229.
87. Moricca, S.; Ragazzi, A.; Mitchelson, K.R.; Assante, G. Antagonism of the Two-Needle Pine Stem Rust Fungi *Cronartium flaccidum* and *Peridermium pini* by *Cladosporium tenuissimum* In Vitro and In Planta. *Phytopathology* **2001**, *91*, 457–468. [CrossRef] [PubMed]
88. Zurlo, D.; Leone, C.; Assante, G.; Salzano, S.; Renzone, G.; Scaloni, A.; Foresta, C.; Colantuoni, V.; Lupo, A. Cladosporol A stimulates G1-phase arrest of the cell cycle by up-regulation of $p21^{waf1/cip1}$ expression in human colon carcinoma HT-29 cells. *Mol. Carcinog.* **2013**, *52*, 1–17. [CrossRef] [PubMed]
89. Zurlo, D.; Assante, G.; Moricca, S.; Colantuoni, V.; Lupo, A. Cladosporol A, a new Peroxisome Proliferator-Activated Receptor γ (PPARγ) ligand, inhibits colorectal cancer cells proliferation through β-catenin/TCF pathway inactivation. *Biochim. Biophys. Acta* **2014**, *1840*, 2361–2372. [CrossRef]
90. Zurlo, D.; Ziccardi, P.; Votino, C.; Colangelo, T.; Cerchia, C.; Dal Piaz, F.; Dallavalle, S.; Moricca, S.; Novellino, E.; Lavecchia, A.; et al. The antiproliferative and proapoptotic effects of cladosporols A and B are related to their different binding mode as PPARγ ligands. *Biochem. Pharmacol.* **2016**, *108*, 22–35. [CrossRef]
91. Fracchiolla, G.; Laghezza, A.; Piemontese, L.; Parente, M.; Lavecchia, A.; Pochetti, G.; Montanari, R.; Di Giovanni, C.; Carbonara, G.; Tortorella, P.; et al. Synthesis, biological evaluation and molecular investigation of fluorinated peroxisome proliferator-activated receptors alpha/gamma dual agonists. *Bioorg. Med. Chem.* **2012**, *20*, 2141–2151. [CrossRef]
92. Porcelli, L.; Gilardi, F.; Laghezza, A.; Piemontese, L.; Mitro, N.; Azzariti, A.; Altieri, F.; Cervoni, L.; Fracchiolla, G.; Giudici, U.; et al. Synthesis, characterization and biological evaluation of ureidofibrate-like derivatives endowed with peroxisome proliferator-activated receptor activity. *J. Med. Chem.* **2012**, *55*, 37–54. [CrossRef] [PubMed]
93. Vasaturo, M.; Fiengo, L.; De Tommasi, N.; Sabatino, L.; Ziccardi, P.; Colantuoni, V.; Bruno, M.; Cerchia, C.; Novellino, E.; Lupo, A.; et al. A compound-based proteomic approach discloses 15-ketoatractyligenin methyl ester as a new PPARγ partial agonist with anti-proliferative ability. *Sci. Rep.* **2017**, *7*, 41273. [CrossRef] [PubMed]
94. Sabatino, L.; Ziccardi, P.; Cerchia, L.; Muccillo, L.; Piemontese, F.; Loiodice, F.; Colantuoni, V.; Lupo, A.; Lavecchia, A. Chiral phenoxyacetic acid analogue inhibit colon cancer cell proliferation acting as PPARγ partial agonists. *Sci. Rep.* **2019**, *9*, 5434. [CrossRef] [PubMed]
95. Rapuano, R.; Ziccardi, P.; Cioffi, V.; Dallavalle, S.; Moricca, S.; Lupo, A. Cladosporol A and B, two natural peroxisome proliferator-activated receptor gamma (PPARγ) agonists, inhibit adipogenesis in 3T3-L1 cells preadipocytes and cause a conditioned–culture-medium-dependent arrest of HT-29 cell proliferation. *BBA-Gen. Subj.* **2021**, *1865*, 129973. [CrossRef] [PubMed]
96. Rapuano, R.; Riccio, A.; Mercuri, A.; Madera, J.R.; Dallavalle, S.; Moricca, S.; Lupo, A. Proliferation and migration of PC-3 prostate cancer cells is counteracted by PPARγ-cladosporol binding-mediated apoptosis and a decreased lipid biosynthesis and accumulation. *Biochem. Pharmacol.* **2024**, *222*, 116097. [CrossRef] [PubMed]
97. Li, F.; Lu, T.; Liu, D.; Zhang, C.; Zhang, Y.; Dong, F. Upregulated *PPARG2* facilitates interaction with demethylated *AKAP12* gene promoter and suppresses proliferation in prostate cancer. *Cell Death Dis.* **2021**, *12*, 528. [CrossRef] [PubMed]
98. Dong, F.; Liu, D.; Lu, T.; Li, F.; Zhang, C.; Qun, E.; Zhang, Y. PPARγ2 functions as a tumor suppressor in a translational mouse model of human prostate cancer. *Asian J. Androl.* **2022**, *24*, 90–96. [PubMed]
99. Medes, G.; Thomas, A.; Weinhouse, S. Metabolism of neoplastic tissue. IV. A study of lipid synthesis in neoplastic tissue slices in vitro. *Cancer Res.* **1953**, *13*, 27–29. [PubMed]
100. Hanahan, D.; Weinberg, R.A. Hallmarks of Cancer: The Next Generation. *Cell* **2011**, *144*, 646–674. [CrossRef] [PubMed]
101. De Berardinis, R.J.; Chandel, N.S. Fundamentals of cancer metabolism. *Sci. Adv.* **2016**, *2*, e1600200.
102. Mohler, J.L.; Titus, M.A.; Bai, S.; Kennerley, B.J.; Lih, F.B.; Tomer, K.B.; Wilson, E.M. Activation of the androgen receptor by intratumoral bioconversion of androstanediol to dihydrotestosterone in prostate cancer. *Cancer Res.* **2011**, *15*, 1486–1496. [CrossRef]
103. Montgomery, R.B.; Mostaghel, E.A.; Vessella, R.; Hess, D.L.; Kalhorn, T.F.; Higano, C.S.; True, L.D.; Nelson, P.S. Maintenance of intratumoral androgens in metastatic prostate cancer: A mechanism for castration-resistant tumor growth. *Cancer Res.* **2008**, *68*, 4447–4454. [CrossRef] [PubMed]
104. Shang, J.; Koietin, D.J. Structural mechanism underlying ligand binding and activation of PPARγ. *Structure* **2021**, *29*, 940–950. [CrossRef] [PubMed]

biomolecules

MDPI

Review

Exploring PPAR Gamma and PPAR Alpha's Regulation Role in Metabolism via Epigenetics Mechanism

Małgorzata Małodobra-Mazur *, Monika Ołdakowska and Tadeusz Dobosz

Department of Forensic Science, Division of Molecular Techniques, Wroclaw Medical University, Sklodowskiej-Curie 52, 51-367 Wroclaw, Poland; monika.oldakowska@umw.edu.pl (M.O.); tadeusz.dobosz@umw.edu.pl (T.D.)

* Correspondence: malgorzata.malodobra-mazur@umw.edu.pl; Tel.: +48-71-784-15-87

Abstract: Peroxisome proliferator-activated receptors (PPARs) belong to a family of nuclear receptors. To date, three types of PPARs, namely PPARα, PPARδ, and PPARγ, have been identified, demonstrating co-expression across numerous tissues. PPARγ is primarily distributed in adipose tissue, the colon, the immune system, and the retina, while PPARα is predominantly expressed in metabolic tissues such as brown adipose tissue, the liver, and the kidneys. Both PPARγ and PPARα play crucial roles in various cellular processes. Recent data suggest that the PPAR family, among other mechanisms, might also be regulated by epigenetic mechanisms. Our recent studies, alongside numerous others, have highlighted the pivotal roles of DNA methylation and histone modifications in the regulation of PPARγ and PPARα, implicating them in the deterioration of metabolic disorders via epigenetic mechanisms. This still not fully understood mechanism of regulation in the nuclear receptors family has been summarized and described in the present paper. The present review summarizes the available data on PPARγ and PPARα regulation via epigenetic mechanisms, elucidating the link between the development of metabolic disorders and the dysregulation of PPARγ and PPARα resulting from these mechanisms.

Keywords: PPARA; PPARG; metabolic syndrome; obesity; DNA methylation; histone modifications

1. Introduction

Peroxisome proliferator-activated receptors (PPARs) belong to the family of nuclear receptors classified as intranuclear receptors, acting as transcription factors when activated [1]. To date, three types of PPARs have been found to be co-expressed in numerous tissues, but with various distributions throughout the organism. The identified PPAR nuclear receptors, namely PPARα, PPARδ, and PPARγ, are similar in structure and function [2]. The most conservative domain across the three types of receptors is the DNA-binding domain (DBD), which contains two zinc-binding sites. The ligand-binding domain (LBD) is the largest domain and has four main features: (1) the dimerization interface, (2) ligand-binding pocket, (3) coregulator-binding surface, and (4) activation function 2 (AF2) [3].

In general, PPARs, after being activated by a specific ligand, bind to the RXR receptor to create a heterodimer and further regulate the expression of numerous genes. PPARγ ligands lead to the activation of insulin sensitization genes, mainly involved in adipogenesis, macrophage metabolism, and inflammatory genes [3,4]. The activation of PPARα leads to the upregulation of enzymes involved in fatty acid uptake, transport into mitochondria, and subsequent oxidation.

Insulin resistance, as well as type 2 diabetes, are classified as disorders in which an epigenetic component is strongly emphasized [5]. Epigenetics is defined as changes in gene function that are inherited by mitotic or meiotic cells and are not related to changes in the DNA sequence [6]. These changes might either enhance or reduce gene expression [7]. Epigenetics is linked with numerous diseases and disorders including cancer, neurodegenerative diseases, and metabolic disorders. Environmental factors have a significant impact on the DNA methylation profile and histone modifications, leading to the dysregulation of the expression of numerous genes, including insulin signaling and lipid metabolism genes [7]. Numerous studies have emphasized the impact of nutrition on human health, mainly via the epigenetic regulation of numerous processes required for maintaining homeostasis. The mechanism linking nutrition with epigenetic modifications is considered as a factor initiating or leading to numerous disorders, especially metabolic disorders [8].

Epidemiological studies consistently demonstrate a positive association between high-fat and carbohydrate-rich diets and the incidence of insulin resistance and type 2 diabetes. Furthermore, sedentary lifestyles exacerbate these effects, highlighting the intricate interplay between genetic predispositions and environmental influences in the pathogenesis of metabolic diseases [9,10]. Understanding the complex interplay between genetic predispositions and environmental factors is essential for elucidating the pathogenesis of metabolic diseases such as insulin resistance and type 2 diabetes.

The interplay between environmental factors and genetic predispositions is identified as one of the many factors contributing to the development of insulin resistance and type 2 diabetes mellitus (T2DM). Epigenetic modifications, such as DNA methylation and histone modifications, are increasingly recognized as crucial mediators in this relationship [7]. Numerous environmental conditions, as mentioned above, are known to induce epigenetic changes, thereby potentially predisposing individuals to metabolic disorders. Based on numerous studies, insulin resistance, obesity, and T2D have been shown to exhibit distinct alterations in the epigenome that result in the dysregulation of the key gene expression patterns involved in insulin signaling and/or lipid metabolism. However, the complexity of genetic and environmental interaction might also result from the inherited patterns of epigenetic changes, as these modifications can be transmitted via the placenta or sperm, influencing the offspring's health and predisposing them to the development of metabolic disorders [11].

Epigenetic modifications are closely related to numerous diseases and disorders, including metabolic disorders. The present study describes and unifies available data regarding the importance of PPARγ and PPARα in proper insulin signaling and glucose and lipid metabolism via mechanisms connecting these nuclear receptors with epigenetic modifications. The present review describes the role of nuclear receptors in the pathogenesis and development of metabolic syndrome.

2. The Role of PPARγ and PPARα in Insulin Signaling and Glucose and Lipid Metabolism

PPARG is predominantly distributed in adipose tissue, the colon, the immune system, and the retina [12]. Four various mRNA transcript variants (*PPARG*1–4) are generated through alternative splicing. PPARγ plays numerous biological roles, including in the development, distribution and metabolism of adipose tissue [4]. PPARγ is the primary regulatory factor that controls the insulin signaling pathway and overall insulin sensitivity, and is necessary for the proper function of mature adipocytes [13–15]. Two main isoforms of PPARγ are distributed: PPARγ1 and PPARγ2, where the latter is mostly restricted to adipose tissue; however, the expression can be induced elsewhere by HFD [16]. It also plays an essential role in cell differentiation, and the regulation of apoptosis. Moreover, PPARγ inhibits inflammatory processes, exhibits anti-atherosclerosis activity, and improves heart performance [4,12,17]. First of all, PPARγ is the main agent that regulates adipogenesis by interaction with other genes that are necessary for the proper maturation of adipocytes (SRBP, FABP4). In terms of metabolic pathway regulation, the heterodimer PPARγ: RXR, and particularly PPARγ2, has been detected on the following target genes of the glucose metabolism pathway: *H6PD*, *PGD*, *GPI1*, *RPIA*, *PFKL*, *PTI1*, *GPD1*, *PDK1*, and *PCK1*. It also regulates lipid metabolism genes, including *GPAT3*, *LPN*, *LPL*, *CD36*, *ACSL1*, *LIPE*, *PNPLA2*, and others [17–19]. Additionally, it also has been shown to regulate the expression of the adiponectin gene (*ADIPOQ*), adiponectin receptor (*ADIPOR2*), and uncoupling protein 1 (*UCP-1*), and suppress the expression of inflammatory genes [20,21].

Numerous compounds act as PPARγ ligands, including both natural and synthetic substances. The natural agonists include docosahexaenoic acid, eicosatetraenoic acid, other polyunsaturated fatty acids, and some monounsaturated fatty acids. The most well-known group of synthetic ligands of PPARγ are the thiazolidinediones, such as troglitazone, rosiglitazone, and pioglitazone [22]. Numerous agonists of PPARγ have been shown to exhibit positive effects in type 2 diabetic patients, increasing insulin sensitivity, lowering blood glucose levels, and regulating lipid metabolism. Thus, several synthetic PPARγ ligands are successfully used for the treatment of metabolic disorders, including type 2 diabetes [2,4].

PPARα is distributed in numerous metabolically active tissues, mainly in the liver and tissues with an increased degree of mitochondrial oxidation and fatty acid catabolism, such as brown adipocytes, heart muscle, skeletal muscle, and the kidneys [12]. The role of PPARα in glucose homeostasis is not fully understood. PPARα plays a central role in regulating the expression of genes involved in fatty acid oxidation, lipid transport, and lipoprotein metabolism. It mainly promotes fatty acid utilization [12,23]. Through its effects on lipid metabolism, PPARα indirectly influences insulin sensitivity. The excessive accumulation of lipid intermediates, such as diacylglycerols and ceramides, in tissues like skeletal muscle and liver can impair insulin signaling, leading to insulin resistance [24]. Furthermore, it has been shown that PPARα agonists such as fenofibrate and Wy14643 can affect glucose homeostasis by increasing insulin sensitivity in adipocytes and muscle cells, which may be related to reduced lipid accumulation in cells through improved fatty acid β-oxidation [25,26]. It has been also suggested that PPARα impacts glucose homeostasis and indirectly affects pancreatic function. On the other hand, no relationship between fibrates and glucose homeostasis in humans has been demonstrated. Further research is needed to fully understand the role of PPARα in regulating blood glucose levels [26,27].

The significance of PPARα in lipid metabolism extends beyond its hepatic functions. In skeletal muscle, PPARα activation enhances fatty acid oxidation, providing an essential energy source during prolonged exercise or fasting states [28]. Moreover, in adipose tissue, it regulates adipocyte differentiation and lipid storage, impacting the overall energy balance [28,29]. PPARα plays a pivotal role in orchestrating the expression of genes involved in lipid uptake, oxidation, and synthesis [30]. The activation of PPARα leads to the upregulation of fatty acid oxidation enzymes such as acyl-CoA oxidase and carnitine palmitoyltransferase-1 [31], facilitating the breakdown of fatty acids for energy production [32]. Furthermore, PPARα is involved in the regulation of lipoprotein metabolism, particularly in the liver [33]. It enhances the expression of *ApoA-I* and *ApoA-II*, key components of high-density lipoprotein (HDL), contributing to the reverse cholesterol transport process. This function of PPARα aids in reducing the levels of low-density lipoprotein (LDL) cholesterol, thus playing a protective role against atherosclerosis [34]. The activation of this receptor also results in the induction of lipoprotein lipase (LPL), an enzyme crucial for the hydrolysis of triglycerides in circulating lipoproteins. Additionally, PPARα activation promotes the expression of APOC3, an inhibitor of LPL, thereby regulating the availability of free fatty acids for storage [35]. A summary of both PPARγ and PPARα expression and metabolic activity in various tissues of human body is presented in Figure 1.

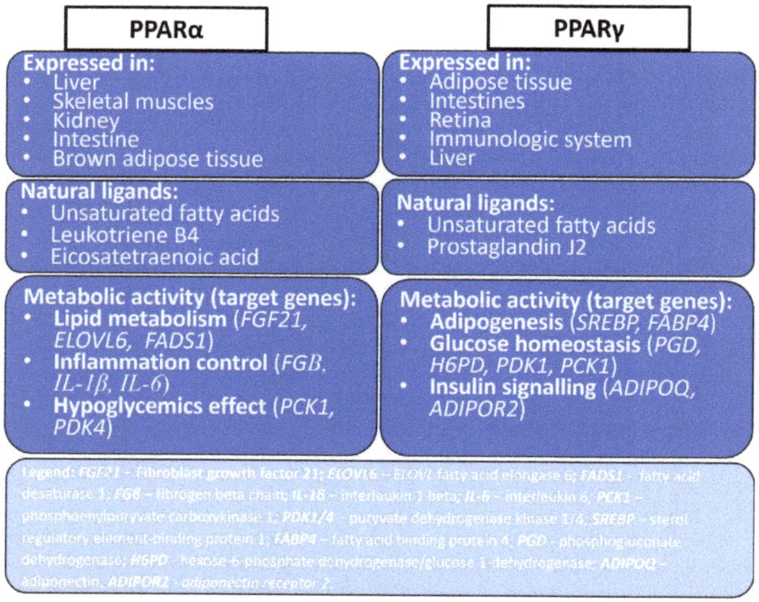

Figure 1. Main characteristic of two isoforms: PPARγ and PPARα. Their expression, function and the list of main ligands that activate the particular peroxisome proliferator are shown [18,36].

3. The Relationship Between Nuclear Receptors and Epigenetic Mechanisms Driving Metabolic Diseases

The interaction between genetic predispositions and environmental influences plays a pivotal role in the pathogenesis of these disorders. Notably, dietary patterns rich in carbohydrates and fats, as well as processed foods, coupled with low physical activity, have been implicated in exacerbating the risk of developing insulin resistance and type 2 diabetes [37].

Peroxisome proliferator-activated receptors are crucial for proper cell metabolism, and any impact on the regulation of these genes substantially influences whole cell homeostasis and metabolism. It has been shown that *PPARG* is among the first genes divergently modified in newly onset insulin resistance [38]. In this context, exploring the complex interplay between epigenetic modifications and metabolic disorders holds promise for uncovering novel therapeutic targets and preventive strategies. A deeper understanding of these processes will enhance our ability to mitigate the burgeoning global burden of insulin resistance and T2D. Specific details of the collected data are presented below and summarized in Table 1.

3.1. Insights from DNA Methylation Studies

Various studies provide evidence for the simultaneous involvement of epigenetic and environmental factors in the development of metabolic diseases. The body of literature has demonstrated a clear relationship between alterations in DNA methylation and the histone modifications affecting various genes implicated in metabolic pathways. Considering the pivotal role of transcription factors in the regulation of the expression of numerous genes, significant attention has been paid to this context.

Numerous studies have indicated the pivotal role of DNA methylation in the regulation of the expression and proper function of *PPARG* in health and homeostasis, influencing many important life processes [17,19,38,39]. PPARγ is also considered to play a significant role in the pathogenesis of many diseases, particularly metabolic disorders and the epigenetic regulation of *PPARG*; in particular, DNA methylation has a significant impact, mainly by regulating *PPARG* expression. Consequently, the disruption of the expression of the *PPARG* gene can lead to various pathologies.

We have recently provided evidence that PPARγ undergoes epigenetic regulation, and any rearrangements lead to numerous metabolic disorders such as obesity or insulin resistance. Firstly, we have shown that the *PPARG* promoter is hypermethylated in obese and type 2 diabetic patients, which correlates with the downregulation of the expression of numerous genes responsible for proper insulin signal transduction in adipocytes [40]. In vivo studies have revealed that the *PPARG* promoter is hypermethylated in the adipose tissue of type 2 diabetic patients, both in visceral and subcutaneous adipose tissues. The hypermethylation positively correlated with the insulin resistance stage (assessed by HOMA-IR) and negatively with the expression of *PPARG*. Our observation has been supported by others who also demonstrated that epigenetic regulation has an impact on *PPARG* expression [41–43]. We and others have observed a distinct promoter methylation pattern in *PPARG* between various human fat depots, especially between subcutaneous and visceral adipose tissue. This observation likely arises from the fact that different fat depots (various types of adipose tissue) perform specific and distinct functions in the human body. We have observed considerable metabolic differences between SAT and VAT [44] concerning various aspects such as lipid metabolism, inflammatory state, insulin resistance induction, and lipid accumulation.

The observations gleaned from the in vivo investigation were subsequently replicated in vitro in a cell culture study, which enabled us to derive congruent conclusions. We showed the hypermethylation of the *PPARG* promoter, which correlated with the downregulation of *PPARG* expression in adipocytes with artificially induced insulin resistance [38]. Additionally, we demonstrated that in adipocytes with newly developed insulin resistance, global DNA methylation was increased, which correlated with the expression of *DNMT1* in those cells. Furthermore, the first gene to respond to changes in the DNA methylation profile due to high-fat diet-induced insulin resistance in adipocytes was *PPARG*. These changes in DNA methylation were observed as early as 72 h after insulin resistance induction by a palmitic acid (16:0), mimicking the high-fat diet. Our results might suggest that *PPARG*, acting as the transcription factor, may be the first response to the changing environmental conditions. No other analyzed genes showed dysregulation in either the expression rate or methylation profile after 72 h of insulin resistance induction [38].

The importance of epigenetic factors in the regulation of PPARs concerning metabolic diseases has been intensively studied by others as well. Volberg et al. observed differentially methylated CpG islands of the *PPARG* promoter in 9-year-old children; these negatively correlated with the birth weight and BMI of the children at the age of 9 years [43]. Another study found that a higher risk of type 2 diabetes is associated with the hypermethylation of the *PPARG* promoter in the pancreatic islets of diabetic patients, which negatively correlated with insulin secretion [45]. Similar results were obtained by Nilsson et al., where the hypermethylation of *PPARG* promoters was shown in type 2 diabetic patients compared to non-diabetes individuals in adipose tissue [46]. Epigenetic regulation has also been shown to impact *PPARG* regulation in non-human subjects. In overweight chickens, the promoter of *Pparg* was differentially methylated at three CpG positions compared to lean chickens [42]. A similar observation was made in db/db mice in terms of the hypermethylation of *Pparg* promoter in epididymal adipose tissue in comparison to wild-type animals; this negatively correlated with *PPARG* expression in the analyzed sample [47]. However, analyzing the above presented data, in vivo studies in both humans and animals should be carefully interpretated, as the analyzed tissues generally are composed of numerous various cells, where the expression profile of genes as well as their methylation status might vary.

The emerging role of epigenetic regulation, particularly DNA methylation, in *PPARG* function and action has been observed in various disorders. The dysregulation of *PPARG* methylation has been documented in idiopathic pulmonary fibrosis (IPF) patients. Wei et al. demonstrated the hypermethylation of *PPARG* in the lungs of IPF patients, which inversely correlated with the expression level and PPARG function [48]. Conversely, demethylation by 5′aza ameliorates the negative effect of IPF and restores the correct expression and function of *PPARG*. Similar results were obtained in the case of liver fibrosis, where the inflammatory state and liver fibrosis strongly correlated with the hypermethylation of *PPARG*, resulting in lower expression [49]. Furthermore, Hardy et al. proposed using the *PPARG* methylation status as a biomarker of liver fibrosis [50].

DNA methylation plays a crucial role in the regulation processes essential for adipogenesis, i.e., the formation of mature adipocytes. Proper epigenetic regulation is essential in this process, which, due to its specificity, is sensitive to external factors. Numerous transcription factors regulate adipogenesis, creating a network that can be easily disturbed [21]. PPARγ is a central regulator of adipogenesis because numerous genes possess PPARγ-binding sites. Thus, the hypermethylation of *PPARG* itself or its target genes might directly or indirectly link *PPARG* with the epigenetic regulation of adipogenesis and the metabolism of mature adipocytes, including a shift towards metabolic disorders. As PPARγ acts as the nuclear transcription factor, changes in the methylation profile of target genes might influence its binding to the specific response elements, thereby regulating adipogenesis. The importance of methylation-specific adipogenesis has been demonstrated, showing the significant role of PPARγ in the differentiation and function of mature adipocytes [21]. We have also previously demonstrated that the methylation of *PPARG* plays a crucial role in adipogenesis [51]. Moreover, we have shown that nutritional factors, especially fatty acids, play a significant role in methylome, including the methylation of *PPARG*. This impacts the differentiation process and the phenotype of mature adipocytes, shifting the adipocyte metabolism toward metabolic disorders [51,52]. There might be several possible mechanisms that affect how nutritional factors, especially fatty acids, influence DNA methylation [53]. First, fatty acids directly influence the expression and action of DNA methyltransferases. Second, ligands of various transcription factors might regulate epigenetic modification. Lastly, it has been proposed that fatty acids interact with MeCP2 (methyl CpG-binding protein (2), mainly in promoter regions regulating the expression of numerous genes [53].

Similar to PPARγ, its isoform PPARα also undergoes epigenetic regulation, including DNA methylation [54]. The promoter of *PPARA* is hypermethylated in type 2 diabetic patients with non-alcoholic fatty liver disease (NAFLD) [55]. Moreover, *PPARA* has been shown to undergo hydroxymethylation modifications that influence its expression, predisposing individuals to NAFLD and the development of metabolic syndrome [56]. Some studies have demonstrated the various DNA methylation patterns of *PPARA* in patients with metabolic syndrome and significant hyperlipidemia [57]. Castellano-Castillo et al. showed global hypermethylation in the visceral adipose tissue of patients with metabolic syndrome by assessing the methylation of LINE-1, which positively correlated with BMI and negatively correlated with insulin sensitivity (assessed by HOMA-IR index).

Furthermore, in addition to global DNA methylation, changes in the site-specific DNA methylation of numerous genes have been observed. These changes also correlated with metabolic dysregulation, including genes important for adipogenesis regulation, lipid metabolism, and inflammation. This suggests that DNA methylation, especially *PPARA*, *LPL*, *SCD* and *TNF-α*, is implicated in metabolism dysregulation and the pathogenesis of metabolic syndrome, involving adipose tissue metabolism dysregulation and the induction of the anti-inflammatory state [57].

3.2. *Insights from the Histones Modifications Studies*

Histone modifications are correlated with both the induction and downregulation of gene expression, depending on the site and type of modification. Generally, histone acetylation is associated with the induction of gene expression, while histone methylation, with some exceptions, is associated with gene expression downregulation. Histone acetylation maintains the negative charge of chromatic by removing the positive charge from the histone tail, neutralizing it, and thereby reducing its interaction with negatively charged

DNA. As a result of chromatin relaxation, DNA becomes more accessible to numerous transcription factors [58]. Histone methylation can involve the mono-, di- or trimethylation of lysines or arginines of histone tails, and its effect on gene expression can be either enhancing or repressing, depending on the site and number of methyl groups added [59]. Our results clearly indicate the emerging role of histone modification in the induction of metabolic disorders, including obesity and insulin resistance. Notably, we have shown a global negative correlation between specific changes (H3K4me3 and H3K9/14ac) and insulin resistance, assessed by HOMA-IR [5]. Additionally, we have shown the downregulation of *SIRT1* and *SIRT7*, a key family of histones deacetylases in adipocytes with insulin resistance.

Lastly, we observed the lower enrichment level of H3K4me3 and H3K9/14ac within the *PPARG* promoter, which are the main markers of chromatin induction, corresponding with lower expression. Similar histone modifications and *SIRT7* downregulation were observed in both visceral and subcutaneously derived adipocytes, indicating a similar mechanism of epigenetic regulation in both fat depots. Furthermore, we previously demonstrated that *SIRT1* and *SIRT7* positively correlated with the expression of numerous genes involved in insulin signaling *(INSR, PIK3R1, AKT, SLC2A4)* and lipid metabolism (*ACC, FASN, SCD-1, LPL*), including *PPARG*. This suggests the emerging role of histone-modifying genes in the regulation of energy metabolism [5]. Indeed, numerous other researchers have shown the regulatory role and impact of Sirtuin family genes on the pathogenesis of metabolic disorders [60,61]. However, some data have indicated a negative correlation between *PPARG* and *SIRT1* [62].

Specific histone modifications have previously been correlated with divergent *PPARG* expression and the development of metabolic disorders. Castellano-Castillo et al. [63] reported the lower H3K4me3 enrichment of the *PPARG* promoter in adipose tissue from obese individuals compared to lean patients, suggesting an association with increased BMI and subsequent metabolic deterioration. Histone acetylation is correlated with chromatin induction and gene expression enhancement. Thus, histone acetylases (HAT) are believed to be key epigenetic players in adipogenesis and the regulation of energy metabolism [64]. According to Lefterova et al., the increased enrichment of H3K9ac marks was observed at *PPARG* binding sites during adipogenesis [65]. In our study, we have shown the lower H3K9/14ac enrichment of the *PPARG* promoter itself, as well as downstream targeted genes [5], in insulin-resistant adipocytes. Among others, Wang et al. [66] demonstrated that the downregulation of *PPARG* decreased the expression of *SLC1A5*, leading to a predisposition to obesity and insulin resistance. Additionally, *PPARG* expression was regulated by H3K27ac or H3K4me3.

The opposing epigenetic-modifying enzymes, specifically the HDAC family, have also been implicated in both adipogenesis and metabolic regulation via PPARs, mainly *PPARG* and *PPARA*. HDACs inhibitors have been shown to be involved in fatty acid metabolism and to play a protective role in the development of diabetic cardiomyopathy [67]. HDAC inhibitors also protect against atherosclerosis by enhancing the expression of *PPARG* [68]. The inhibition of HDACs maintains a higher rate of chromatic acetylation, which is a primary chromatin activation marker. According to Fu et al., the possible mechanism linking the activity of HDACs with the PPAR nuclear family (both *PPARG* and *PPARA)* is Cyclin D1 [69]. However, other mechanisms of interaction between *PPARG* and histone-modifying enzymes are possible and remain to be elucidated.

Table 1. Epigenetic regulation of *PPARG* and *PPARA* and its effect on metabolism regulation, mainly in relation to glucose and lipid metabolism as well as insulin sensitivity.

Epigenetic Modification	Gene of Interest	Sample Type	Observed Effects	Pathway	References
Promoter hypermethylation	*PPARG*	Human adipose tissue SAT and VAT—in vivo study	Downregulation of *PPARG* expression as well as other insulin signaling and lipid metabolism genes. Insulin resistance and dysregulation of lipid levels	Insulin signaling Lipid metabolism	[40,46]
		Adipose tissue—animal model in vivo study	Metabolic syndrome development	Lipid metabolism	[47]
Promoter hypermethylation	*PPARG*	Human adipose tissue: surface and deep—in vivo study	Hypermethylation in deep adipose tissue correlated with lower PPARγ protein content in fat depot, lower adipogenicity properties and sensitivity to adipogenic agents	Adipogenesis	[41]
Various promoter methylation pattern	*PPARG*	Chicken adipose tissue—animal model in vivo study	Lower methylation in adipose tissue of fat chicken Age-related promoter methylation	Adipogenesis Lipid accumulation in adipose tissue	[42]
Promoter methylation pattern	*PPARG*	Human blood—in vivo study	Methylation status of some CpG was negatively correlated with birth weight and increased risk of obesity. Impact of birth weight on metabolism	Association with perinatal factors Hypothesis of programming metabolism	[43]
Promoter and gene body hypomethylation	*PPARG*	Human adipocytes—in vitro study	Hypomethylation of *PPARG* promoter during adipogenesis influences the rate of adipogenesis, lipid accumulation and phenotype of mature adipocytes. Hypomethylation is promoted by fatty acids supplementation	Adipogenesis	[21,51]
Promoter hypermethylation	*PPARA*	Hepatocytes—animal model in vitro study	Downregulation of mRNA and protein level of PPARα. Relationship with the pathogenesis of non-alcoholic fatty liver disease	Disruption in lipid accumulation	[55]
Global and gene-specific hypermethylation	*PPARA*	Human VAT—in vivo study	Dysregulation of gene expression including PPARα and downstream genes that strongly positively correlated with TG level	Metabolic syndrome development	[57]
Epigenetic changes related to circadian clock by H3K27ac and H3K4me3	*PPARG*	Adipose tissue—animal model in vivo study	Downregulation of PPARγ was a consequence of histone modification at H3K27ac or H3K4me3 leading to downregulation of further genes, including *SLC1A5*	Downregulation of SLC1A5 and further reduction in glutamine and methionine uptake	[66]
Changes at H3K4me3 and H3K9/14ac	*PPARG*	Human VAT and SAT—in vivo study	Dysregulation of *PPARG* and correlation with insulin resistance. H3K4me3 enrichment of *PPARG* directly correlated with BMI	Insulin signaling Lipids metabolism	[5,63]
Sirt-1-dependent action	*PPARG*	Adipocytes—in vivo and in vitro animal model study	Downregulation of PPARγ, lower mRNA and protein content	Lipolysis	[65]

The potential mechanism linking epigenetics (both DNA methylation and histone modification) with PPARs and the regulation of their expression is presented in Figure 2.

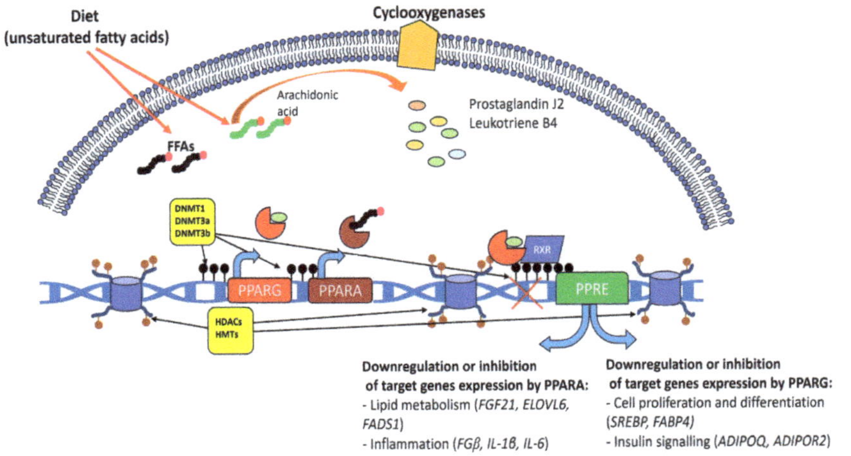

Figure 2. Schematic mechanism of epigenetic regulation involved in PPARγ and PPARγ expression and action, and their effects on cell metabolism. DNMT—DNA methyltransferases; HDAC—histone deacetylases; HMT—histone methylases, FGF21—Fibroblast growth factor 21; ELOVL6—ELOVL fatty acid elongase 6; FADS1—fatty acid desaturase 1; FGβ—fibrogen beta chain; IL-1β—interleukin 1 beta; IL-6—interleukin 6; SREBP—sterol regulatory element-binding protein 1; FABP4—fatty acid-binding protein 4; ADIPOQ—adiponectin; ADIPOR2—adiponectin receptor 2 (graph was prepared using PowerPoint software 2021).

4. Conclusions

To conclude, the above data provide evidence that epigenetic regulation is one of the main mechanisms controlling PPAR action, which directly or indirectly affects the downstream genes responsible for metabolism regulation. Discrepancies in epigenetic regulation, such as dysregulation in the DNA methylation of histones modifications, might lead to disruptions in homeostasis and consequently contribute to the pathogenesis of metabolic disorders.

Understanding epigenetic regulation is crucial for the prevention, prediction, and future treatment of metabolic disorders. It is very likely that epigenetic-modifying agents could be effective in managing metabolic disorders.

Author Contributions: Conceptualization, T.D. and M.M.-M.; methodology, M.M.-M. and M.O.; resources, M.M.-M. and M.O.; data curation, M.M.-M. and M.O.; writing—original draft preparation, M.M.-M. and M.O.; writing—review and editing, T.D. and M.M.-M.; visualization, M.M.-M. All authors have read and agreed to the published version of the manuscript.

Funding: This research received no external funding.

Institutional Review Board Statement: Not applicable.

Informed Consent Statement: Not applicable.

Data Availability Statement: Not applicable.

Conflicts of Interest: The authors declare no conflicts of interest.

References

1. Li, J.; Guo, C.; Wu, J. The Agonists of Peroxisome Proliferator-Activated Receptor-γ for Liver Fibrosis. *Drug Des. Devel. Ther.* **2021**, *15*, 2619–2628. [CrossRef] [PubMed]
2. Yousefnia, S.; Momenzadeh, S.; Seyed Forootan, F.; Ghaedi, K.; Nasr Esfahani, M.H. The Influence of Peroxisome Proliferator-Activated Receptor γ (PPARγ) Ligands on Cancer Cell Tumorigenicity. *Gene* **2018**, *649*, 14–22. [CrossRef] [PubMed]
3. Chandra, V.; Huang, P.; Hamuro, Y.; Raghuram, S.; Wang, Y.; Burris, T.P.; Rastinejad, F. Structure of the Intact PPAR-γ–RXR-α Nuclear Receptor Complex on DNA. *Nature* **2008**, *456*, 350–356. [CrossRef] [PubMed]
4. Mirza, A.Z.; Althagafi, I.I.; Shamshad, H. Role of PPAR Receptor in Different Diseases and Their Ligands: Physiological Importance and Clinical Implications. *Eur. J. Med. Chem.* **2019**, *166*, 502–513. [CrossRef]
5. Małodobra-Mazur, M.; Cierzniak, A.; Myszczyszyn, A.; Kaliszewski, K.; Dobosz, T. Histone Modifications Influence the Insulin-Signaling Genes and Are Related to Insulin Resistance in Human Adipocytes. *Int. J. Biochem. Cell Biol.* **2021**, *137*, 106031. [CrossRef]
6. Deans, C.; Maggert, K.A. What Do You Mean, "Epigenetic"? *Genetics* **2015**, *199*, 887–896. [CrossRef]
7. Zhang, L.; Lu, Q.; Chang, C. Epigenetics in Health and Disease. *Adv. Exp. Med. Biol.* **2020**, *1253*, 3–55. [CrossRef]
8. Jiménez-Chillarón, J.C.; Díaz, R.; Martínez, D.; Pentinat, T.; Ramón-Krauel, M.; Ribó, S.; Plösch, T. The Role of Nutrition on Epigenetic Modifications and Their Implications on Health. *Biochimie* **2012**, *94*, 2242–2263. [CrossRef]
9. Roberts, C.K.; Hevener, A.L.; Barnard, R.J. Metabolic Syndrome and Insulin Resistance: Underlying Causes and Modification by Exercise Training. *Compr. Physiol.* **2013**, *3*, 1–58. [CrossRef]
10. Swarup, S.; Ahmed, I.; Grigorova, Y.; Zeltser, R. Metabolic Syndrome. In *StatPearls*; StatPearls Publishing: Treasure Island, FL, USA, 2024.
11. Bhadsavle, S.S.; Golding, M.C. Paternal Epigenetic Influences on Placental Health and Their Impacts on Offspring Development and Disease. *Front. Genet.* **2022**, *13*, 1068408. [CrossRef]
12. Wang, Y.-X. PPARs: Diverse Regulators in Energy Metabolism and Metabolic Diseases. *Cell Res.* **2010**, *20*, 124–137. [CrossRef] [PubMed]
13. Ahmadian, M.; Suh, J.M.; Hah, N.; Liddle, C.; Atkins, A.R.; Downes, M.; Evans, R.M. PPARγ Signaling and Metabolism: The Good, the Bad and the Future. *Nat. Med.* **2013**, *19*, 557–566. [CrossRef] [PubMed]
14. Imai, T.; Takakuwa, R.; Marchand, S.; Dentz, E.; Bornert, J.-M.; Messaddeq, N.; Wendling, O.; Mark, M.; Desvergne, B.; Wahli, W.; et al. Peroxisome Proliferator-Activated Receptor Gamma Is Required in Mature White and Brown Adipocytes for Their Survival in the Mouse. *Proc. Natl. Acad. Sci. USA* **2004**, *101*, 4543–4547. [CrossRef] [PubMed]
15. He, W.; Barak, Y.; Hevener, A.; Olson, P.; Liao, D.; Le, J.; Nelson, M.; Ong, E.; Olefsky, J.M.; Evans, R.M. Adipose-Specific Peroxisome Proliferator-Activated Receptor Gamma Knockout Causes Insulin Resistance in Fat and Liver but Not in Muscle. *Proc. Natl. Acad. Sci. USA* **2003**, *100*, 15712–15717. [CrossRef]
16. Medina-Gomez, G.; Gray, S.L.; Yetukuri, L.; Shimomura, K.; Virtue, S.; Campbell, M.; Curtis, R.K.; Jimenez-Linan, M.; Blount, M.; Yeo, G.S.H.; et al. PPAR Gamma 2 Prevents Lipotoxicity by Controlling Adipose Tissue Expandability and Peripheral Lipid Metabolism. *PLoS Genet.* **2007**, *3*, e64. [CrossRef]
17. Medina-Gomez, G.; Gray, S.; Vidal-Puig, A. Adipogenesis and Lipotoxicity: Role of Peroxisome Proliferator-Activated Receptor γ (PPARγ) and PPARγcoactivator-1 (PGC1). *Public Health Nutr.* **2007**, *10*, 1132–1137. [CrossRef]
18. Nielsen, R.; Pedersen, T.Å.; Hagenbeek, D.; Moulos, P.; Siersbæk, R.; Megens, E.; Denissov, S.; Børgesen, M.; Francoijs, K.-J.; Mandrup, S.; et al. Genome-Wide Profiling of PPARγ:RXR and RNA Polymerase II Occupancy Reveals Temporal Activation of Distinct Metabolic Pathways and Changes in RXR Dimer Composition during Adipogenesis. *Genes Dev.* **2008**, *22*, 2953. [CrossRef]
19. Lehrke, M.; Lazar, M.A. The Many Faces of PPARγ. *Cell* **2005**, *123*, 993–999. [CrossRef]
20. Wang, S.; Dougherty, E.J.; Danner, R.L. PPARγ Signaling and Emerging Opportunities for Improved Therapeutics. *Pharmacol. Res.* **2016**, *111*, 76–85. [CrossRef]
21. Takada, H.; Saito, Y.; Mituyama, T.; Wei, Z.; Yoshihara, E.; Jacinto, S.; Downes, M.; Evans, R.M.; Kida, Y.S. Methylome, Transcriptome, and PPARγ Cistrome Analyses Reveal Two Epigenetic Transitions in Fat Cells. *Epigenetics* **2014**, *9*, 1195–1206. [CrossRef]
22. Dumasia, R.; Eagle, K.A.; Kline-Rogers, E.; May, N.; Cho, L.; Mukherjee, D. Role of PPAR- Gamma Agonist Thiazolidinediones in Treatment of Pre-Diabetic and Diabetic Individuals: A Cardiovascular Perspective. *Curr. Drug Targets Cardiovasc. Haematol. Disord.* **2005**, *5*, 377–386. [CrossRef] [PubMed]
23. Kersten, S. Peroxisome Proliferator Activated Receptors and Lipoprotein Metabolism. *PPAR Res.* **2007**, *2008*, e132960. [CrossRef] [PubMed]
24. Park, S.S.; Seo, Y.-K. Excess Accumulation of Lipid Impairs Insulin Sensitivity in Skeletal Muscle. *Int. J. Mol. Sci.* **2020**, *21*, 1949. [CrossRef] [PubMed]
25. Lefere, S.; Puengel, T.; Hundertmark, J.; Penners, C.; Frank, A.K.; Guillot, A.; de Muynck, K.; Heymann, F.; Adarbes, V.; Defrêne, E.; et al. Differential Effects of Selective- and Pan-PPAR Agonists on Experimental Steatohepatitis and Hepatic Macrophages ☆. *J. Hepatol.* **2020**, *73*, 757–770. [CrossRef]
26. Lin, Y.; Wang, Y.; Li, P. PPARα: An Emerging Target of Metabolic Syndrome, Neurodegenerative and Cardiovascular Diseases. *Front. Endocrinol.* **2022**, *13*, 1074911. [CrossRef]

27. Black, R.N.A.; Ennis, C.N.; Young, I.S.; Hunter, S.J.; Atkinson, A.B.; Bell, P.M. The Peroxisome Proliferator-Activated Receptor Alpha Agonist Fenofibrate Has No Effect on Insulin Sensitivity Compared to Atorvastatin in Type 2 Diabetes Mellitus; a Randomised, Double-Blind Controlled Trial. *J. Diabetes Complicat.* **2014**, *28*, 323–327. [CrossRef]
28. Burri, L.; Thoresen, G.H.; Berge, R.K. The Role of PPAR α Activation in Liver and Muscle. *PPAR Res.* **2010**, *2010*, 542359. [CrossRef]
29. Petersen, K.F.; Dufour, S.; Savage, D.B.; Bilz, S.; Solomon, G.; Yonemitsu, S.; Cline, G.W.; Befroy, D.; Zemany, L.; Kahn, B.B.; et al. The Role of Skeletal Muscle Insulin Resistance in the Pathogenesis of the Metabolic Syndrome. *Proc. Natl. Acad. Sci. USA* **2007**, *104*, 12587–12594. [CrossRef]
30. Morris, D.L.; Cho, K.W.; Zhou, Y.; Rui, L. SH2B1 Enhances Insulin Sensitivity by Both Stimulating the Insulin Receptor and Inhibiting Tyrosine Dephosphorylation of Insulin Receptor Substrate Proteins. *Diabetes* **2009**, *58*, 2039–2047. [CrossRef]
31. Pawlak, M.; Lefebvre, P.; Staels, B. Molecular Mechanism of PPARα Action and Its Impact on Lipid Metabolism, Inflammation and Fibrosis in Non-Alcoholic Fatty Liver Disease. *J. Hepatol.* **2015**, *62*, 720–733. [CrossRef]
32. Montagner, A.; Polizzi, A.; Fouché, E.; Ducheix, S.; Lippi, Y.; Lasserre, F.; Barquissau, V.; Régnier, M.; Lukowicz, C.; Benhamed, F.; et al. Liver PPARα Is Crucial for Whole-Body Fatty Acid Homeostasis and Is Protective against NAFLD. *Gut* **2016**, *65*, 1202–1214. [CrossRef] [PubMed]
33. Todisco, S.; Santarsiero, A.; Convertini, P.; De Stefano, G.; Gilio, M.; Iacobazzi, V.; Infantino, V. PPAR Alpha as a Metabolic Modulator of the Liver: Role in the Pathogenesis of Nonalcoholic Steatohepatitis (NASH). *Biology* **2022**, *11*, 792. [CrossRef] [PubMed]
34. Zandbergen, F.; Plutzky, J. PPARα in Atherosclerosis and Inflammation. *Biochim. Biophys. Acta (BBA)-Mol. Cell Biol. Lipids* **2007**, *1771*, 972–982. [CrossRef] [PubMed]
35. Fuior, E.V.; Zvintzou, E.; Filippatos, T.; Giannatou, K.; Mparnia, V.; Simionescu, M.; Gafencu, A.V.; Kypreos, K.E. Peroxisome Proliferator-Activated Receptor α in Lipoprotein Metabolism and Atherosclerotic Cardiovascular Disease. *Biomedicines* **2023**, *11*, 2696. [CrossRef]
36. Rakhshandehroo, M.; Knoch, B.; Müller, M.; Kersten, S. Peroxisome Proliferator-Activated Receptor Alpha Target Genes. *PPAR Res.* **2010**, *2010*, 612089. [CrossRef]
37. Fahed, G.; Aoun, L.; Bou Zerdan, M.; Allam, S.; Bou Zerdan, M.; Bouferraa, Y.; Assi, H.I. Metabolic Syndrome: Updates on Pathophysiology and Management in 2021. *Int. J. Mol. Sci.* **2022**, *23*, 786. [CrossRef]
38. Małodobra-Mazur, M.; Cierzniak, A.; Kaliszewski, K.; Dobosz, T. PPARG Hypermethylation as the First Epigenetic Modification in Newly Onset Insulin Resistance in Human Adipocytes. *Genes* **2021**, *12*, 889. [CrossRef]
39. Shen, Y.; Su, Y.; Silva, F.J.; Weller, A.H.; Sostre-Colón, J.; Titchenell, P.M.; Steger, D.J.; Seale, P.; Soccio, R.E. Shared PPARα/γ Target Genes Regulate Brown Adipocyte Thermogenic Function. *Cell Rep.* **2020**, *30*, 3079–3091.e5. [CrossRef]
40. Cierzniak, A.; Pawelka, D.; Kaliszewski, K.; Rudnicki, J.; Dobosz, T.; Malodobra-Mazur, M. DNA Methylation in Adipocytes from Visceral and Subcutaneous Adipose Tissue Influences Insulin-Signaling Gene Expression in Obese Individuals. *Int. J. Obes.* **2021**, *45*, 650–658. [CrossRef]
41. Kosaka, K.; Kubota, Y.; Adachi, N.; Akita, S.; Sasahara, Y.; Kira, T.; Kuroda, M.; Mitsukawa, N.; Bujo, H.; Satoh, K. Human Adipocytes from the Subcutaneous Superficial Layer Have Greater Adipogenic Potential and Lower PPAR-γ DNA Methylation Levels than Deep Layer Adipocytes. *Am. J. Physiol. Cell Physiol.* **2016**, *311*, C322–C329. [CrossRef]
42. Sun, Y.N.; Gao, Y.; Qiao, S.P.; Wang, S.Z.; Duan, K.; Wang, Y.X.; Li, H.; Wang, N. Epigenetic DNA Methylation in the Promoters of Peroxisome Proliferator-Activated Receptor γ in Chicken Lines Divergently Selected for Fatness. *J. Anim. Sci.* **2014**, *92*, 48–53. [CrossRef] [PubMed]
43. Volberg, V.; Yousefi, P.; Huen, K.; Harley, K.; Eskenazi, B.; Holland, N. CpG Methylation across the Adipogenic PPARγ Gene and Its Relationship with Birthweight and Child BMI at 9 Years. *BMC Med. Genet.* **2017**, *18*, 7. [CrossRef] [PubMed]
44. Małodobra-Mazur, M.; Cierzniak, A.; Pawełka, D.; Kaliszewski, K.; Rudnicki, J.; Dobosz, T. Metabolic Differences between Subcutaneous and Visceral Adipocytes Differentiated with an Excess of Saturated and Monounsaturated Fatty Acids. *Genes* **2020**, *11*, 1092. [CrossRef] [PubMed]
45. Ling, C.; Del Guerra, S.; Lupi, R.; Rönn, T.; Granhall, C.; Luthman, H.; Masiello, P.; Marchetti, P.; Groop, L.; Del Prato, S. Epigenetic Regulation of PPARGC1A in Human Type 2 Diabetic Islets and Effect on Insulin Secretion. *Diabetologia* **2008**, *51*, 615–622. [CrossRef] [PubMed]
46. Nilsson, E.; Jansson, P.A.; Perfilyev, A.; Volkov, P.; Pedersen, M.; Svensson, M.K.; Poulsen, P.; Ribel-Madsen, R.; Pedersen, N.L.; Almgren, P.; et al. Altered DNA Methylation and Differential Expression of Genes Influencing Metabolism and Inflammation in Adipose Tissue From Subjects With Type 2 Diabetes. *Diabetes* **2014**, *63*, 2962–2976. [CrossRef]
47. Fujiki, K.; Kano, F.; Shiota, K.; Murata, M. Expression of the Peroxisome Proliferator Activated Receptor Gamma Gene Is Repressed by DNA Methylation in Visceral Adipose Tissue of Mouse Models of Diabetes. *BMC Biol.* **2009**, *7*, 38. [CrossRef]
48. Wei, A.; Gao, Q.; Chen, F.; Zhu, X.; Chen, X.; Zhang, L.; Su, X.; Dai, J.; Shi, Y.; Cao, W. Inhibition of DNA Methylation De-Represses Peroxisome Proliferator-Activated Receptor-γ and Attenuates Pulmonary Fibrosis. *Br. J. Pharmacol.* **2022**, *179*, 1304–1318. [CrossRef]
49. Zhao, Q.; Fan, Y.-C.; Zhao, J.; Gao, S.; Zhao, Z.-H.; Wang, K. DNA Methylation Patterns of Peroxisome Proliferator-Activated Receptor Gamma Gene Associated with Liver Fibrosis and Inflammation in Chronic Hepatitis B. *J. Viral Hepat.* **2013**, *20*, 430–437. [CrossRef]

50. Hardy, T.; Zeybel, M.; Day, C.P.; Dipper, C.; Masson, S.; McPherson, S.; Henderson, E.; Tiniakos, D.; White, S.; French, J.; et al. Plasma DNA Methylation: A Potential Biomarker for Stratification of Liver Fibrosis in Non-Alcoholic Fatty Liver Disease. *Gut* **2017**, *66*, 1321–1328. [CrossRef]
51. Malodobra-Mazur, M.; Cierzniak, A.; Dobosz, T. Oleic Acid Influences the Adipogenesis of 3T3-L1 Cells via DNA Methylation and May Predispose to Obesity and Obesity-Related Disorders. *Lipids Health Dis.* **2019**, *18*, 230. [CrossRef]
52. Gomez-Alonso, M.d.C.; Kretschmer, A.; Wilson, R.; Pfeiffer, L.; Karhunen, V.; Seppälä, I.; Zhang, W.; Mittelstraß, K.; Wahl, S.; Matias-Garcia, P.R.; et al. DNA Methylation and Lipid Metabolism: An EWAS of 226 Metabolic Measures. *Clin. Epigenetics* **2021**, *13*, 7. [CrossRef] [PubMed]
53. González-Becerra, K.; Ramos-Lopez, O.; Barrón-Cabrera, E.; Riezu-Boj, J.I.; Milagro, F.I.; Martínez-López, E.; Martínez, J.A. Fatty Acids, Epigenetic Mechanisms and Chronic Diseases: A Systematic Review. *Lipids Health Dis.* **2019**, *18*, 178. [CrossRef] [PubMed]
54. Theys, C.; Lauwers, D.; Perez-Novo, C.; Vanden Berghe, W. PPARα in the Epigenetic Driver Seat of NAFLD: New Therapeutic Opportunities for Epigenetic Drugs? *Biomedicines* **2022**, *10*, 3041. [CrossRef] [PubMed]
55. Li, Y.Y.; Tang, D.; Du, Y.L.; Cao, C.Y.; Nie, Y.Q.; Cao, J.; Zhou, Y.J. Fatty Liver Mediated by Peroxisome Proliferator-Activated Receptor-α DNA Methylation Can Be Reversed by a Methylation Inhibitor and Curcumin. *J. Dig. Dis.* **2018**, *19*, 421–430. [CrossRef]
56. Wang, J.; Zhang, Y.; Zhuo, Q.; Tseng, Y.; Wang, J.; Ma, Y.; Zhang, J.; Liu, J. TET1 Promotes Fatty Acid Oxidation and Inhibits NAFLD Progression by Hydroxymethylation of PPARα Promoter. *Nutr. Metab.* **2020**, *17*, 46. [CrossRef]
57. Castellano-Castillo, D.; Moreno-Indias, I.; Sanchez-Alcoholado, L.; Ramos-Molina, B.; Alcaide-Torres, J.; Morcillo, S.; Ocaña-Wilhelmi, L.; Tinahones, F.; Queipo-Ortuño, M.I.; Cardona, F. Altered Adipose Tissue DNA Methylation Status in Metabolic Syndrome: Relationships Between Global DNA Methylation and Specific Methylation at Adipogenic, Lipid Metabolism and Inflammatory Candidate Genes and Metabolic Variables. *J. Clin. Med.* **2019**, *8*, 87. [CrossRef]
58. Struhl, K. Histone Acetylation and Transcriptional Regulatory Mechanisms. *Genes Dev.* **1998**, *12*, 599–606. [CrossRef]
59. Miller, J.L.; Grant, P.A. The Role of DNA Methylation and Histone Modifications in Transcriptional Regulation in Humans. *Subcell. Biochem.* **2013**, *61*, 289–317. [CrossRef]
60. Strycharz, J.; Rygielska, Z.; Swiderska, E.; Drzewoski, J.; Szemraj, J.; Szmigiero, L.; Sliwinska, A. SIRT1 as a Therapeutic Target in Diabetic Complications. *Curr. Med. Chem.* **2018**, *25*, 1002–1035. [CrossRef]
61. Arunachalam, G.; Samuel, S.M.; Marei, I.; Ding, H.; Triggle, C.R. Metformin Modulates Hyperglycaemia-Induced Endothelial Senescence and Apoptosis through SIRT1. *Br. J. Pharmacol.* **2014**, *171*, 523–535. [CrossRef]
62. Picard, F.; Kurtev, M.; Chung, N.; Topark-Ngarm, A.; Senawong, T.; de Oliveira, R.M.; Leid, M.; McBurney, M.W.; Guarente, L. Sirt1 Promotes Fat Mobilization in White Adipocytes by Repressing PPAR-γ. *Nature* **2004**, *429*, 771. [CrossRef] [PubMed]
63. Castellano-Castillo, D.; Denechaud, P.-D.; Fajas, L.; Moreno-Indias, I.; Oliva-Olivera, W.; Tinahones, F.; Queipo-Ortuño, M.I.; Cardona, F. Human Adipose Tissue H3K4me3 Histone Mark in Adipogenic, Lipid Metabolism and Inflammatory Genes Is Positively Associated with BMI and HOMA-IR. *PLoS ONE* **2019**, *14*, e0215083. [CrossRef] [PubMed]
64. Sugii, S.; Evans, R.M. Epigenetic Codes of PPARγ in Metabolic Disease. *FEBS Lett.* **2011**, *585*, 2121–2128. [CrossRef] [PubMed]
65. Lefterova, M.I.; Zhang, Y.; Steger, D.J.; Schupp, M.; Schug, J.; Cristancho, A.; Feng, D.; Zhuo, D.; Stoeckert, C.J.; Liu, X.S.; et al. PPARgamma and C/EBP Factors Orchestrate Adipocyte Biology via Adjacent Binding on a Genome-Wide Scale. *Genes Dev.* **2008**, *22*, 2941–2952. [CrossRef] [PubMed]
66. Wang, S.; Lin, Y.; Gao, L.; Yang, Z.; Lin, J.; Ren, S.; Li, F.; Chen, J.; Wang, Z.; Dong, Z.; et al. PPAR-γ Integrates Obesity and Adipocyte Clock through Epigenetic Regulation of Bmal1. *Theranostics* **2022**, *12*, 1589–1606. [CrossRef]
67. Lee, T.-I.; Kao, Y.-H.; Tsai, W.-C.; Chung, C.-C.; Chen, Y.-C.; Chen, Y.-J. HDAC Inhibition Modulates Cardiac PPARs and Fatty Acid Metabolism in Diabetic Cardiomyopathy. *PPAR Res.* **2016**, *2016*, 5938740. [CrossRef]
68. Gao, Q.; Wei, A.; Chen, F.; Chen, X.; Ding, W.; Ding, Z.; Wu, Z.; Du, R.; Cao, W. Enhancing PPARγ by HDAC Inhibition Reduces Foam Cell Formation and Atherosclerosis in ApoE Deficient Mice. *Pharmacol. Res.* **2020**, *160*, 105059. [CrossRef]
69. Fu, M.; Rao, M.; Bouras, T.; Wang, C.; Wu, K.; Zhang, X.; Li, Z.; Yao, T.-P.; Pestell, R.G. Cyclin D1 Inhibits Peroxisome Proliferator-Activated Receptor Gamma-Mediated Adipogenesis through Histone Deacetylase Recruitment. *J. Biol. Chem.* **2005**, *280*, 16934–16941. [CrossRef]

biomolecules

MDPI

Review

New Insights into the Role of PPARγ in Skin Physiopathology

Stefania Briganti, Sarah Mosca, Anna Di Nardo, Enrica Flori * and Monica Ottaviani

Laboratory of Cutaneous Physiopathology and Integrated Center of Metabolomics Research, San Gallicano Dermatological Institute, IRCCS, 00144 Rome, Italy; stefania.briganti@ifo.it (S.B.); sarah.mosca@ifo.it (S.M.); anna.dinardo@ifo.it (A.D.N.); monica.ottaviani@ifo.it (M.O.)
* Correspondence: enrica.flori@ifo.it

Abstract: Peroxisome proliferator-activated receptor gamma (PPARγ) is a transcription factor expressed in many tissues, including skin, where it is essential for maintaining skin barrier permeability, regulating cell proliferation/differentiation, and modulating antioxidant and inflammatory responses upon ligand binding. Therefore, PPARγ activation has important implications for skin homeostasis. Over the past 20 years, with increasing interest in the role of PPARs in skin physiopathology, considerable effort has been devoted to the development of PPARγ ligands as a therapeutic option for skin inflammatory disorders. In addition, PPARγ also regulates sebocyte differentiation and lipid production, making it a potential target for inflammatory sebaceous disorders such as acne. A large number of studies suggest that PPARγ also acts as a skin tumor suppressor in both melanoma and non-melanoma skin cancers, but its role in tumorigenesis remains controversial. In this review, we have summarized the current state of research into the role of PPARγ in skin health and disease and how this may provide a starting point for the development of more potent and selective PPARγ ligands with a low toxicity profile, thereby reducing unwanted side effects.

Keywords: PPARs; skin physiology; inflammatory skin disease; skin cancer; sebaceous gland; lipids

1. Introduction

Peroxisome proliferator-activated receptors (PPARs) are a class of nuclear receptor proteins that act as transcription factors regulating target gene expression. Since their identification in the 1990s [1], PPARs have been the focus of many studies in the literature due to their role in several critical biological processes, such as energy homeostasis and inflammation [2–5].

PPARs are known to be classically activated by binding to specific ligands, as well as other members of the nuclear receptor superfamily. Following ligand interaction, PPAR undergoes a conformational change that allows it to form heterodimers with the 9-cis retinoid X receptor (RXR) (Figure 1A). The nature of the ligand is important in driving the genomic PPAR response, leading to the recruitment of several co-activators or co-repressors [6].

The resulting complex is able to bind specific DNA sequences in their promoter regions, known as PPAR response elements (PPREs), thereby regulating the expression of target genes involved in a range of physiological processes related to inflammation, metabolism, and cell differentiation [5,7]. In addition to the transcriptional activation and repression, an additional mechanism described in the literature is that PPARs can also act as transrepressors of genes primarily associated with pro-inflammatory pathways through a protein–protein PPRE-independent interaction with other transcription factors, such as NF-kB and AP1 [8,9].

There are three PPAR isoforms, PPARα, PPARβ/δ, and PPARγ, which share structural and sequence homology but differ in tissue distribution, ligand selectivity, and receptor responsiveness, demonstrating their ability to regulate distinct sets of genes [2,10–13]. The size of the ligand-binding cavities of PPARs is comparable, with about 80% of the amino acids conserved. Nevertheless, amino acid sequence variations between PPARs affect the specificity of the ligands (Figure 2) [14].

A

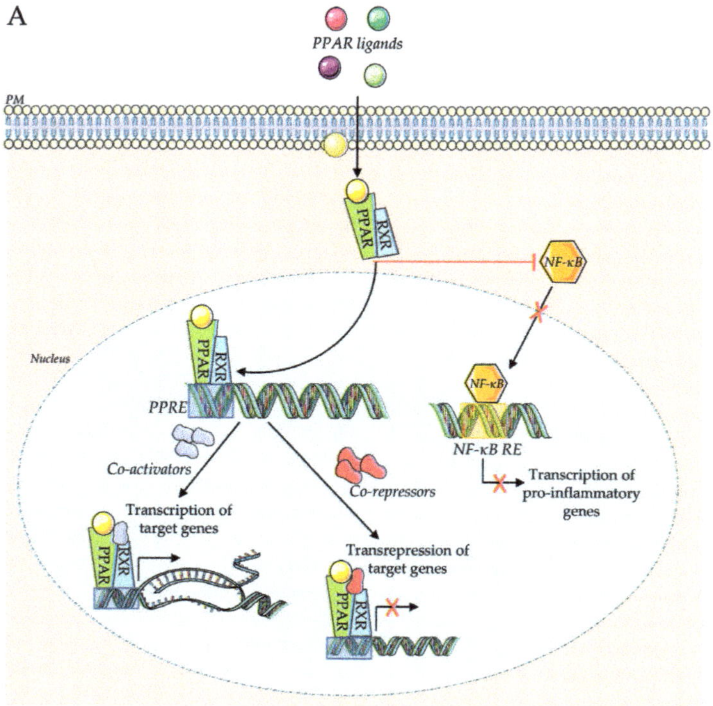

B

PPARα	PPARβ/δ	PPARγ
-Cell proliferation/differentiation -Skin epidermal barrier -SG homeostasis -Anti-inflammatory processes -Melanogenesis -Wound healing	-Cell survival, migration and differentiation -Skin epidermal barrier -Wound healing	-Cell proliferation/differentiation -Skin epidermal barrier -SG homeostasis -Lipidogenesis -Anti-inflammatory processes -Melanogenesis -Fibrosis inhibition -Oxidative stress containment

Figure 1. PPARs signaling and functions. (**A**) Schematic representation of the ligand-dependent PPARs activation. The ligand nature leads to the recruitment of several co-activators or co-repressors, important in driving the PPAR genomic actions. Cytoplasmic PPARs, by interacting with other transcription factors, such as NF-kB, elicit non-genomic effects by negatively regulating pro-inflammatory gene expression (**B**) Principal roles exerted by the different PPARs isoforms in skin homeostasis.

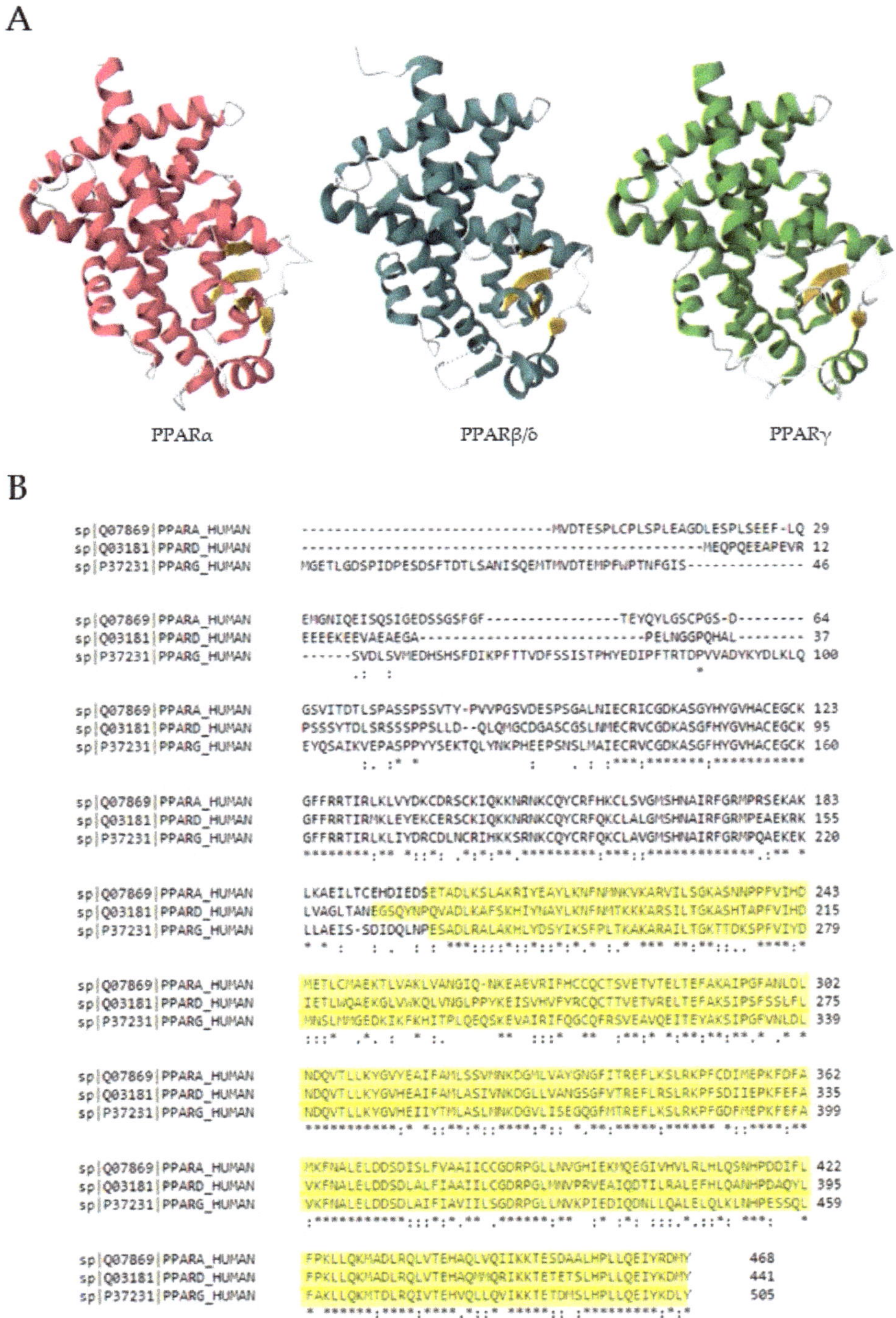

Figure 2. Three-dimensional structure and ligand binding pocket aminoacid sequences of three PPAR isoforms. (**A**) 3D structure of PPARα (1I7G [15]), PPARβ/δ (3TKM [16]), and PPARγ (2F4B [17]). (**B**) Structure-based sequence alignment of hPPARα (UniProt ID Q07869), hPPARβ/δ (UniProt ID Q03181),

and hPPARγ (UniProt ID P37231). For each aligned residue pair, ":" is for conserved substitutions, "." denotes semi-conserved substitutions, and "*" denotes identical residues. Residues binding ligands are highlighted in yellow.

All PPAR isoforms are expressed in the human skin and are involved in various cellular functions, including cell proliferation and differentiation [4,18–20] (Figure 1B). PPARα and PPARγ are known to induce sebocyte differentiation and are involved in keratinocyte differentiation and epidermal lipid synthesis, with a critical role in epidermal barrier repair [21,22]. In particular, the activation of PPARγ leads to the inhibition of keratinocyte proliferation, promoting terminal epidermal differentiation instead [23,24]. PPARβ/δ is involved in skin wound healing, lipid synthesis, keratinocyte survival, migration, and sebocyte maturation [25–28]. Several studies in the literature show that activation of PPARα and PPARγ is associated with the inhibition of melanocyte proliferation and stimulation of pigmentation, whereas PPARβ/δ does not yet appear to affect melanocytes [29–32]. Furthermore, cutaneous immune cells, such as Langerhans and T cells, express all three PPARs isoforms, highlighting their importance in exerting anti-inflammatory effects [4,33,34]. Therefore, the dysregulation of PPAR expression is associated with inflammatory skin disorders and skin malignancies, suggesting their potentiality as therapeutic targets using specific ligands [20].

Typically, PPARs are activated by binding specific ligands that can be natural or synthetic compounds. Fatty acids and their derivatives are examples of endogenous ligands, while synthetic ligands include therapeutic drugs such as thiazolidinediones (TZDs) (PPARγ agonists) and fibrates (PPARα agonists) [2,6,35,36]. In particular, TZDs, such as pioglitazone, and WY14643 are some of the receptor ligands that have demonstrated promise in the management of a range of skin conditions, including psoriasis, atopic dermatitis, and skin cancers [20,28,37,38]. Moreover, it has been discovered that PPARγ agonists decrease skin fibrosis and accelerate wound healing [39,40]. However, PPAR modulators should be carefully evaluated to determine their efficacy and potential side effects [6,37]. Data in the literature show concerns about possible cardiovascular hazards and carcinogenicity associated with PPAR agonist use, especially PPARα/γ dual agonists [41]. TZDs, which are known for their effectiveness in treating type 2 diabetes due to their ability to improve glycaemic control and induce insulin sensitivity, are not good candidates for long-term therapies, such as chemotherapy for skin cancer, due to their lack of specificity and harmful side effects [42–47]. Much attention is therefore being paid to the development of new classes of selective and potent PPARγ modulators retaining the beneficial activity of PPARγ agonists while eliminating many of the undesirable side effects [6].

In this review, we have attempted to describe current knowledge of PPARγ in skin physiopathology as the main cutaneous therapeutic target in skin inflammatory disorders and cancer.

2. Methods

2.1. Three-Dimensional Crystal Structure of PPARs and Sequence Alignment

The 3D structures of hPPARα, hPPARβ/δ, and hPPARγ were obtained by using the US RCSB PDB (RCSB.org) data center for the global Protein Data Bank (PDB) archive of 3D structure data for large biological molecules, with PDBentries of 1I7G [15], 3TKM [16], and 2F4B [17], respectively. Ligands and water in crystal structures were removed (Figure 2A).

The amino acids sequence alignment of the three receptors was obtained by using UniProt Align of hPPARα (Q07869), hPPARβ/δ (Q03181), and hPPARγ (P37231) (UniProt: the Universal Protein Knowledgebase 2002–2024; current release 2024).

2.2. Chemical Structures Drawing

The chemical structures of the most representative PPARγ activators (Figure 3) were created using ACD/ChemSketch (Freeware) Version 2022.2.3 Software (Advanced Chemistry Development, Inc., Toronto, Canada).

3. PPARγ Modulation and Skin Barrier Homeostasis

The structure and composition of the outermost layer of the epidermis, the stratum corneum (SC), are related to the proper functioning of the skin barrier, mediating the maintenance of water and electrolyte balance and preventing dehydration [48,49]. The SC consists of corneocytes, which are anucleated, terminally differentiated keratinocytes embedded in a lamellar lipid matrix, and provide a barrier to the movement of water and electrolytes [50,51]. The process of differentiation of keratinocytes into corneocytes is also associated with the expression of specific proteins, such as keratin, fillagrin, transglutaminase, involucrin, and loricrin, which have the function of forming the structural scaffolds of the extracellular lipid matrix [50]. Moreover, these proteins are also targets of PPAR activation [52]. In addition, differentiating keratinocytes secrete ceramides, the predominant lipid species in the extracellular matrix, which play an important role in the water-holding and barrier functions of the SC [51]. As epidermal lipid metabolism and skin barrier function are closely linked, modulation of lipid metabolism by PPAR ligands was tested for its effects on the skin barrier. During keratinocyte differentiation induced by high calcium concentration, the levels of PPARγ increase along with the terminal differentiation [18]. In terminally differentiated keratinocytes, PPARγ levels increase 5-fold, peaking in the suprabasal layer [53].

Several studies in epidermal keratinocytes or mice have shown that either the overexpression of PPARγ or its activation by agonists can have beneficial effects on skin barrier function. PPARγ induces a shift in the balance between differentiation and proliferation towards differentiation, leading to the normalization of terminal differentiation of epidermal keratinocytes and a reduction in their proliferation rate [24]. The treatment of animals with PPARγ agonists also reduces the proliferation rate of epidermal keratinocytes, and this anti-proliferative effect is more rapid in recovering epidermis with a disrupted skin barrier [23,54,55]. In mice knockout for epidermal PPARγ, a significant reduction in the transcription of several genes associated with lipid barrier formation has been reported [21,56]. In human epidermal equivalents, barrier disruption mediated by chemical agents such as SDS or acetone resulted in reduced *PPARγ* gene expression levels, suggesting that PPARγ signaling is closely correlated with epidermal barrier fitness [57]. PPARγ ligands may therefore represent useful therapies for various inflammatory skin conditions characterized by epidermal barrier impairment.

4. PPARγ and the Sebaceous Gland

The Sebaceous Gland (SG) is an epithelial appendage with an acinar structure consisting of sebocytes; SG is part of the pilosebaceous unit, and its development is closely associated with the formation of the hair follicle (HF) during skin morphogenesis [58,59]. SG expresses all PPAR isoforms. However, PPARγ seems to be a crucial and necessary requirement for SG genesis and homeostasis [60–63]. Mice with a targeted deletion of PPARγ show the dramatic downregulation of transcripts encoding for genes related to gland development and sebum production, leading to an asebia phenotype. The almost complete ablation of SGs is also associated with a slow hair cycle, altered HF morphology, and skin inflammation, indicating the importance of PPARγ not only for the SG but likely for the whole pilosebaceous unit [21,63–65]. Due to its role in SG specification, PPARγ can also be considered a fate determinant for SG cell reprogramming; in fact, the presence of a PPARγ agonist in the culture medium of epidermal stem cells is necessary to induce their differentiation into sebocytes [66]. More recently, it has been proven that the stable overexpression of PPARγ alone is sufficient to drive human keratinocyte conversion into culture-expandible SG cells, with gene expression patterns and functional properties mimicking sebocytes, particularly in terms of lipid synthesis. Furthermore, in mice, the intradermal injection of these "induced SG cells" gives rise to a de novo SG-like structure capable of producing a lipidomic profile closely related to those expected for native SG [63,66].

In addition to its relevance in SG morphogenesis, PPARγ also plays a key role in the maintenance of the SG primary function, namely sebum production, since its expression is associated with sebocyte differentiation. Sebocytes undergo multiple cell state transformations moving toward the sebaceous duct and reach their maturation at the distal end of the SG which constitutes about 20–50% of the SG volume [58,63,67] Variation in the differentiation stage of sebocytes thus characterizes the layered structure of the SG. The peripheral zone is made up of undifferentiated and proliferative sebocytes, whereas differentiating sebocytes, which synthesize and accumulate lipids in cytoplasmic droplets that notably increase their size, constitute the SG maturation zone. Finally, in the necrotic zone, mature and terminally differentiated sebocytes undergo a specialized form of cell death called holocrine secretion, releasing sebum on the skin surface through the HF infundibulum and the sebaceous duct [58,68–70]. Together with the increase in lipid production, more differentiated sebocytes of the central zone of the gland highly express PPARγ in contrast to sebocytes present in the SG peripheral zone, underlining the importance of this transcription factor in lipid synthesis and metabolism [62,71,72]. PPARγ activation resulted, in fact, in lipidogenic induction and increased sebum secretion in sebocyte cell lines (SZ95, SEB-1) and human skin, respectively [62,72]. Moreover, PPARγ expression in the different SG zones correlates with the expression of androgen receptors and androgen-metabolizing enzymes, indicating the cooperative role of PPARγ towards androgenic activity. This interplay is also observed in the SZ95 experimental model, where PPARγ induction produced increased sensitivity to testosterone [62,67,73–75]. Further investigation of the role of PPARγ in the SG may help to better define the still underestimated and not fully explored SG activity and function.

5. PPARγ and Skin Inflammation

Inflammatory responses promote acute or chronic diseases characterized by excessive production of arachidonic acid-derived eicosanoids, inflammatory cytokines, and adhesion molecules [76]. Skin inflammation is a strictly regulated process. PPARs represent one of the main regulators, and their activation by nitrated fatty acid derivatives and cyclopentenone prostaglandins, which are products formed in the late stages of the inflammatory reaction, results in the inhibition of core elements of the inflammatory reaction [9]. Multiple direct and indirect mechanisms promote the anti-inflammatory effects of PPARs, which are involved in the regulation of immune cells and the resolution of skin inflammation [3]. Numerous inflammatory mediators and cytokines are inhibited by PPARγ ligands in various cell types, including monocytes/macrophages, epithelial cells, smooth muscle cells, endothelial cells, dendritic cells, and lymphocytes [77–79]. Additionally, PPARγ regulates the biological activities of Langerhans cells [80,81] and decreases the expression of adhesion molecules [82]. PPARγ agonists promote the differentiation of hemopoietic progenitor cells in Langherans cells [83], whereas the pro-inflammatory cytokines inhibit this process. Increased expression of PPARγ affects the maturation and function of Langerhans cells, mainly by accelerating lipid metabolism [84] and fatty acid oxidation [85]. In addition, the PPAR-signaling pathway enhances immunogenicity and T cell priming by Langerhans cells [86]. PPARγ is involved in the differentiation and proliferation of T cells and plays a crucial role in their activation following antigen recognition, causing rapid changes in their phenotype. A shift from a resting state to a state with a much higher metabolic demand follows stimulation of the T cell receptor, activating, in turn, PPARγ and inducing the genes involved in glucose and fatty acid uptake [87,88]. The activated CD4+ T cells differentiate into several subpopulations with different inflammatory and metabolic phenotypes, namely Th1, Th2, Th17, and Treg [89]. Various PPARγ agonists reduce proliferation, inhibit Th1 and Th17 differentiation, and promote Treg differentiation of CD4+ T cells [33,90,91].

In macrophages, PPARγ regulates polarization, maturation, epigenetics, and metabolism. In particular, PPARγ modulates the polarization of pro-inflammatory M1 macrophages into anti-inflammatory M2 macrophages [92]. In addition, after stimulation with LPS, a condition that drives M1 polarization, macrophages knockout for PPARγ produce high levels of pro-inflammatory cytokines, such as IL-1, which are considered markers of M1 polarization [79]. PPARγ also plays a role in CD36 expression, oxLDL uptake, and foam cell formation by macrophages [93]. PPARγ deficient macrophages have, in fact, diminished capacity to uptake and degrade OxLDL [94]. Notably, not only cholesterol uptake but also its efflux is under the transcriptional control of PPARγ through the cassettes of the ABCA1 family [94]. Furthermore, the overexpression of dominant-negative PPARgamma in macrophages results in the upregulation of pro-inflammatory cytokines/chemokines and expansion of myeloid-derived suppressor cells (MDSCs), leading to inflammation, immunosuppression, and tumorigenesis [95]. These data suggest that the PPARgamma signaling pathway and its downstream gene products are essential for controlling chronic inflammation (especially MDSC homeostasis) and tumorigenesis.

PPARγ is a key regulator of the functional maturation of dendritic cells (DC), driving the induction of immunogenic T cell responses versus immune tolerance. Klotz et al. demonstrated that pharmacological modulation of PPARγ signaling in murine DC reduced the expression of costimulatory molecules and IL-12 associated with the maturation process and significantly inhibited the ability of DC to prime naive CD4+ T cells in vitro [96]. In addition, CD4+ T cells primed by PPARγ-activated DCs failed to express Th1 and Th2 cytokines and did not respond to further T cell receptor-mediated stimulation with secondary clonal expansion [96]. The authors proposed that PPARγ controls DC function in a complex manner that allows the survival of Ag-reactive CD4+ T cells but induces CD4+ T cell anergy instead of immunity upon secondary Ag encounter. Conversely, PPARγ ablation increased DC immunogenicity, suggesting that PPARγ may act as a constitutive regulator of DC suppression [96]. The activation of PPARγ with 15d-PGJ2 or troglitazone inhibits Toll Like Receptor-mediated activation of the MAP kinase and NF-kB pathways and results in a reduced capacity of DC to stimulate T cell proliferation. This highlights the inhibitory effect of PPARγ activation on DC maturation [97]. Through its ability to reduce the activity of transcription factors, such as AP-1, STAT, NF-kB, and NFAT, PPARγ negatively regulates inflammatory gene expression in skin immune cells [98,99]. The formation of inhibitory complexes between PPARγ and the aforementioned transcription factors is called trans-repression [100] and leads to the suppression of pro-inflammatory cytokine genes and the downregulation of cyclooxygenase-2/COX-2/PTGS2 expression, thereby reducing the production of prostaglandins [101]. Due to its anti-inflammatory effects, PPARγ has been the subject of numerous studies in recent years to elucidate pathogenetic mechanisms and develop new therapies for inflammatory skin diseases.

6. PPARγ Agonists/Modulators in Skin Disease Management

6.1. *Atopic Dermatitis and Contact Dermatitis*

Atopic dermatitis (AD) is a chronic relapsing inflammatory skin disease, which may be associated with IgE-dependent diseases like allergic asthma, food allergy, and allergic rhinitis. AD patients have skin barrier impairment within crusted erythematous areas, associated with epidermal hyperplasia, scaling, and lichenification [102]. Loss-of-function mutations in the filaggrin gene are a major predisposing factor for the manifestation of AD, increasing the permeability of the skin [103]. However, not only do patients with mutations show decreased filaggrin levels in the skin but inflammatory processes in general also decrease filaggrin expression in people not carrying the respective mutation [104]. Skin injury often triggers AD, allowing increased penetration of environmental allergens and inducing a cytokine milieu that favors a Th2 response [52]. Although information on PPARγ in AD is conflicting, the systemic administration of PPARγ ligands improved experimental skin allergy in mice [105] and the intensity of clinical symptoms in patients with severe AD [106]. Despite the evidence that AD patients may benefit from PPARγ agonists, the clinical observations derive from a small number of patients, so the potential role of PPARγ modulation in the treatment of AD should be evaluated in future controlled trials.

PPARγ agonists may maintain mast cell homeostasis by inhibiting the maturation of their precursors, reducing mast cell phenotypic markers and viability, inhibiting degranulation, and inducing cell apoptosis [107]. These data suggest that PPARγ ligands may serve as effective anti-inflammatory reagents in the treatment of mast cell-related diseases, such as allergic reactions and contact dermatitis. However, recent data report that PPARγ is essential for the promotion of 'type 2' immune responses that are typically associated with allergic diseases, highlighting the contrasting role of PPARγ in allergic inflammation [108]. The high availability of PPARγ ligands in the environment is thought to play a key role in the increase in allergic diseases worldwide [108]. This effect is mainly observed with PPARγ ligands from environmental pollutants. The discovery and testing of compounds that can bind PPARγ but primarily activate anti-inflammatory responses may be a valid strategy to counteract or mitigate the negative effects of environmental PPARγ activators. Remarkably, 50% of children with AD develop asthma later in life, and the risk of developing hay fever is up to 75% [109]. This typical sequence of allergic manifestations is called the "atopic march" [110,111]. Glucocorticoids (GCs) are the standard anti-inflammatory drugs used to treat AD and attempt to halt the atopic march. However, GCs have detrimental side effects on the integrity of the epidermal barrier [112–115]. Therefore, the combination of GCs with therapies, such as PPARγ ligands, that have an anti-inflammatory effect while maintaining the integrity of the skin barrier is an interesting topic for study [115]. The activation of the glucocorticoid receptor (GR) has been shown to contribute to the anti-inflammatory effects of PPARγ activity [116]. Deckers et al. developed a murine model that mimics the atopic march. In this model, skin inflammation is associated with a mixed Th2/Th17 phenotype and severe airway inflammation induced by dust mite challenges. Combined activation of GR/PPARγ suppresses local skin inflammation to a greater extent than single activation but appears insufficient to prevent the allergic immune response in the lung, even though the severity of asthma is effectively reduced by counteracting the Th17 response [117]. Thus, GR/PPARγ co-activation represents a potent remedy against allergic skin inflammation and worsening of atopic march.

The beneficial effects of PPARγ modulators in AD therapy are, therefore, mostly related to in vitro/in vivo studies with cell cultures and animal models, whereas conflicting data are reported in patients. Therefore, controlled clinical studies are needed to test the therapeutic efficacy of PPARγ modulators.

6.2. *Psoriasis*

Psoriasis is a multifactorial inflammatory skin disorder, generally considered a polygenic, immune activation-dominated disease characterized by keratinocyte hyperplasia and disturbed differentiation [118]. The interaction between psoriatic keratinocytes and cells of the immune system is a crucial element in the development of psoriatic lesions since keratinocytes produce chemokines to attract T cells to the inflammatory site [119]. Due to their pro-differentiating, anti-proliferative, and immunomodulatory effects, PPARγ may play a central role in the pathogenesis and treatment of psoriasis. In psoriatic lesional skin, the expression of PPARγ is decreased [98,120]. In a 3D tissue-engineered psoriatic skin model characterized by the reduced expression of PPARγ, the addition of DHA, a long-chain n-3 polyunsaturated fatty acid, regulates the differentiation of psoriatic keratinocytes and promotes the synthesis of anti-inflammatory lipid mediators through its ability to activate the PPARγ pathway [121]. According to the studies performed on mice, either the overexpression of PPARγ or its activation by agonists may potentially produce variable beneficial effects on the skin. In skin hyperproliferative disease murine models, the treatment of animals with PPARγ agonists (troglitazone, rosiglitazone, pioglitazone, and BP-1107) suppresses inflammatory signals, reduces the proliferation rate, and normalizes terminal differentiation of epidermal keratinocytes [23,122,123]. A disruption of fatty acid metabolism and associated PPARγ signaling was revealed by whole transcriptome sequencing [124] and mRNA expression profiling [125] of psoriatic lesions and adjacent normal skin. The PPARγ signaling pathway is proposed as a key regulatory factor in the pathogenesis and a potential therapeutic target of psoriasis. Sobolev et al. combined experimental results and network functional analysis to reconstruct the model of down-regulated PPARγ signaling in psoriasis and tested the hypothesis that low PPARγ levels may alter the activity of cellular signaling pathways in the skin, facilitating the chronic inflammatory and immune response in human psoriatic lesions [120]. They found higher expression of *STAT3, RORC, FOXP3,* and *IL17A* genes overall with low *PPARγ* mRNA levels in psoriatic lesions. Their data support the hypothesis that PPARγ attenuates the expression of genes involved in the development of psoriatic lesions and show that the regulation of PPARγ expression by FOSL1 and by the STAT3/FOSL1 feedback loop may be central in psoriatic skin and T cells [126]. Based on this evidence, the modulation of PPARγ may have therapeutic potential in psoriasis. Several clinical studies show that PPARγ agonists are effective in normalizing the key morphological features of psoriasis. Patients receiving systemic treatment with TDZs troglitazone or pioglitazone showed reduced hyperplasia and normalized histological features of lesional skin [23,127–131]. However, oral rosiglitazone was not more effective than placebo in patients with moderate to severe chronic plaque psoriasis [132]. Topical rosiglitazone was also ineffective [133], suggesting that differences in the ability to modulate PPARγ may explain the different therapeutic responses. In patients with psoriasis, PPARγ agonists may also have an additive effect in combination with typical antipsoriatic drugs. Pioglitazone improved the effect of acitritin [134], methotrexate [135,136], or other antipsoriatic drugs [137] and increased the effectiveness of phototherapy in eligible psoriasis patients [138]. However, like many other drugs, TZD therapies have unwanted side effects due to their simultaneous action on different cells and non-specific off-target effects. TZDs promote adipogenesis and adipocyte maturation, leading to adipocyte hypertrophy, and induce edema, weight gain, and increased risk of cardiovascular events and bone fractures in patients [98]. The new generation of PPARγ agonists is expected to retain the beneficial properties of TZDs while significantly reducing the disruption of lipid and glucose metabolism, which is actively involved in the side effects observed in patients. A recently discovered non-TZD agonist of PPARγ, GED-0507-34L, binds to PPARγ with relatively high affinity and reduces the inflammatory response by inhibiting NF-kB and upregulating the expression of IkBα, as well as suppressing the prostaglandin-endoperoxide synthase 2 (PTGS2) in normal human epidermal keratinocytes and lymphocytes [101]. In IL21-induced psoriasis-like skin lesions in mice, topical application of GED-0507-34L reduces the accumulation of cellular infiltrate and epidermal hyperplasia, suggesting that this PPARγ modulator may potentially benefit patients with psoriasis [101]. A Phase I clinical trial in psoriasis patients (Reg. No.: GED0507-PSO-01-12, USA) showed that, unlike TZDs, GED-0507-34L had no serious side effects other than redness and itching in the treated areas (in 1 of 24 treated patients) [98]. However, despite the increasing development of new agonists, the potential benefits of new PPARγ modulators in the treatment of psoriasis need to be confirmed in clinical trials to ensure that the increase in efficacy is balanced by a reduction in potential side effects in long-term therapy.

6.3. Vitiligo

Vitiligo is a relatively common skin disorder characterized by a reduction in the number and function of melanocytes, resulting in hypopigmented or depigmented skin lesions affecting 0.5–1% of the world's population. Several factors including genetics, autoimmunity, oxidative stress, melanocyte apoptosis, and neurological mechanisms have been implicated in the pathogenesis of vitiligo [139]. The destruction of melanocytes within vitiligo lesions is mediated by the differentiation of DCs, leading to the transformation of CD4+ T cells into Th1 or Th17 lymphocytes, which secrete various pro-inflammatory cytokines such as IFN-γ, TNF-α, and IL-17, further damaging melanocytes and activating B lymphocytes to produce antibodies against autoantigens such as tyrosinase, tyrosin hydroxylase, and Sox10 [73,140–142]. Simultaneously, CD8+ T cells are activated and exert their destructive effect on melanocytes [143]. Improving the understanding of the inflammatory pathways in the pathogenesis of vitiligo and identifying more therapeutic methods to promote the proliferation and function of melanocytes has been a hot topic in the field of vitiligo treatment. In a recent study, a comprehensive bioinformatics analysis was conducted to explore potential genes and signaling pathways associated with vitiligo and metabolic diseases [144]. PPARγ pathway genes were significantly downregulated in vitiligo samples, and the drug–gene interaction analysis suggested that the PPARγ agonist rosiglitazone may activate the *EDNRB* gene expression to enhance melanogenesis. Thus, exploring the role of the PPARγ pathway may provide new insights into the pathogenesis and discovery of potential therapeutic targets of vitiligo. Treatment of vitiligo melanocytes and fibroblasts with pioglitazone ameliorates mitochondrial alterations by reducing ROS generation, restoring mitochondrial membrane potential, increasing ATP levels, and increasing anaerobic glycolytic enzyme expression and protein levels [145]. In the same study, pioglitazone reverses premature melanocyte senescence in vitiligo. Recently, pioglitazone was found to ameliorate defective differentiation/barrier constitution and reduce inflammatory mediator production in vitiligo keratinocytes [146]. This suggests that PPARγ activation may preserve and enhance skin barrier homeostasis in vitiligo patients, potentially counteracting the effects of environmental stressors and subsequent triggering of inflammation. Further clinical studies are needed to investigate the efficacy of selective modulators of PPARγ as a therapy to treat and/or slow the clinical manifestations of vitiligo and its spread.

6.4. Acne

Acne is one of the most common SG-related inflammatory disorders, in which the altered SG activity, characterized by both quantitative and qualitative sebum alterations, plays a pivotal role. This abnormal sebum production contributes to and affects the other pathogenetic factors, namely follicular hyper-keratinization, inflammation, and *C. acnes* colonization, representing the driving multifactorial process of the disease. Due to its role in sebocyte differentiation and lipid production, PPARγ has been considered a possible target for the treatment of acne [70,147,148]. In addition to PPARγ, SG activity is also regulated by insulin and insulin-like growth factor (IGF1), indicated by growing evidence as important hormonal triggers in acne development [149–155]. In SZ95 sebocytes, insulin stimulates lipogenesis, as well as inflammation, through PI3K-Akt-mTOR signaling. As far as lipid induction is concerned, insulin challenge gives rise to qualitative lipid composition mimicking acne sebum in terms of eicosanoid production, and thus in terms of lipo-inflammatory process driven by eicosanoid-producing enzymes, such as lipoxygenase (LOX) [22,67,148,156,157]. The magnitude of the insulin effect depends on the sebocyte differentiation stage. In fact, the reduced expression of PPARγ in low differentiated SZ95 sebocytes is associated with a higher level of both insulin and insulin-like growth factor-1 receptors (IR, IGF1R), and with the consequent up-modulation of Akt-mTOR signaling, leading to higher responsiveness of these cells to insulin stimulus in comparison with more differentiated sebocytes. Furthermore, the reduced level of PPARγ may also be associated with a higher inflammatory reaction after insulin stimulation due to the increase in LOX activity, which can also result in the production of IL6 and IL8 cytokines, beyond eicosanoid increase, in sebocytes [22,148,156]. Accordingly, in vivo data in acne skin demonstrated lower levels of PPARγ together with increased AkT-mTOR signaling as well as higher 5-LOX expression and eicosanoid levels [158–160]. The PPARγ modulation by the selective and newly designed agonist N-Acetyl-Ged-0507-34-Levo (NAC-GED0507; GMG-43AC) is able to counteract the acne-like altered lipogenesis and inflammatory process produced by insulin treatment in SZ95 sebocytes. The induction of sebocyte differentiation and thus the increase in PPARγ expression by NAC-GED is considered the basis for the correct sebocyte response to insulin stimulation, leading to improved sebum composition. In addition, parallel anti-inflammatory effects of PPARγ were observed through the reduction in LOX activity and antagonism of the NF-kB pathway [22,156]. The clinical efficacy of topical NAC-GED0507 gel was demonstrated and then confirmed by a phase 1 and phase

2 trial, respectively. The knowledge gained so far identifies PPARγ modulation as a key element to be taken into account in novel and effective therapeutic strategies for acne treatment [156,161].

6.5. PPARγ and Skin Aging

Aged skin is characterized by structural and functional changes in the dermis, which include the formation of lines and wrinkles, increased pigmentation, loss of elasticity and firmness, and gray skin [162,163]. Skin aging is the hallmark of prolonged UV exposure, which causes photoaging, and intrinsic aging-related factors, such as collagen breakdown, by increasing the expression levels of metalloproteinase (MMP) enzymes [164–166]. Endogenous systems for photoprotection include epidermal thickening, pigmentation, and complex antioxidant and DNA repair systems [167–169]. Excessive production of ROS is known to be an initiator and promoter of intrinsic aging and photoaging [170]. UV-induced skin inflammation is associated with both intrinsic and extrinsic skin aging processes through the release of pro-inflammatory cytokines by the accumulation of ROS [171–173]. Since intracellular oxidative stress is mainly related to the decline of endogenous antioxidants, such as reduced glutathione and catalase with UV exposure and aging, treatment with exogenous antioxidants would be an effective approach to prevent the progression of skin aging [174,175]. The PPAR family, through its ability to modulate inflammatory response and antioxidant defense, can modulate the balance between MMP activity and collagen expression to maintain skin homeostasis and counteract skin aging. Several studies have suggested that PPARγ represents a therapeutic target for the photoaging process, as it is able to inhibit the expression of MMPs and AP-1 and modulate oxidative stress-sensitive pathways [176,177]. Intensive studies of PPARα/γ dual modulators have revealed their importance in age-related inflammation and photoaging as regulators of cytokines, MMPs, and NF-kB [178–180]. Decreased PPARγ activity is closely related to chronic inflammation associated with the aging process in vivo. Pioglitazone treatment contributes to attenuating several age-related disorders in aged apoE-/- mice, thereby representing a promising protective therapy against aging and age-related diseases [181]. Aging impairs epidermal re-growth during wound healing and results in lower expression of peroxisome proliferator-activated receptor gamma coactivator-1 alpha (Pgc-1a) [182]. In addition, it has been reported that PPARγ modulates UVB-induced COX-2 expression and subsequent synthesis of PGE2, a lipid mediator that suppresses collagen synthesis in dermal fibroblasts by partially interfering with TGF-β signaling [183,184]. In UVB-exposed human keratinocytes, chimyl alcohol reduces cellular damage by suppressing *COX-2* mRNA expression and PGE2 synthesis and stimulates the intracellular defense system against ROS through Nrf2 signaling activated by a PPAR-γ-dependent mechanism [185]. 3-O-laurylglyceryl ascorbate, an amphipathic derivative of ascorbic acid, scavenges intracellular ROS and stimulates intracellular antioxidants such as GSH through the PPARγ and Nrf2 signaling pathways [186].

The ability of PPARγ modulators to counteract the photo-aging process has been extensively studied in experimental models of cell senescence. Due to the key role of oxidative stress in the photoaging process, the transformation of proliferating skin cells into photoaged cells under conditions of artificially elevated ROS levels resembles premature senescence. Therefore, stress-induced premature senescence (SIPS) models can be useful tools to investigate the biological and biochemical mechanisms involved in photo-induced skin damage. PPARγ activation has been demonstrated to reduce age-related inflammation and aging progression in an H_2O_2-induced SIPS cell phenotype [187]. In human dermal fibroblasts induced to SIPS by PUVA (a single sub-cytotoxic exposure to UVA in the presence of activated 8-methoxypsoralen), a decrease in PPARγ expression and activity is an early event associated with the induction of a senescence-like phenotype [188,189]. Azelaic acid, a nine-carbon linear saturated dicarboxylic acid, and 2,4,6-octatrienoic acid, a member of the parrodiene family, increase the transactivation of PPARγ and counteract PUVA-SIPS markers, including long-term growth arrest, flattened morphology, increased synthesis of MMPs, increased degradation of ECM, and senescence-associated beta-galactosidase staining, and also interfere with ROS-dependent cellular signaling mechanisms [188,189]. In a reproducible photo-damaged model of mouse dermal fibroblasts treated with repeated UVB exposures, rosiglitazone counteracts the photoaging process [190]. Taken together, PPARγ modulation may counteract both intrinsic and extrinsic skin aging processes. However, further investigation is needed to develop new therapeutic or preventive approaches with oral or topical administration of PPARγ modulators while limiting potential side effects such as lipid accumulation due to increased adipocyte differentiation.

7. PPARγ and Skin Cancer

Over the past twenty years, considerable progress has been made in understanding the role of PPARγ in skin cancer. UV exposure is the major cause of skin carcinogenesis. This process involves DNA damage and alterations in signaling pathways essential for cell proliferation, survival, oxidative stress, inflammation, immunosuppression, differentiation, remodeling, and apoptosis [191–193].

7.1. Non-Melanoma Skin Cancer (NMSC)

The ability of PPARγ activation to induce terminal differentiation, afford photoprotective effects, and inhibit cell growth and inflammation has been demonstrated in several in vitro studies using malignant skin cell types and in animal studies using murine skin tumor models, suggesting that PPARγ agonists may act as tumor suppressors.

The study by Sahu et al. provides evidence that epidermal PPARγ plays a protective role in suppressing UVB-induced NMSC formation and progression in mice. Indeed, mice lacking epidermal PPAR (PPARγ $-/-^{epi}$ mice) show a marked increase in photocarcinogenesis associated with UVB-induced apoptosis, inflammation, barrier dysfunction, and epidermal hyperplasia [56]. A marked increase in TNF-α expression has been demonstrated in the same mouse model. In addition, the PPARγ agonist rosiglitazone reverses the systemic immunosuppression induced by chronic UVB [21]. Park et al. show the protective effects of a PPARα/γ dual agonist on inflammatory responses, epidermal thickness, and lipid peroxidation associated with chronic UVB exposure [194]. Furthermore, mice lacking the PPARγ heterodimerization partner RXRα in epidermal keratinocytes show increased apoptosis, altered epidermal proliferation, and augmented DNA damage in response to UVB exposure [195]. The studies by Ren and Konger and Balupillai et al. support the idea that PPARγ activation may be a potential pharmacological target for the prevention or treatment of skin cancer by suppressing tumor-promoting chronic inflammation associated with UV exposure, as well as by enhancing antitumor immune responses in the skin [196,197]. More recently, reports on human epidermoid carcinoma cells demonstrate the ability of PPARγ ligands to inhibit cell growth by regulating the expression of cell cycle-associated proteins and inducing terminal differentiation [198–200]. These results are consistent with the observed reduced anchorage-dependent clonogenicity and marked tumor volume inhibition in ectopic xenografts [199].

The activation of PPARγ by the parrodiene derivative 2,4,6-octatrienoic acid (Octa) prevents cutaneous UV damage by acting on the different skin cell populations [189,201]. Most interestingly, Octa promotes antioxidant defense, greater cell survival, and the effective removal of UV-induced DNA damage in human keratinocytes and human epidermal skin equivalents [202]. In line with our results, Babino et al. proved the good tolerability and long-term efficacy of a medical device containing Octa and urea in the reduction in grade III actinic keratoses (AKs) [203], skin lesions resulting from chronic and excessive UV exposure with a certain risk of becoming cancerous [204,205]. In addition, topical application of sunscreen containing inorganic filters and Octa has been shown to protect against the formation of sunburn cells, reduce the number of apoptotic keratinocytes, and prevent UV-induced molecular alterations [206]. More recently, Flori et al. demonstrated the ability of Octa and its derivative A02 to antagonize, through PPARγ activation, the TGF-β1-mediated changes associated with the epithelial–mesenchymal transition (EMT) process in a human squamous cell carcinoma cell line. Most interestingly, both compounds counteracted the TGF-β1-mediated activation of EMT-related signaling pathways, cell migration capacity, cell membrane lipid remodeling, and the release of bioactive lipids involved in EMT. In addition, in vivo experiments in a mouse model with cutaneous papillomas showed that the same PPARγ ligands, applied topically, were able to reduce the number of lesions and tumor area [200].

In terms of human epidemiological evidence, a large population-based retrospective cohort study in Taiwan suggests that the long-term use of the PPARγ agonist rosiglitazone in patients with type 2 diabetes mellitus may be associated with a lower risk of NMSC [207]. Unfortunately, there is little direct evidence from clinical trials on the feasibility of using PPARγ modulators for the prevention and/or treatment of keratinocyte-derived cancers. However, data from the literature strongly support the hypothesis that targeting PPARγ may represent a potential pharmacological target for the prevention or treatment of skin cancer. Current knowledge could be considered as a starting point for developing more potent and selective chemopreventive or chemotherapeutic PPARγ ligands with a low toxicity profile, thus reducing unwanted side effects.

7.2. Melanoma

Some evidence in the literature suggests that PPARγ may be involved in protecting against melanoma, although current knowledge is limited and controversial. Similar to non-melanoma skin cancers, growth inhibition via PPARγ activation has been described in melanoma. Most of these studies used melanoma cell lines and treatment with PPARγ agonists in vitro [193]. The study by Botton et al. shows that the PPARγ ligand ciglitazone inhibits melanoma cell growth by inducing both cell cycle arrest via a PPARγ-dependent pathway at low concentrations of the compound and apoptosis at higher concentrations and independently of PPARγ activation. Furthermore, in vivo treatment of nude mice with ciglitazone significantly impairs the development of human melanoma xenografts [208]. In addition, the PPARγ ligand 15-deoxy-D12,14 prostaglandin J2 (15d-PGJ2) has been shown to induce cell cycle arrest in several melanoma cell lines and to inhibit tumor cell migration, tumor-associated fibroblast proliferation, and endothelial cell tube formation [209]. A connection between the α-melanocyte stimulating hormone (α-MSH) and PPARγ in human melanocytes and melanoma cells has been identified, resulting in reduced melanoma cell proliferation associated with cell cycle withdrawal [32,210,211]. In line with these findings, Borland et al. demonstrated that overexpression and/or activation of PPARγ by rosiglitazone inhibits cell proliferation and anchorage-dependent clonogenicity in a human melanoma cell line. This effect is reflected in the observed inhibition of ectopic xenograft tumorigenicity [199]. Moreover, Konger et al. demonstrate that B16F10 melanoma tumor growth is enhanced in syngeneic PPARγ $-/-^{epi}$ mice, indicating that loss of epidermal PPARγ acts through indirect mechanisms to regulate tumor growth. Moreover, the pretreatment of mice with rosiglitazone counteracts B16F10 tumor growth induced by UVB [212]. A clinical study on patients with late metastatic stage melanoma shows a significantly prolonged progression-free survival in the group of patients that received chemotherapy in combination with pioglitazone compared to the group that received chemotherapy alone [213]. In contrast to these literature data showing a potential efficacy of PPARγ activation in the prevention or treatment of melanoma, a cohort study conducted in diabetic patients treated with pioglitazone suggests that the use of the TZD class of PPARγ ligands is associated with an increased risk of melanoma [214]. In line with this evidence, Pich et al. show that rosiglitazone induces a tumorigenic paracrine communication program in a subset of human metastatic melanoma cells involving the secretion of cytokines, chemokines, and angiogenic factors, leading to the activation of non-malignant stromal cells of the tumor microenvironment. Moreover, in vivo data show that a rosiglitazone-supplemented diet promotes melanoma tumor growth in xenografts, accompanied by increased inflammation and angiogenesis [215]. The role of PPARγ agonists in tumourigenicity is complex [216]. The co-occurrence of data suggesting both beneficial and tumorigenic effects of PPARγ ligands indicates a great complexity in the mechanisms of PPARγ action in relation to the tumor context and microenvironment. Therefore, the role of PPARγ in melanoma remains to be elucidated. Further clinical studies are needed to determine whether PPARγ modulators may have a beneficial role in the prevention and/or treatment of melanoma. Therapeutic modulation of PPARγ should, therefore, be considered with caution.

8. Conclusions

Over the past 20 years, considerable progress has been made in understanding the role of PPARγ in skin physiopathology. Keratinocyte differentiation, barrier permeability maintenance, sebaceous gland development and function, and inflammatory response are all processes regulated by PPARγ transcriptional activity. Alteration and/or disruption of these physiological functions are associated with the onset of skin disorders. The development of drugs targeting PPARγ is an area of active research, with PPARγ agonists being considered potential future dermatology therapeutics (Table 1; Figure 3).

Table 1. Therapeutic targeting of PPARγ in skin diseases. Beneficial effects of PPARγ activation in various skin disorders, listing the most tested molecules and the corresponding references.

Skin Condition	PPARγ Beneficial Effects	PPARγ Activators	References
Atopic Dermatitis, Contact Dermatitis	- Promotion of anti-inflammatory responses - Maintenance of mast cell homeostasis - Maintenance of barrier integrity and homeostasis	pioglitazone, ciglitazone	[107,115,116]
Psoriasis	- Reduction in the expression of genes involved in the development of psoriatic lesions - Normalisation of lesional skin - Reduction in epidermal hyperplasia	troglitazone, rosiglitazone, pioglitazone, BP-1107, GED-050734L	[23,98,101,122–126,134–138]
Vitiligo	- Enhancement of melanogenesis - Reduction in mitochondrial alterations - Reduction in ROS generation - Reversal of premature melanocyte senescence - Improvement of defective differentiation/barrier constitution	rosiglitazone, pioglitazone	[145,146]
Acne	- Reduction in altered lipogenesis - Anti-inflammatory activity - Improvement of sebum composition	NAC-GED0507	[22,156,161]
Skin Aging	- Modulation of aged-related inflammation by inhibition of pro-inflammatory gene expression - Reduction in aging progression and age-related disorders	pioglitazone, azelaic acid, kojyl cinnamate ester derivatives, octa, rosiglitazone	[177,181,188–190]
Non-melanoma skin cancer	- Antioxidant defense - Reduction in UV-dependent DNA damage - Inhibition of cell growth and induction of terminal differentiation - Antagonization of the EMT process	rosiglitazone, octa, A02	[21,198–200,202]
Melanoma	- Inhibition of cell growth - Induction of cell cycle arrest - Inhibition of tumor cell migration - Reduction in melanoma cell proliferation	ciglitazone,15d-PGJ2, rosiglitazone, pioglitazone	[193,199,209,212,213]

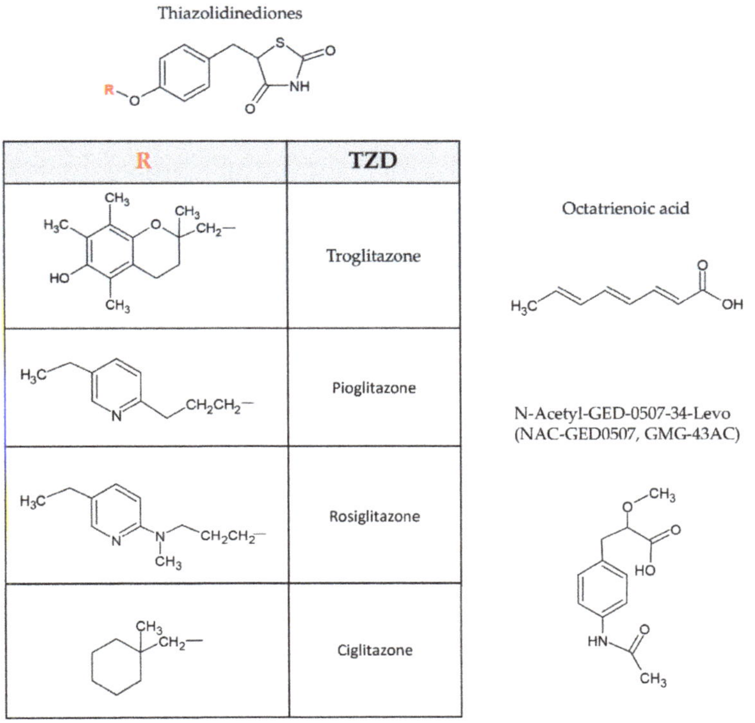

Figure 3. Representative PPARγ activators. Chemical structures of the most tested PPARγ activators related to skin pathophysiology. Thiazolidinedione (TZD).

A major challenge for future research will be the development of novel PPARγ ligands capable of selectively activating the different patterns of genes involved in the desired beneficial activity while reducing unwanted side effects through the design of ligands with a low toxicity profile. Much work is needed to fully characterize the pharmacological activity and specificity of these compounds.

Author Contributions: Conceptualization: S.B., S.M., E.F. and M.O.; writing—original draft preparation: S.B., S.M., E.F. and M.O.; writing—review and editing: S.B., S.M., E.F., A.D.N. and M.O. All authors have read and agreed to the published version of the manuscript.

Funding: This work was supported by public funds from the Italian Ministry of Health (RC2024).

Acknowledgments: Selected artworks (cell membranes, nucleus, nucleic acids, ligands, and receptors) shown in Figure 1 were used from or adapted from pictures provided by Servier Medical Art (Servier; https://smart.servier.com/) (accessed on 18 April 2024), licensed under a Creative Commons Attribution 4.0 Unported License.

Conflicts of Interest: The authors declare no conflicts of interest.

References

1. Issemann, I.; Green, S. Activation of a member of the steroid hormone receptor superfamily by peroxisome proliferators. *Nature* **1990**, *347*, 645–650. [CrossRef]
2. Varga, T.; Czimmerer, Z.; Nagy, L. PPARs are a unique set of fatty acid regulated transcription factors controlling both lipid metabolism and inflammation. *Biochim. Biophys. Acta (BBA) Mol. Basis Dis.* **2011**, *1812*, 1007–1022. [CrossRef]
3. Wahli, W.; Michalik, L. PPARs at the crossroads of lipid signaling and inflammation. *Trends Endocrinol. Metab.* **2012**, *23*, 351–363. [CrossRef] [PubMed]

4. Schmuth, M.; Moosbrugger-Martinz, V.; Blunder, S.; Dubrac, S. Role of PPAR, LXR, and PXR in epidermal homeostasis and inflammation. *Biochim. Biophys. Acta (BBA) Mol. Cell Biol. Lipids* **2014**, *1841*, 463–473. [CrossRef] [PubMed]
5. Vázquez-Carrera, M.; Wahli, W. PPARs as Key Mediators in the Regulation of Metabolism and Inflammation. *Int. J. Mol. Sci.* **2022**, *23*, 5025. [CrossRef] [PubMed]
6. Ramot, Y.; Mastrofrancesco, A.; Camera, E.; Desreumaux, P.; Paus, R.; Picardo, M. The role of PPARgamma-mediated signalling in skin biology and pathology: New targets and opportunities for clinical dermatology. *Exp. Dermatol.* **2015**, *24*, 245–251. [CrossRef]
7. Viswakarma, N.; Jia, Y.; Bai, L.; Vluggens, A.; Borensztajn, J.; Xu, J.; Reddy, J.K. Coactivators in PPAR-Regulated Gene Expression. *PPAR Res.* **2010**, *2010*, 250126. [CrossRef]
8. Peraza, M.A.; Burdick, A.D.; Marin, H.E.; Gonzalez, F.J.; Peters, J.M. The toxicology of ligands for peroxisome proliferator-activated receptors (PPAR). *Toxicol. Sci.* **2006**, *90*, 269–295. [CrossRef]
9. Korbecki, J.; Bobiński, R.; Dutka, M. Self-regulation of the inflammatory response by peroxisome proliferator-activated receptors. *Inflamm. Res.* **2019**, *68*, 443–458. [CrossRef]
10. Michalik, L.; Wahli, W. Peroxisome proliferator-activated receptors (PPARs) in skin health, repair and disease. *Biochim. Biophys. Acta (BBA) Mol. Cell Biol. Lipids* **2007**, *1771*, 991–998. [CrossRef]
11. Michalik, L.; Auwerx, J.; Berger, J.P.; Chatterjee, V.K.; Glass, C.K.; Gonzalez, F.J.; Grimaldi, P.A.; Kadowaki, T.; Lazar, M.A.; O'Rahilly, S.; et al. International Union of Pharmacology. LXI. Peroxisome Proliferator-Activated Receptors. *Pharmacol. Rev.* **2006**, *58*, 726–741. [CrossRef] [PubMed]
12. Kliewer, S.A.; Forman, B.M.; Blumberg, B.; Ong, E.S.; Borgmeyer, U.; Mangelsdorf, D.J.; Umesono, K.; Evans, R.M. Differential expression and activation of a family of murine peroxisome proliferator-activated receptors. *Proc. Natl. Acad. Sci. USA* **1994**, *91*, 7355–7359. [CrossRef] [PubMed]
13. Tanaka, N.; Aoyama, T.; Kimura, S.; Gonzalez, F.J. Targeting nuclear receptors for the treatment of fatty liver disease. *Pharmacol. Ther.* **2017**, *179*, 142–157. [CrossRef] [PubMed]
14. Zoete, V.; Grosdidier, A.; Michielin, O. Peroxisome proliferator-activated receptor structures: Ligand specificity, molecular switch and interactions with regulators. *Biochim. Biophys. Acta (BBA) Mol. Cell Biol. Lipids* **2007**, *1771*, 915–925. [CrossRef] [PubMed]
15. Cronet, P.; Petersen, J.F.; Folmer, R.; Blomberg, N.; Sjöblom, K.; Karlsson, U.; Lindstedt, E.-L.; Bamberg, K. Structure of the PPARα and -γ Ligand Binding Domain in Complex with AZ 242; Ligand Selectivity and Agonist Activation in the PPAR Family. *Structure* **2001**, *9*, 699–706. [CrossRef] [PubMed]
16. Batista, F.A.H.; Trivella, D.B.B.; Bernardes, A.; Gratieri, J.; Oliveira, P.S.L.; Figueira, A.C.M.; Webb, P.; Polikarpov, I. Structural insights into human peroxisome proliferator activated receptor delta (PPAR-delta) selective ligand binding. *PLoS ONE* **2012**, *7*, e33643. [CrossRef] [PubMed]
17. Mahindroo, N.; Wang, C.-C.; Liao, C.-C.; Huang, C.-F.; Lu, I.-L.; Lien, T.-W.; Peng, Y.-H.; Huang, W.-J.; Lin, Y.-T.; Hsu, M.-C.; et al. Indol-1-yl acetic acids as peroxisome proliferator-activated receptor agonists: Design, synthesis, structural biology, and molecular docking studies. *J. Med. Chem.* **2006**, *49*, 1212–1216. [CrossRef] [PubMed]
18. Westergaard, M.; Henningsen, J.; Kratchmarova, I.; Kristiansen, K.; Svendsen, M.L.; Johansen, C.; Jensen, U.B.; Schrøder, H.D.; Berge, R.K.; Iversen, L.; et al. Modulation of keratinocyte gene expression and differentiation by PPAR-selective ligands and tetradecylthioacetic acid. *J. Investig. Dermatol.* **2001**, *116*, 702–712. [CrossRef] [PubMed]
19. Friedmann, P.S.; Cooper, H.L.; Healy, E. Peroxisome proliferator-activated receptors and their relevance to dermatology. *Acta Derm. Venereol.* **2005**, *85*, 194–202. [CrossRef]
20. Sertznig, P.; Seifert, M.; Tilgen, W.; Reichrath, J. Peroxisome proliferator-activated receptors (PPARs) and the human skin: Importance of PPARs in skin physiology and dermatologic diseases. *Am. J. Clin. Dermatol.* **2008**, *9*, 15–31. [CrossRef]
21. Konger, R.L.; Derr-Yellin, E.; Zimmers, T.A.; Katona, T.; Xuei, X.; Liu, Y.; Zhou, H.-M.; Simpson, E.R.; Turner, M.J. Epidermal PPARγ Is a Key Homeostatic Regulator of Cutaneous Inflammation and Barrier Function in Mouse Skin. *Int. J. Mol. Sci.* **2021**, *22*, 8634. [CrossRef] [PubMed]
22. Mastrofrancesco, A.; Ottaviani, M.; Cardinali, G.; Flori, E.; Briganti, S.; Ludovici, M.; Zouboulis, C.; Lora, V.; Camera, E.; Picardo, M. Pharmacological PPARγ modulation regulates sebogenesis and inflammation in SZ95 human sebocytes. *Biochem. Pharmacol.* **2017**, *138*, 96–106. [CrossRef] [PubMed]
23. Ellis, C.N.; Varani, J.; Fisher, G.J.; Zeigler, M.E.; Pershadsingh, H.A.; Benson, S.C.; Chi, Y.; Kurtz, T.W. Troglitazone improves psoriasis and normalizes models of proliferative skin disease: Ligands for peroxisome proliferator-activated receptor-gamma inhibit keratinocyte proliferation. *Arch. Dermatol.* **2000**, *136*, 609–616. [CrossRef] [PubMed]
24. Mao-Qiang, M.; Fowler, A.J.; Schmuth, M.; Lau, P.; Chang, S.; Brown, B.E.; Moser, A.H.; Michalik, L.; Desvergne, B.; Wahli, W.; et al. Peroxisome-proliferator-activated receptor (PPAR)-gamma activation stimulates keratinocyte differentiation. *J. Investig. Dermatol.* **2004**, *123*, 305–312. [CrossRef] [PubMed]
25. Montagner, A.; Rando, G.; Degueurce, G.; Leuenberger, N.; Michalik, L.; Wahli, W. New insights into the role of PPARs. *Prostaglandins Leukot. Essent. Fat. Acids* **2011**, *85*, 235–243. [CrossRef] [PubMed]
26. Icre, G.; Wahli, W.; Michalik, L. Functions of the peroxisome proliferator-activated receptor (PPAR) alpha and beta in skin homeostasis, epithelial repair, and morphogenesis. *J. Investig. Dermatol. Symp. Proc.* **2006**, *11*, 30–35. [CrossRef] [PubMed]
27. Sertznig, P.; Reichrath, J. Peroxisome proliferator-activated receptors (PPARs) in dermatology: Challenge and promise. *Dermato-Endocrinology* **2011**, *3*, 130–135. [CrossRef] [PubMed]

28. Gupta, M.; Mahajan, V.K.; Mehta, K.S.; Chauhan, P.S.; Rawat, R. Peroxisome proliferator-activated receptors (PPARs) and PPAR agonists: The 'future' in dermatology therapeutics? *Arch. Dermatol. Res.* **2015**, *307*, 767–780. [CrossRef] [PubMed]
29. Kang, H.Y.; Chung, E.; Lee, M.; Cho, Y.; Kang, W.H. Expression and function of peroxisome proliferator-activated receptors in human melanocytes. *Br. J. Dermatol.* **2004**, *150*, 462–468. [CrossRef]
30. Lee, J.S.; Choi, Y.M.; Kang, H.Y. PPAR-gamma agonist, ciglitazone, increases pigmentation and migration of human melanocytes. *Exp. Dermatol.* **2007**, *16*, 118–123. [CrossRef]
31. Michalik, L.; Wahli, W. PPARs Mediate Lipid Signaling in Inflammation and Cancer. *PPAR Res.* **2008**, *2008*, 134059. [CrossRef]
32. Flori, E.; Rosati, E.; Cardinali, G.; Kovacs, D.; Bellei, B.; Picardo, M.; Maresca, V. The α-melanocyte stimulating hormone/peroxisome proliferator activated receptor-γ pathway down-regulates proliferation in melanoma cell lines. *J. Exp. Clin. Cancer Res.* **2017**, *36*, 142. [CrossRef] [PubMed]
33. Clark, R.B.; Bishop-Bailey, D.; Estrada-Hernandez, T.; Hla, T.; Puddington, L.; Padula, S.J. The nuclear receptor PPAR gamma and immunoregulation: PPAR gamma mediates inhibition of helper T cell responses. *J. Immunol.* **2000**, *164*, 1364–1371. [CrossRef] [PubMed]
34. al Yacoub, N.; Romanowska, M.; Krauss, S.; Schweiger, S.; Foerster, J. PPARδ Is a Type 1 IFN Target Gene and Inhibits Apoptosis in T Cells. *J. Investig. Dermatol.* **2008**, *128*, 1940–1949. [CrossRef]
35. Grygiel-Górniak, B. Peroxisome proliferator-activated receptors and their ligands: Nutritional and clinical implications—A review. *Nutr. J.* **2014**, *13*, 17. [CrossRef] [PubMed]
36. Virendra, S.A.; Kumar, A.; Chawla, P.A.; Mamidi, N. Development of Heterocyclic PPAR Ligands for Potential Therapeutic Applications. *Pharmaceutics* **2022**, *14*, 2139. [CrossRef]
37. Weindl, G.; Schäfer-Korting, M.; Schaller, M.; Korting, H.C. Peroxisome proliferator-activated receptors and their ligands: Entry into the post-glucocorticoid era of skin treatment? *Drugs* **2005**, *65*, 1919–1934. [CrossRef]
38. El-Jamal, N.; Dubuquoy, L.; Auwerx, J.; Bertin, B.; Desreumaux, P. In vivo imaging reveals selective PPAR activity in the skin of peroxisome proliferator-activated receptor responsive element-luciferase reporter mice. *Exp. Dermatol.* **2013**, *22*, 137–140. [CrossRef]
39. Chen, H.; Shi, R.; Luo, B.; Yang, X.; Qiu, L.; Xiong, J.; Jiang, M.; Liu, Y.; Zhang, Z.; Wu, Y. Macrophage peroxisome proliferator-activated receptor γ deficiency delays skin wound healing through impairing apoptotic cell clearance in mice. *Cell Death Dis.* **2015**, *6*, e1597. [CrossRef]
40. Ruzehaji, N.; Frantz, C.; Ponsoye, M.; Avouac, J.; Pezet, S.; Guilbert, T.; Luccarini, J.-M.; Broqua, P.; Junien, J.-L.; Allanore, Y. Pan PPAR agonist IVA337 is effective in prevention and treatment of experimental skin fibrosis. *Ann. Rheum. Dis.* **2016**, *75*, 2175–2183. [CrossRef]
41. Balakumar, P.; Rose, M.; Ganti, S.S.; Krishan, P.; Singh, M. PPAR dual agonists: Are they opening Pandora's Box? *Pharmacol. Res.* **2007**, *56*, 91–98. [CrossRef] [PubMed]
42. Schmidt, M.V.; Brüne, B.; von Knethen, A. The nuclear hormone receptor PPARγ as a therapeutic target in major diseases. *Sci. World J.* **2010**, *10*, 2181–2197. [CrossRef]
43. Ahmadian, M.; Suh, J.M.; Hah, N.; Liddle, C.; Atkins, A.R.; Downes, M.; Evans, R.M. PPARγ signaling and metabolism: The good, the bad and the future. *Nat. Med.* **2013**, *19*, 557–566. [CrossRef] [PubMed]
44. Jia, Z.; Sun, Y.; Yang, G.; Zhang, A.; Huang, S.; Heiney, K.M.; Zhang, Y. New Insights into the PPAR γ Agonists for the Treatment of Diabetic Nephropathy. *PPAR Res.* **2014**, *2014*, 818530. [CrossRef] [PubMed]
45. Kung, J.; Henry, R.R. Thiazolidinedione safety. *Expert Opin. Drug Saf.* **2012**, *11*, 565–579. [CrossRef] [PubMed]
46. Wang, S.; Dougherty, E.J.; Danner, R.L. PPARγ signaling and emerging opportunities for improved therapeutics. *Pharmacol. Res.* **2016**, *111*, 76–85. [CrossRef] [PubMed]
47. Lu, P.; Zhao, Z. Advances on PPARγ Research in the Emerging Era of Precision Medicine. *Curr. Drug Targets* **2018**, *19*, 663–673. [CrossRef] [PubMed]
48. Lefèvre-Utile, A.; Braun, C.; Haftek, M.; Aubin, F. Five Functional Aspects of the Epidermal Barrier. *Int. J. Mol. Sci.* **2021**, *22*, 11676. [CrossRef] [PubMed]
49. Rawlings, A.V.; Harding, C.R. Moisturization and skin barrier function. *Dermatol. Ther.* **2004**, *17* (Suppl. S1), 43–48. [CrossRef]
50. Elias, P.M.; Gruber, R.; Crumrine, D.; Menon, G.; Williams, M.L.; Wakefield, J.S.; Holleran, W.M.; Uchida, Y. Formation and functions of the corneocyte lipid envelope (CLE). *Biochim. Biophys. Acta (BBA) Mol. Cell Biol. Lipids* **2014**, *1841*, 314–318. [CrossRef]
51. Moore, D.J.; Rawlings, A.V. The chemistry, function and (patho)physiology of stratum corneum barrier ceramides. *Int. J. Cosmet. Sci.* **2017**, *39*, 366–372. [CrossRef] [PubMed]
52. Hatano, Y.; Elias, P.M. "Outside-to-inside", "inside-to-outside", and "intrinsic" endogenous pathogenic mechanisms in atopic dermatitis: Keratinocytes as the key functional cells involved in both permeability barrier dysfunction and immunological alterations. *Front. Immunol.* **2023**, *14*, 1239251. [CrossRef] [PubMed]
53. Hegazy, R.A.; Hay, R.M.A.; Shaker, O.; Sayed, S.S.; Halim, D.A.A. Psoriasis and metabolic syndrome: Is peroxisome proliferator-activated receptor-γ part of the missing link? *Eur. J. Dermatol.* **2012**, *22*, 622–628. [CrossRef] [PubMed]
54. Demerjian, M.; Man, M.; Choi, E.; Brown, B.E.; Crumrine, D.; Chang, S.; Mauro, T.; Elias, P.M.; Feingold, K.R. Topical treatment with thiazolidinediones, activators of peroxisome proliferator-activated receptor-γ, normalizes epidermal homeostasis in a murine hyperproliferative disease model. *Exp. Dermatol.* **2006**, *15*, 154–160. [CrossRef] [PubMed]

55. Man, M.-Q.; Choi, E.-H.; Schmuth, M.; Crumrine, D.; Uchida, Y.; Elias, P.M.; Holleran, W.M.; Feingold, K.R. Basis for improved permeability barrier homeostasis induced by PPAR and LXR activators: Liposensors stimulate lipid synthesis, lamellar body secretion, and post-secretory lipid processing. *J. Investig. Dermatol.* **2006**, *126*, 386–392. [CrossRef]
56. Sahu, R.P.; DaSilva, S.C.; Rashid, B.; Martel, K.C.; Jernigan, D.; Mehta, S.R.; Mohamed, D.R.; Rezania, S.; Bradish, J.R.; Armstrong, A.B.; et al. Mice lacking epidermal PPARgamma exhibit a marked augmentation in photocarcinogenesis associated with increased UVB-induced apoptosis, inflammation and barrier dysfunction. *Int. J. Cancer* **2012**, *131*, 1055. [CrossRef] [PubMed]
57. Blunder, S.; Krimbacher, T.; Moosbrugger-Martinz, V.; Gruber, R.; Schmuth, M.; Dubrac, S. Keratinocyte-derived IL-1β induces *PPARG* downregulation and *PPARD* upregulation in human reconstructed epidermis following barrier impairment. *Exp. Dermatol.* **2021**, *30*, 1298–1308. [CrossRef] [PubMed]
58. Clayton, R.W.; Langan, E.A.; Ansell, D.M.; de Vos, I.J.H.M.; Göbel, K.; Schneider, M.R.; Picardo, M.; Lim, X.; van Steensel, M.A.M.; Paus, R. Neuroendocrinology and neurobiology of sebaceous glands. *Biol. Rev.* **2020**, *95*, 592–624. [CrossRef] [PubMed]
59. Zouboulis, C.C.; Coenye, T.; He, L.; Kabashima, K.; Kobayashi, T.; Niemann, C.; Nomura, T.; Oláh, A.; Picardo, M.; Quist, S.R.; et al. Sebaceous immunobiology-skin homeostasis, pathophysiology, coordination of innate immunity and inflammatory response and disease associations. *Front. Immunol.* **2022**, *13*, 1029818. [CrossRef] [PubMed]
60. Rosenfield, R.L.; Kentsis, A.; Deplewski, D.; Ciletti, N. Rat preputial sebocyte differentiation involves peroxisome proliferator-activated receptors. *J. Investig. Dermatol.* **1999**, *112*, 226–232. [CrossRef]
61. Schmuth, M.; Jiang, Y.J.; Dubrac, S.; Elias, P.M.; Feingold, K.R. Thematic review series: Skin lipids. Peroxisome proliferator-activated receptors and liver X receptors in epidermal biology. *J. Lipid Res.* **2008**, *49*, 499–509. [CrossRef] [PubMed]
62. Dozsa, A.; Dezso, B.; Toth, B.I.; Bacsi, A.; Poliska, S.; Camera, E.; Picardo, M.; Zouboulis, C.C.; Bíró, T.; Schmitz, G.; et al. PPARγ-mediated and arachidonic acid-dependent signaling is involved in differentiation and lipid production of human sebocytes. *J. Investig. Dermatol.* **2014**, *134*, 910–920. [CrossRef] [PubMed]
63. Veniaminova, N.A.; Jia, Y.Y.; Hartigan, A.M.; Huyge, T.J.; Tsai, S.-Y.; Grachtchouk, M.; Nakagawa, S.; Dlugosz, A.A.; Atwood, S.X.; Wong, S.Y. Distinct mechanisms for sebaceous gland self-renewal and regeneration provide durability in response to injury. *Cell Rep.* **2023**, *42*, 113121. [CrossRef] [PubMed]
64. Karnik, P.; Tekeste, Z.; McCormick, T.S.; Gilliam, A.C.; Price, V.H.; Cooper, K.D.; Mirmirani, P. Hair follicle stem cell-specific PPARgamma deletion causes scarring alopecia. *J. Investig. Dermatol.* **2009**, *129*, 1243–1257. [CrossRef] [PubMed]
65. Sardella, C.; Winkler, C.; Quignodon, L.; Hardman, J.A.; Toffoli, B.; Attianese, G.M.P.G.; Hundt, J.E.; Michalik, L.; Vinson, C.R.; Paus, R.; et al. Delayed Hair Follicle Morphogenesis and Hair Follicle Dystrophy in a Lipoatrophy Mouse Model of Pparg Total Deletion. *J. Investig. Dermatol.* **2018**, *138*, 500–510. [CrossRef] [PubMed]
66. Wang, X.; Wang, X.; Liu, J.; Cai, T.; Guo, L.; Wang, S.; Wang, J.; Cao, Y.; Ge, J.; Jiang, Y.; et al. Hair Follicle and Sebaceous Gland De Novo Regeneration with Cultured Epidermal Stem Cells and Skin-Derived Precursors. *Stem Cells Transl. Med.* **2016**, *5*, 1695–1706. [CrossRef] [PubMed]
67. Smith, R.N.; Braue, A.; Varigos, G.A.; Mann, N.J. The effect of a low glycemic load diet on acne vulgaris and the fatty acid composition of skin surface triglycerides. *J. Dermatol. Sci.* **2008**, *50*, 41–52. [CrossRef]
68. Fischer, H.; Fumicz, J.; Rossiter, H.; Napirei, M.; Buchberger, M.; Tschachler, E.; Eckhart, L. Holocrine Secretion of Sebum Is a Unique DNase2-Dependent Mode of Programmed Cell Death. *J. Investig. Dermatol.* **2017**, *137*, 587–594. [CrossRef] [PubMed]
69. Koenig, U.; Robenek, H.; Barresi, C.; Brandstetter, M.; Resch, G.P.; Gröger, M.; Pap, T.; Hartmann, C. Cell death induced autophagy contributes to terminal differentiation of skin and skin appendages. *Autophagy* **2020**, *16*, 932–945. [CrossRef]
70. Del Rosso, J.Q.; Kircik, L. The primary role of sebum in the pathophysiology of acne vulgaris and its therapeutic relevance in acne management. *J. Dermatol. Treat.* **2024**, *35*, 2296855. [CrossRef]
71. Furue, M.; Takemura, M.; Nishio, K.; Sato, Y.; Nagata, S.; Kan, N.; Suenaga, A.; Furue, K.; Yoshida, M.; Konishi, S.; et al. IImmunohistological Localization of Peroxisome Proliferator-Activated Receptor α and γ in Human Sebaceous Glands. *Fukuoka Igaku Zasshi* **2016**, *107*, 199–203. [PubMed]
72. Trivedi, N.R.; Cong, Z.; Nelson, A.M.; Albert, A.J.; Rosamilia, L.L.; Sivarajah, S.; Gilliland, K.L.; Liu, W.; Mauger, D.T.; Gabbay, R.A.; et al. Peroxisome Proliferator-Activated Receptors Increase Human Sebum Production. *J. Investig. Dermatol.* **2006**, *126*, 2002–2009. [CrossRef] [PubMed]
73. Thiboutot, D.; Harris, G.; Iles, V.; Cimis, G.; Gilliland, K.; Hagari, S. Activity of the type 1 5 alpha-reductase exhibits regional differences in isolated sebaceous glands and whole skin. *J. Investig. Dermatol.* **1995**, *105*, 209–214. [CrossRef] [PubMed]
74. Makrantonaki, E.; Zouboulis, C. Testosterone metabolism to 5?-dihydrotestosterone and synthesis of sebaceous lipids is regulated by the peroxisome proliferator-activated receptor ligand linoleic acid in human sebocytes. *Br. J. Dermatol.* **2007**, *156*, 428–432. [CrossRef] [PubMed]
75. Inoue, T.; Miki, Y.; Kakuo, S.; Hachiya, A.; Kitahara, T.; Aiba, S.; Zouboulis, C.C.; Sasano, H. Expression of steroidogenic enzymes in human sebaceous glands. *J. Endocrinol.* **2014**, *222*, 301–312. [CrossRef] [PubMed]
76. Calder, P.C. n − 3 Polyunsaturated Fatty Acids, Inflammation, and Inflammatory Diseases. *Am. J. Clin. Nutr.* **2006**, *83* (Suppl. S6), 1505S–1519S. [CrossRef] [PubMed]
77. Henson, P. Suppression of macrophage inflammatory responses by PPARs. *Proc. Natl. Acad. Sci. USA* **2003**, *100*, 6295–6296. [CrossRef]
78. Rizzo, G.; Fiorucci, S. PPARs and other nuclear receptors in inflammation. *Curr. Opin. Pharmacol.* **2006**, *6*, 421–427. [CrossRef] [PubMed]

79. Heming, M.; Gran, S.; Jauch, N.L.; Fischer-Riepe, L.; Russo, A.; Klotz, L.; Hermann, S.; Schäfers, M.; Roth, J.; Barczyk-Kahlert, K. Peroxisome Proliferator-Activated Receptor-γ Modulates the Response of Macrophages to Lipopolysaccharide and Glucocorticoids. *Front. Immunol.* **2018**, *9*, 893. [CrossRef]
80. Dubrac, S.; Stoitzner, P.; Pirkebner, D.; Elentner, A.; Schoonjans, K.; Auwerx, J.; Saeland, S.; Hengster, P.; Fritsch, P.; Romani, N.; et al. Peroxisome proliferator-activated receptor-alpha activation inhibits Langerhans cell function. *J. Immunol.* **2007**, *178*, 4362–4372. [CrossRef]
81. Wculek, S.K.; Khouili, S.C.; Priego, E.; Heras-Murillo, I.; Sancho, D. Metabolic Control of Dendritic Cell Functions: Digesting Information. *Front. Immunol.* **2019**, *10*, 775. [CrossRef]
82. Dubrac, S.; Schmuth, M. PPAR-alpha in cutaneous inflammation. *Derm. Endocrinol.* **2011**, *3*, 23–26. [CrossRef]
83. Bonetti, M.I.; Bacci, S.; Santosuosso, M.; Mazzanti, B.; Aldinucci, A.; Ballerini, C.; Guasti, D.; Calosi, L.; Bosi, A.; Romagnoli, P. Rosiglitazone promotes the differentiation of Langerhans cells and inhibits that of other dendritic cell types from CD133 positive hematopoietic precursors. *Histol. Histopathol.* **2014**, *29*, 323–332. [CrossRef]
84. Gogolak, P.; Rethi, B.; Szatmari, I.; Lanyi, A.; Dezso, B.; Nagy, L.; Rajnavolgyi, E. Differentiation of CD1a− and CD1a+ monocyte-derived dendritic cells is biased by lipid environment and PPARγ. *Blood* **2007**, *109*, 643–652. [CrossRef]
85. Zhao, F.; Xiao, C.; Evans, K.S.; Theivanthiran, T.; DeVito, N.; Holtzhausen, A.; Liu, J.; Liu, X.; Boczkowski, D.; Nair, S.; et al. Paracrine Wnt5a-β-Catenin Signaling Triggers a Metabolic Program that Drives Dendritic Cell Tolerization. *Immunity* **2018**, *48*, 147.e7–160.e7. [CrossRef] [PubMed]
86. Angeli, V.; Hammad, H.; Staels, B.; Capron, M.; Lambrecht, B.N.; Trottein, F. Peroxisome proliferator-activated receptor gamma inhibits the migration of dendritic cells: Consequences for the immune response. *J. Immunol.* **2003**, *170*, 5295–5301. [CrossRef]
87. Geltink, R.I.K.; Kyle, R.L.; Pearce, E.L. Unraveling the Complex Interplay Between T Cell Metabolism and Function. *Annu. Rev. Immunol.* **2018**, *36*, 461–488. [CrossRef] [PubMed]
88. Angela, M.; Endo, Y.; Asou, H.K.; Yamamoto, T.; Tumes, D.J.; Tokuyama, H.; Yokote, K.; Nakayama, T. Fatty acid metabolic reprogramming via mTOR-mediated inductions of PPARγ directs early activation of T cells. *Nat. Commun.* **2016**, *7*, 13683. [CrossRef] [PubMed]
89. Choi, J.-M.; Bothwell, A.L.M. The nuclear receptor PPARs as important regulators of T-cell functions and autoimmune diseases. *Mol. Cells* **2012**, *33*, 217–222. [CrossRef]
90. Stark, J.M.; Tibbitt, C.A.; Coquet, J.M. The Metabolic Requirements of Th2 Cell Differentiation. *Front. Immunol.* **2019**, *10*, 2318. [CrossRef]
91. Klotz, L.; Burgdorf, S.; Dani, I.; Saijo, K.; Flossdorf, J.; Hucke, S.; Alferink, J.; Novak, N.; Beyer, M.; Mayer, G.; et al. The nuclear receptor PPARγ selectively inhibits Th17 differentiation in a T cell–intrinsic fashion and suppresses CNS autoimmunity. *J. Exp. Med.* **2009**, *206*, 2079–2089. [CrossRef] [PubMed]
92. Daniel, B.; Nagy, G.; Czimmerer, Z.; Horvath, A.; Hammers, D.W.; Cuaranta-Monroy, I.; Poliska, S.; Tzerpos, P.; Kolostyak, Z.; Hays, T.T.; et al. The Nuclear Receptor PPARγ Controls Progressive Macrophage Polarization as a Ligand-Insensitive Epigenomic Ratchet of Transcriptional Memory. *Immunity* **2018**, *49*, 615–626.e6. [CrossRef] [PubMed]
93. Kotla, S.; Singh, N.K.; Rao, G.N. ROS via BTK-p300-STAT1-PPARγ signaling activation mediates cholesterol crystals-induced CD36 expression and foam cell formation. *Redox Biol.* **2016**, *11*, 350–364. [CrossRef] [PubMed]
94. Moore, K.J.; Rosen, E.D.; Fitzgerald, M.L.; Randow, F.; Andersson, L.P.; Altshuler, D.; Milstone, D.S.; Mortensen, R.M.; Spiegelman, B.M.; Freeman, M.W. The role of PPAR-γ in macrophage differentiation and cholesterol uptake. *Nat. Med.* **2001**, *7*, 41–47. [CrossRef] [PubMed]
95. Wu, L.; Yan, C.; Czader, M.; Foreman, O.; Blum, J.S.; Kapur, R.; Du, H. Inhibition of PPARγ in myeloid-lineage cells induces systemic inflammation, immunosuppression, and tumorigenesis. *Blood* **2012**, *119*, 115–126. [CrossRef] [PubMed]
96. Klotz, L.; Dani, I.; Edenhofer, F.; Nolden, L.; Evert, B.; Paul, B.; Kolanus, W.; Klockgether, T.; Knolle, P.; Diehl, L. Peroxisome Proliferator-Activated Receptor γ Control of Dendritic Cell Function Contributes to Development of CD4+ T Cell Anergy. *J. Immunol.* **2007**, *178*, 2122–2131. [CrossRef] [PubMed]
97. Appel, S.; Mirakaj, V.; Bringmann, A.; Weck, M.M.; Grünebach, F.; Brossart, P. PPAR-γ agonists inhibit toll-like receptor-mediated activation of dendritic cells via the MAP kinase and NF-κB pathways. *Blood* **2005**, *106*, 3888–3894. [CrossRef]
98. Sobolev, V.V.; Tchepourina, E.; Korsunskaya, I.M.; Geppe, N.A.; Chebysheva, S.N.; Soboleva, A.G.; Mezentsev, A. The Role of Transcription Factor PPAR-γ in the Pathogenesis of Psoriasis, Skin Cells, and Immune Cells. *Int. J. Mol. Sci.* **2022**, *23*, 9708. [CrossRef] [PubMed]
99. Kim, I.S.; Silwal, P.; Jo, E.-K. Peroxisome Proliferator-Activated Receptor-Targeted Therapies: Challenges upon Infectious Diseases. *Cells* **2023**, *12*, 650. [CrossRef]
100. Ricote, M.; Glass, C.K. PPARs and molecular mechanisms of transrepression. *Biochim. Biophys. Acta (BBA) Mol. Cell Biol. Lipids* **2007**, *1771*, 926–935. [CrossRef]
101. Mastrofrancesco, A.; Kovacs, D.; Sarra, M.; Bastonini, E.; Cardinali, G.; Aspite, N.; Camera, E.; Chavatte, P.; Desreumaux, P.; Monteleone, G.; et al. Preclinical Studies of a Specific PPARγ Modulator in the Control of Skin Inflammation. *J. Investig. Dermatol.* **2013**, *134*, 1001–1011. [CrossRef] [PubMed]
102. Bieber, T. Atopic dermatitis. *N. Engl. J. Med.* **2008**, *358*, 1483–1494. [CrossRef] [PubMed]

103. Palmer, C.N.A.; Irvine, A.D.; Terron-Kwiatkowski, A.; Zhao, Y.; Liao, H.; Lee, S.P.; Goudie, D.R.; Sandilands, A.; Campbell, L.E.; Smith, F.J.D.; et al. Common loss-of-function variants of the epidermal barrier protein filaggrin are a major predisposing factor for atopic dermatitis. *Nat. Genet.* **2006**, *38*, 441–446. [CrossRef] [PubMed]
104. Howell, M.D.; Kim, B.E.; Gao, P.; Grant, A.V.; Boguniewicz, M.; DeBenedetto, A.; Schneider, L.; Beck, L.A.; Barnes, K.C.; Leung, D.Y. Cytokine modulation of atopic dermatitis filaggrin skin expression. *J. Allergy Clin. Immunol.* **2009**, *124*, R7–R12. [CrossRef] [PubMed]
105. Dahten, A.; Mergemeier, S.; Worm, M. PPARγ expression profile and its cytokine driven regulation in atopic dermatitis. *Allergy* **2007**, *62*, 926–933. [CrossRef]
106. Behshad, R.; Cooper, K.D.; Korman, N.J. A Retrospective Case Series Review of the Peroxisome Proliferator-Activated Receptor Ligand Rosiglitazone in the Treatment of Atopic Dermatitis. *Arch. Dermatol.* **2008**, *144*, 84–88. [CrossRef] [PubMed]
107. Zhang, Y.; Li, X.; Fang, S.; Zhu, Z.; Yao, M.; Ying, L.; Zhu, L.; Ma, Z.; Wang, W. Peroxisome proliferator-activated receptor γ agonist suppresses mast cell maturation and induces apoptosis. *Mol. Med. Rep.* **2017**, *16*, 1793–1800. [CrossRef]
108. Stark, J.M.; Coquet, J.M.; Tibbitt, C.A. The Role of PPAR-γ in Allergic Disease. *Curr. Allergy Asthma Rep.* **2021**, *21*, 45. [CrossRef] [PubMed]
109. Thomsen, S.F. Epidemiology and natural history of atopic diseases. *Eur. Clin. Respir. J.* **2015**, *2*, 24642. [CrossRef]
110. Bantz, S.K.; Zhu, Z.; Zheng, T. The Atopic March: Progression from Atopic Dermatitis to Allergic Rhinitis and Asthma. *J. Clin. Cell. Immunol.* **2014**, *5*, 202. [CrossRef]
111. Dharmage, S.C.; Lowe, A.J.; Matheson, M.C.; Burgess, J.A.; Allen, K.J.; Abramson, M.J. Atopic dermatitis and the atopic march revisited. *Allergy* **2014**, *69*, 17–27. [CrossRef] [PubMed]
112. Schäcke, H.; Döcke, W.-D.; Asadullah, K. Mechanisms involved in the side effects of glucocorticoids. *Pharmacol. Ther.* **2002**, *96*, 23–43. [CrossRef] [PubMed]
113. Kao, J.S.; Fluhr, J.W.; Man, M.-Q.; Fowler, A.J.; Hachem, J.-P.; Crumrine, D.; Ahn, S.K.; Brown, B.E.; Elias, P.M.; Feingold, K.R. Short-Term Glucocorticoid Treatment Compromises Both Permeability Barrier Homeostasis and Stratum Corneum Integrity: Inhibition of Epidermal Lipid Synthesis Accounts for Functional Abnormalities. *J. Investig. Dermatol.* **2003**, *120*, 456–464. [CrossRef] [PubMed]
114. Aberg, K.M.; Radek, K.A.; Choi, E.-H.; Kim, D.-K.; Demerjian, M.; Hupe, M.; Kerbleski, J.; Gallo, R.L.; Ganz, T.; Mauro, T.; et al. Psychological stress downregulates epidermal antimicrobial peptide expression and increases severity of cutaneous infections in mice. *J. Clin. Investig.* **2007**, *117*, 3339–3349. [CrossRef] [PubMed]
115. Demerjian, M.; Choi, E.; Man, M.; Chang, S.; Elias, P.M.; Feingold, K.R. Activators of PPARs and LXR decrease the adverse effects of exogenous glucocorticoids on the epidermis. *Exp. Dermatol.* **2009**, *18*, 643–649. [CrossRef] [PubMed]
116. Yamamoto, A.; Kakuta, H.; Sugimoto, Y. Involvement of glucocorticoid receptor activation on anti-inflammatory effect induced by peroxisome proliferator-activated receptor γ agonist in mice. *Int. Immunopharmacol.* **2014**, *22*, 204–208. [CrossRef] [PubMed]
117. Deckers, J.; Bougarne, N.; Mylka, V.; Desmet, S.; Luypaert, A.; Devos, M.; Tanghe, G.; Van Moorleghem, J.; Vanheerswynghels, M.; De Cauwer, L.; et al. Co-Activation of Glucocorticoid Receptor and Peroxisome Proliferator–Activated Receptor-γ in Murine Skin Prevents Worsening of Atopic March. *J. Investig. Dermatol.* **2018**, *138*, 1360–1370. [CrossRef] [PubMed]
118. Greb, J.E.; Goldminz, A.M.; Elder, J.T.; Lebwohl, M.G.; Gladman, D.D.; Wu, J.J.; Mehta, N.N.; Finlay, A.Y.; Gottlieb, A.B. Psoriasis. *Nat. Rev. Dis. Prim.* **2016**, *2*, 16082. [CrossRef] [PubMed]
119. Lowes, M.A.; Suárez-Fariñas, M.; Krueger, J.G. Immunology of Psoriasis. *Annu. Rev. Immunol.* **2014**, *32*, 227–255. [CrossRef]
120. Sobolev, V.; Nesterova, A.; Soboleva, A.; Dvoriankova, E.; Piruzyan, A.; Mildzikhova, D.; Korsunskaya, I.; Svitich, O. The Model of *PPARγ*-Downregulated Signaling in Psoriasis. *PPAR Res.* **2020**, *2020*, 6529057. [CrossRef]
121. Morin, S.; Simard, M.; Flamand, N.; Pouliot, R. Biological action of docosahexaenoic acid in a 3D tissue-engineered psoriatic skin model: Focus on the PPAR signaling pathway. *Biochim. Biophys. Acta (BBA) Mol. Cell Biol. Lipids* **2021**, *1866*, 159032. [CrossRef] [PubMed]
122. Varani, J.; Bhagavathula, N.; Ellis, C.N.; Pershadsingh, H.A. Thiazolidinediones: Potential as therapeutics for psoriasis and perhaps other hyperproliferative skin disease. *Expert Opin. Investig. Drugs* **2006**, *15*, 1453–1468. [CrossRef] [PubMed]
123. Bhagavathula, N.; Nerusu, K.C.; Reddy, M.; Ellis, C.N.; Chittiboyina, A.; Avery, M.; Pershadsingh, H.A.; Kurtz, T.W.; Varani, J. BP-1107 [{2-[4-(2,4-dioxo-thiazolidin-5-ylmethyl)-phenoxy]-ethyl}-methyl-amide]: A novel synthetic thiazolidinedione that inhibits epidermal hyperplasia in psoriatic skin-severe-combined immunodeficient mouse transplants after topical application. *J. Pharmacol. Exp. Ther.* **2005**, *315*, 996–1004. [CrossRef] [PubMed]
124. Yu, Z.; Gong, Y.; Cui, L.; Hu, Y.; Zhou, Q.; Chen, Z.; Yu, Y.; Chen, Y.; Xu, P.; Zhang, X.; et al. High-throughput transcriptome and pathogenesis analysis of clinical psoriasis. *J. Dermatol. Sci.* **2020**, *98*, 109–118. [CrossRef] [PubMed]
125. Li, X.; Xing, J.; Wang, F.; Li, J.; Li, J.; Hou, R.; Zhang, K. The mRNA Expression Profile of Psoriatic Lesion Distinct from Non-Lesion. *Clin. Cosmet. Investig. Dermatol.* **2022**, *15*, 2035–2043. [CrossRef] [PubMed]
126. Sobolev, V.; Nesterova, A.; Soboleva, A.; Mezentsev, A.; Dvoriankova, E.; Piruzyan, A.; Denisova, E.; Melnichenko, O.; Korsunskaya, I. Analysis of *PPARγ* Signaling Activity in Psoriasis. *Int. J. Mol. Sci.* **2021**, *22*, 8603. [CrossRef] [PubMed]
127. Pershadsingh, H.A.; Sproul, J.A.; Benjamin, E.; Finnegan, J.; Amin, N.M. Treatment of Psoriasis with Troglitazone Therapy. *Arch. Dermatol.* **1998**, *134*, 1304–1305. [CrossRef] [PubMed]
128. Bongartz, T.; Coras, B.; Vogt, T.; Schölmerich, J.; Müller-Ladner, U. Treatment of active psoriatic arthritis with the PPAR ligand pioglitazone: An open-label pilot study. *Rheumatology* **2005**, *44*, 126–129. [CrossRef]

129. Robertshaw, H.; Friedmann, P.S. Pioglitazone: A promising therapy for psoriasis. *Br. J. Dermatol.* **2005**, *152*, 189–191. [CrossRef]
130. Shafiq, N.; Malhotra, S.; Pandhi, P.; Gupta, M.; Kumar, B.; Sandhu, K. Pilot trial: Pioglitazone versus placebo in patients with plaque psoriasis (the P6). *Int. J. Dermatol.* **2005**, *44*, 328–333. [CrossRef]
131. Hafez, V.G.; Bosseila, M.; Halim, M.R.E.A.; Shaker, O.G.; Kamal, M.; Kareem, H.S. Clinical effects of "pioglitazone", an insulin sensitizing drug, on psoriasis vulgaris and its co-morbidities, a double blinded randomized controlled Trialxj. *Dermatol. Treat.* **2015**, *26*, 208–214. [CrossRef]
132. Ellis, C.N.; Barker, J.N.; Haig, A.E.; Parker, C.A.; Daly, S.; Jayawardene, D.A. Placebo response in two long-term randomized psoriasis studies that were negative for rosiglitazone. *Am. J. Clin. Dermatol.* **2007**, *8*, 93–102. [CrossRef]
133. Kuenzli, S.; Saurat, J.-H. Effect of topical PPARbeta/delta and PPARgamma agonists on plaque psoriasis. A pilot study. *Dermatology* **2003**, *206*, 252–256. [CrossRef] [PubMed]
134. Mittal, R.; Malhotra, S.; Pandhi, P.; Kaur, I.; Dogra, S. Efficacy and safety of combination Acitretin and Pioglitazone therapy in patients with moderate to severe chronic plaque-type psoriasis: A randomized, double-blind, placebo-controlled clinical trial. *Arch. Dermatol.* **2009**, *145*, 387–393. [CrossRef]
135. Lajevardi, V.; Hallaji, Z.; Daklan, S.; Abedini, R.; Goodarzi, A.; Abdolreza, M. The efficacy of methotrexate plus pioglitazone vs. methotrexate alone in the management of patients with plaque-type psoriasis: A single-blinded randomized controlled trial. *Int. J. Dermatol.* **2015**, *54*, 95–101. [CrossRef] [PubMed]
136. Abidi, A.; Rizvi, D.A.; Saxena, K.; Chaudhary, S.; Ahmad, A. The evaluation of efficacy and safety of methotrexate and pioglitazone in psoriasis patients: A randomized, open-labeled, active-controlled clinical trial. *Indian J. Pharmacol.* **2020**, *52*, 16–22. [CrossRef]
137. El-Gharabawy, R.M.; Ahmed, A.S.; Al-Najjar, A.H. Mechanism of action and effect of immune-modulating agents in the treatment of psoriasis. *Biomed. Pharmacother.* **2017**, *85*, 141–147. [CrossRef] [PubMed]
138. Ghiasi, M.; Ebrahimi, S.; Lajevardi, V.; Taraz, M.; Azizpour, A. Efficacy and safety of pioglitazone plus phototherapy versus phototherapy in patients with plaque type psoriasis: A Double Blinded Randomized Controlled Trial. *J. Dermatol. Treat.* **2019**, *30*, 664–667. [CrossRef]
139. Bergqvist, C.; Ezzedine, K. Vitiligo: A Review. *Dermatology* **2020**, *236*, 571–592. [CrossRef]
140. Zhen, Y.; Yao, L.; Zhong, S.; Song, Y.; Cui, Y.; Li, S. Enhanced Th1 and Th17 responses in peripheral blood in active non-segmental vitiligo. *Arch. Dermatol. Res.* **2016**, *308*, 703–710. [CrossRef]
141. Singh, R.K.; Lee, K.M.; Vujkovic-Cvijin, I.; Ucmak, D.; Farahnik, B.; Abrouk, M.; Nakamura, M.; Zhu, T.H.; Bhutani, T.; Wei, M.; et al. The role of IL-17 in vitiligo: A review. *Autoimmun. Rev.* **2016**, *15*, 397–404. [CrossRef] [PubMed]
142. Kemp, E.H.; Emhemad, S.; Akhtar, S.; Watson, P.F.; Gawkrodger, D.J.; Weetman, A.P. Autoantibodies against tyrosine hydroxylase in patients with non-segmental (generalised) vitiligo. *Exp. Dermatol.* **2011**, *20*, 35–40. [CrossRef] [PubMed]
143. Jimbo, H.; Nagai, H.; Fujiwara, S.; Shimoura, N.; Nishigori, C. Fas-FasL interaction in cytotoxic T cell-mediated vitiligo: The role of lesional expression of tumor necrosis factor-α and interferon-γ in Fas-mediated melanocyte apoptosis. *Exp. Dermatol.* **2020**, *29*, 61–70. [CrossRef] [PubMed]
144. Zhao, S.; Chen, X.; Dutta, K.; Chen, J.; Wang, J.; Zhang, Q.; Jia, H.; Sun, J.; Lai, Y. Multiple gene-drug prediction tool reveals Rosiglitazone based treatment pathway for non-segmental vitiligo. *Inflammation* **2023**, *47*, 678–695. [CrossRef] [PubMed]
145. Papaccio, F.; Bellei, B.; Ottaviani, M.; D'arino, A.; Truglio, M.; Caputo, S.; Cigliana, G.; Sciuto, L.; Migliano, E.; Pacifico, A.; et al. A Possible Modulator of Vitiligo Metabolic Impairment: Rethinking a PPARγ Agonist. *Cells* **2022**, *11*, 3583. [CrossRef] [PubMed]
146. Bastonini, E.; Kovacs, D.; Briganti, S.; Ottaviani, M.; D'Arino, A.; Migliano, E.; Pacifico, A.; Iacovelli, P.; Picardo, M. Effects of pioglitazone on the differentiation and inflammation in vitiligo keratinocytes. *J. Eur. Acad. Dermatol. Venereol.* **2024**, *ahead of print*. [CrossRef] [PubMed]
147. Picardo, M.; Ottaviani, M.; Camera, E.; Mastrofrancesco, A. Sebaceous gland lipids. *Dermato-Endocrinology* **2009**, *1*, 68–71. [CrossRef]
148. Ottaviani, M.; Camera, E.; Picardo, M. Lipid Mediators in Acne. *Mediat. Inflamm.* **2010**, *2010*, 858176. [CrossRef]
149. Schneider, M.R.; Paus, R. Sebocytes, multifaceted epithelial cells: Lipid production and holocrine secretion. *Int. J. Biochem. Cell Biol.* **2010**, *42*, 181–185. [CrossRef]
150. Melnik, B.C.; Schmitz, G. Role of insulin, insulin-like growth factor-1, hyperglycaemic food and milk consumption in the pathogenesis of acne vulgaris. *Exp. Dermatol.* **2009**, *18*, 833–841. [CrossRef]
151. Melnik, B. Linking diet to acne metabolomics, inflammation, and comedogenesis: An update. *Clin. Cosmet. Investig. Dermatol.* **2015**, *8*, 371–388. [CrossRef] [PubMed]
152. Clatici, V.G.; Voicu, C.; Voaides, C.; Roseanu, A.; Icriverzi, M.; Jurcoane, S. Diseases of Civilization—Cancer, Diabetes, Obesity and Acne—The Implication of Milk, IGF-1 and mTORC1. *Maedica* **2018**, *13*, 273–281. [CrossRef] [PubMed]
153. Cong, T.-X.; Hao, D.; Wen, X.; Li, X.-H.; He, G.; Jiang, X. From pathogenesis of acne vulgaris to anti-acne agents. *Arch. Dermatol. Res.* **2019**, *311*, 337–349. [CrossRef] [PubMed]
154. Briganti, S.; Flori, E.; Mastrofrancesco, A.; Ottaviani, M. Acne as an altered dermato-endocrine response problem. *Exp. Dermatol.* **2020**, *29*, 833–839. [CrossRef] [PubMed]
155. Okoro, O.E.; Camera, E.; Flori, E.; Ottaviani, M. Insulin and the sebaceous gland function. *Front. Physiol.* **2023**, *14*, 1252972. [CrossRef] [PubMed]

156. Ottaviani, M.; Flori, E.; Mastrofrancesco, A.; Briganti, S.; Lora, V.; Capitanio, B.; Zouboulis, C.C.; Picardo, M. Sebocyte differentiation as a new target for acne therapy: An in vivo experience. *J. Eur. Acad. Dermatol. Venereol.* **2020**, *34*, 1803–1814. [CrossRef]
157. Camera, E.; Ludovici, M.; Tortorella, S.; Sinagra, J.-L.; Capitanio, B.; Goracci, L.; Picardo, M. Use of lipidomics to investigate sebum dysfunction in juvenile acne. *J. Lipid Res.* **2016**, *57*, 1051–1058. [CrossRef] [PubMed]
158. Alestas, T.; Ganceviciene, R.; Fimmel, S.; Müller-Decker, K.; Zouboulis, C.C. Enzymes involved in the biosynthesis of leukotriene B4 and prostaglandin E2 are active in sebaceous glands. *J. Mol. Med.* **2006**, *84*, 75–87. [CrossRef] [PubMed]
159. Dozsa, A.; Mihaly, J.; Dezso, B.; Csizmadia, E.; Keresztessy, T.; Marko, L.; Rühl, R.; Remenyik, E.; Nagy, L. Decreased peroxisome proliferator-activated receptor γ level and signalling in sebaceous glands of patients with acne vulgaris. *Clin. Exp. Dermatol.* **2016**, *41*, 547–551. [CrossRef]
160. Monfrecola, G.; Lembo, S.; Caiazzo, G.; De Vita, V.; Di Caprio, R.; Balato, A.; Fabbrocini, G. Mechanistic target of rapamycin (mTOR) expression is increased in acne patients' skin. *Exp. Dermatol.* **2016**, *25*, 153–155. [CrossRef]
161. Picardo, M.; Cardinali, C.; La Placa, M.; Lewartowska-Białek, A.; Lora, V.; Micali, G.; Montisci, R.; Morbelli, L.; Nova, A.; Parodi, A.; et al. Efficacy and safety of N-acetyl-GED-0507-34-LEVO gel in patients with moderate-to severe facial acne vulgaris: A phase IIb randomized double-blind, vehicle-controlled trial. *Br. J. Dermatol.* **2022**, *187*, 507–514. [CrossRef]
162. Gilchrest, B.A. Photoaging. *J. Investig. Dermatol.* **2013**, *133*, E2–E6. [CrossRef] [PubMed]
163. Lupa, D.M.W.; Kalfalah, F.; Safferling, K.; Boukamp, P.; Poschmann, G.; Volpi, E.; Götz-Rösch, C.; Bernerd, F.; Haag, L.; Huebenthal, U.; et al. Characterization of Skin Aging–Associated Secreted Proteins (SAASP) Produced by Dermal Fibroblasts Isolated from Intrinsically Aged Human Skin. *J. Investig. Dermatol.* **2015**, *135*, 1954–1968. [CrossRef]
164. Jenkins, G. Molecular mechanisms of skin ageing. *Mech. Ageing Dev.* **2002**, *123*, 801–810. [CrossRef]
165. Ma, W.; Wlaschek, M.; Tantcheva-Poór, I.; Schneider, L.A.; Naderi, L.; Razi-Wolf, Z.; Schüller, J.; Scharffetter-Kochanek, K. Chronological ageing and photoageing of the fibroblasts and the dermal connective tissue. *Clin. Exp. Dermatol.* **2001**, *26*, 592–599. [CrossRef] [PubMed]
166. Rittié, L.; Fisher, G.J. UV-light-induced signal cascades and skin aging. *Ageing Res. Rev.* **2002**, *1*, 705–720. [CrossRef]
167. Abdel-Malek, Z.A.; Kadekaro, A.L.; Swope, V.B. Stepping up melanocytes to the challenge of UV exposure. *Pigment. Cell Melanoma Res.* **2010**, *23*, 171–186. [CrossRef] [PubMed]
168. González, S.; Fernández-Lorente, M.; Gilaberte-Calzada, Y. The latest on skin photoprotection. *Clin. Dermatol.* **2008**, *26*, 614–626. [CrossRef]
169. Wang, S.Q.; Balagula, Y.; Osterwalder, U. Photoprotection: A review of the current and future technologies. *Dermatol. Ther.* **2010**, *23*, 31–47. [CrossRef]
170. Yaar, M.; Gilchrest, B.A. Photoageing: Mechanism, prevention and therapy. *Br. J. Dermatol.* **2007**, *157*, 874–887. [CrossRef]
171. Naderi-Hachtroudi, L.; Peters, T.; Brenneisen, P.; Meewes, C.; Hommel, C.; Razi-Wolf, Z.; Schneider, L.A.; Schüller, J.; Wlaschek, M.; Scharffetter-Kochanek, K. Induction of manganese superoxide dismutase in human dermal fibroblasts: A UV-B-mediated paracrine mechanism with the release of epidermal interleukin 1 alpha, interleukin 1 beta, and tumor necrosis factor alpha. *Arch. Dermatol.* **2002**, *138*, 1473–1479. [CrossRef] [PubMed]
172. Fisher, G.J.; Kang, S.; Varani, J.; Bata-Csorgo, Z.; Wan, Y.; Datta, S.; Voorhees, J.J. Mechanisms of photoaging and chronological skin aging. *Arch. Dermatol.* **2002**, *138*, 1462–1470. [CrossRef] [PubMed]
173. Sarkar, D.; Fisher, P.B. Molecular mechanisms of aging-associated inflammation. *Cancer Lett.* **2006**, *236*, 13–23. [CrossRef] [PubMed]
174. Masaki, H.; Okano, Y.; Ochiai, Y.; Obayashi, K.; Akamatsu, H.; Sakurai, H. alpha-tocopherol increases the intracellular glutathione level in HaCaT keratinocytes. *Free. Radic. Res.* **2002**, *36*, 705–709. [CrossRef] [PubMed]
175. Shindo, Y.; Witt, E.; Packer, L. Antioxidant defense mechanisms in murine epidermis and dermis and their responses to ultraviolet light. *J. Investig. Dermatol.* **1993**, *100*, 260–265. [CrossRef] [PubMed]
176. Chung, J.H.; Seo, A.Y.; Chung, S.W.; Kim, M.K.; Leeuwenburgh, C.; Yu, B.P.; Chung, H.Y. Molecular mechanism of PPAR in the regulation of age-related inflammation. *Ageing Res. Rev.* **2008**, *7*, 126–136. [CrossRef] [PubMed]
177. Chen, J.; Mehta, J.L. Angiotensin II-mediated oxidative stress and procollagen-1 expression in cardiac fibroblasts: Blockade by pravastatin and pioglitazone. *Am. J. Physiol. Circ. Physiol.* **2006**, *291*, H1738–H1745. [CrossRef] [PubMed]
178. Kim, J.K.; Mun, S.; Kim, M.; Kim, M.; Sa, B.; Hwang, J. 5,7-Dimethoxyflavone, an activator of PPARα/γ, inhibits UVB-induced MMP expression in human skin fibroblast cells. *Exp. Dermatol.* **2012**, *21*, 211–216. [CrossRef] [PubMed]
179. Kim, S.O.; Han, Y.; Ahn, S.; An, S.; Shin, J.C.; Choi, H.; Kim, H.-J.; Park, N.H.; Kim, Y.-J.; Jin, S.H.; et al. Kojyl cinnamate esters are peroxisome proliferator-activated receptor α/γ dual agonists. *Bioorganic Med. Chem.* **2018**, *26*, 5654–5663. [CrossRef]
180. Shin, M.H.; Lee, S.-R.; Kim, M.-K.; Shin, C.-Y.; Lee, D.H.; Chung, J.H. Activation of Peroxisome Proliferator-Activated Receptor Alpha Improves Aged and UV-Irradiated Skin by Catalase Induction. *PLoS ONE* **2016**, *11*, e0162628. [CrossRef]
181. Shen, D.; Li, H.; Zhou, R.; Liu, M.-J.; Yu, H.; Wu, D.-F. Pioglitazone attenuates aging-related disorders in aged apolipoprotein E deficient mice. *Exp. Gerontol.* **2018**, *102*, 101–108. [CrossRef] [PubMed]
182. Wong, W.; Crane, E.D.; Zhang, H.; Li, J.; Day, T.A.; Green, A.E.; Menzies, K.J.; Crane, J.D. Pgc-1α controls epidermal stem cell fate and skin repair by sustaining NAD^+ homeostasis during aging. *Mol. Metab.* **2022**, *65*, 101575. [CrossRef] [PubMed]

183. Yoshizaki, N.; Fujii, T.; Masaki, H.; Okubo, T.; Shimada, K.; Hashizume, R. Orange peel extract, containing high levels of polymethoxyflavonoid, suppressed UVB-induced COX-2 expression and PGE_2 production in HaCaT cells through PPAR-γ activation. *Exp. Dermatol.* **2014**, *23* (Suppl. S1), 18–22. [CrossRef] [PubMed]
184. Sandulache, V.C.; Parekh, A.; Li-Korotky, H.; Dohar, J.E.; Hebda, P.A. Prostaglandin E2 inhibition of keloid fibroblast migration, contraction, and transforming growth factor (TGF)-β1–induced collagen synthesis. *Wound Repair Regen.* **2007**, *15*, 122–133. [CrossRef] [PubMed]
185. Yokota, M.; Yahagi, S.; Tokudome, Y.; Masaki, H. Chimyl Alcohol Suppresses PGE_2 Synthesis by Human Epidermal Keratinocytes through the Activation of PPAR-γ. *J. Oleo Sci.* **2018**, *67*, 455–462. [CrossRef] [PubMed]
186. Katsuyama, Y.; Tsuboi, T.; Taira, N.; Yoshioka, M.; Masaki, H. 3-O-Laurylglyceryl ascorbate activates the intracellular antioxidant system through the contribution of PPAR-γ and Nrf2. *J. Dermatol. Sci.* **2016**, *82*, 189–196. [CrossRef] [PubMed]
187. Lee, Y.; Lee, N.; Bhattarai, G.; Yun, J.; Kim, T.; Jhee, E.; Yi, H. PPARγ inhibits inflammatory reaction in oxidative stress induced human diploid fibloblast. *Cell Biochem. Funct.* **2010**, *28*, 490–496. [CrossRef]
188. Briganti, S.; Flori, E.; Mastrofrancesco, A.; Kovacs, D.; Camera, E.; Ludovici, M.; Cardinali, G.; Picardo, M. Azelaic acid reduced senescence-like phenotype in photo-irradiated human dermal fibroblasts: Possible implication of PPARγ. *Exp. Dermatol.* **2012**, *22*, 41–47. [CrossRef]
189. Briganti, S.; Flori, E.; Bellei, B.; Picardo, M. Modulation of PPARγ Provides New Insights in a Stress Induced Premature Senescence Model. *PLoS ONE* **2014**, *9*, e104045. [CrossRef]
190. Chen, L.; Bi, B.; Zeng, J.; Zhou, Y.; Yang, P.; Guo, Y.; Zhu, J.; Yang, Q.; Zhu, N.; Liu, T. Rosiglitazone ameliorates senescence-like phenotypes in a cellular photoaging model. *J. Dermatol. Sci.* **2015**, *77*, 173–181. [CrossRef]
191. Pihl, C.; Togsverd-Bo, K.; Andersen, F.; Haedersdal, M.; Bjerring, P.; Lerche, C.M. Keratinocyte Carcinoma and Photoprevention: The Protective Actions of Repurposed Pharmaceuticals, Phytochemicals and Vitamins. *Cancers* **2021**, *13*, 3684. [CrossRef] [PubMed]
192. Piipponen, M.; Riihilä, P.; Nissinen, L.; Kähäri, V.-M. The Role of p53 in Progression of Cutaneous Squamous Cell Carcinoma. *Cancers* **2021**, *13*, 4507. [CrossRef] [PubMed]
193. Wagner, N.; Wagner, K.-D. Peroxisome Proliferator-Activated Receptors and the Hallmarks of Cancer. *Cells* **2022**, *11*, 2432. [CrossRef] [PubMed]
194. Park, M.H.; Park, J.Y.; Lee, H.J.; Kim, D.H.; Chung, K.W.; Park, D.; Jeong, H.O.; Kim, H.R.; Park, C.H.; Kim, S.R.; et al. The novel PPAR alpha/gamma dual agonist MHY 966 modulates UVB-induced skin inflammation by inhibiting NF-kappaB activity. *PLoS ONE* **2013**, *8*, e76820.
195. Wang, Z.; Coleman, D.J.; Bajaj, G.; Liang, X.; Ganguli-Indra, G.; Indra, A.K. RXRalpha ablation in epidermal keratinocytes enhances UVR-induced DNA damage, apoptosis, and proliferation of keratinocytes and melanocytes. *J. Investig. Dermatol.* **2011**, *131*, 177–187. [CrossRef] [PubMed]
196. Ren, L.; Konger, R.L. Evidence that peroxisome proliferator-activated receptor γ suppresses squamous carcinogenesis through anti-inflammatory signaling and regulation of the immune response. *Mol. Carcinog.* **2019**, *58*, 1589–1601. [CrossRef] [PubMed]
197. Balupillai, A.; Prasad, R.N.; Ramasamy, K.; Muthusamy, G.; Shanmugham, M.; Govindasamy, K.; Gunaseelan, S. Caffeic Acid Inhibits UVB-induced Inflammation and Photocarcinogenesis through Activation of Peroxisome Proliferator-activated Receptor-gamma in Mouse Skin. *Photochem. Photobiol.* **2015**, *91*, 1458–1468. [CrossRef]
198. Li, Q.; Peng, Y.S.; Chen, P.J.; Wang, M.L.; Cao, C.; Xiong, H.; Zhang, J.; Chen, M.H.; Peng, X.B.; Zeng, K. Peroxisome proliferator-activated receptor-gamma agonist-mediated inhibition of cell growth is independent of apoptosis in human epidermoid carcinoma A431 cells. *Oncol. Lett.* **2018**, *15*, 6578–6584.
199. Borland, M.G.; Kehres, E.M.; Lee, C.; Wagner, A.L.; Shannon, B.E.; Albrecht, P.P.; Zhu, B.; Gonzalez, F.J.; Peters, J.M. Inhibition of tumorigenesis by peroxisome proliferator-activated receptor (PPAR)-dependent cell cycle blocks in human skin carcinoma cells. *Toxicology* **2018**, *404–405*, 25–32. [CrossRef]
200. Flori, E.; Mosca, S.; Cardinali, G.; Briganti, S.; Ottaviani, M.; Kovacs, D.; Manni, I.; Truglio, M.; Mastrofrancesco, A.; Zaccarini, M.; et al. The Activation of PPARγ by (2Z,4E,6E)-2-methoxyocta-2,4,6-trienoic Acid Counteracts the Epithelial–Mesenchymal Transition Process in Skin Carcinogenesis. *Cells* **2023**, *12*, 1007. [CrossRef]
201. Flori, E.; Mastrofrancesco, A.; Kovacs, D.; Ramot, Y.; Briganti, S.; Bellei, B.; Paus, R.; Picardo, M. 2,4,6-Octatrienoic acid is a novel promoter of melanogenesis and antioxidant defence in normal human melanocytes via PPAR-gamma activation. *Pigment. Cell Melanoma Res.* **2011**, *24*, 618–630. [CrossRef] [PubMed]
202. Flori, E.; Mastrofrancesco, A.; Kovacs, D.; Bellei, B.; Briganti, S.; Maresca, V.; Cardinali, G.; Picardo, M. The activation of PPARγ by 2,4,6-Octatrienoic acid protects human keratinocytes from UVR-induced damages. *Sci. Rep.* **2017**, *7*, 9241–9243. [CrossRef] [PubMed]
203. Babino, G.; Caccavale, S.; Pinto, D.; Trink, A.; Giuliani, G.; Rinaldi, F.; Argenziano, G. A Randomized Double-Blind Parallel-Group Study to Evaluate the Long-Term Effects of a Medical Device Containing 0.3% Octatrienoic Acid in the Treatment of Grade III Actinic Keratosis. *Dermatol. Ther.* **2021**, *11*, 1751–1762. [CrossRef] [PubMed]
204. Filosa, A.; Filosa, G. Actinic keratosis and squamous cell carcinoma: Clinical and pathological features. *G Ital. Dermatol. Venereol.* **2015**, *150*, 379–384. [PubMed]
205. Massone, C.; Cerroni, L. The many clinico-pathologic faces of actinic keratosis: An atlas. *Curr. Probl. Dermatol.* **2015**, *46*, 64–69. [CrossRef] [PubMed]

206. Pinto, D.; Trink, A.; Giuliani, G.; Rinaldi, F. Protective effects of sunscreen (50+) and octatrienoic acid 0.1% in actinic keratosis and UV damages. *J. Investig. Med.* **2022**, *70*, 92–98. [CrossRef] [PubMed]
207. Tseng, C.-H. Rosiglitazone may reduce non-melanoma skin cancer risk in Taiwanese. *BMC Cancer* **2015**, *15*, 41. [CrossRef] [PubMed]
208. Botton, T.; Puissant, A.; Bahadoran, P.; Annicotte, J.-S.; Fajas, L.; Ortonne, J.-P.; Gozzerino, G.; Zamoum, T.; Tartare-Deckert, S.; Bertolotto, C.; et al. In Vitro and In Vivo Anti-Melanoma Effects of Ciglitazone. *J. Investig. Dermatol.* **2009**, *129*, 1208–1218. [CrossRef] [PubMed]
209. Paulitschke, V.; Gruber, S.; Hofstätter, E.; Haudek-Prinz, V.; Klepeisz, P.; Schicher, N.; Jonak, C.; Petzelbauer, P.; Pehamberger, H.; Gerner, C.; et al. Proteome analysis identified the PPARγ ligand 15d-PGJ2 as a novel drug inhibiting melanoma progression and interfering with tumor-stroma interaction. *PLoS ONE* **2012**, *7*, e46103. [CrossRef]
210. Maresca, V.; Flori, E.; Camera, E.; Bellei, B.; Aspite, N.; Ludovici, M.; Catricalà, C.; Cardinali, G.; Picardo, M. Linking αMSH with PPARγ in B16-F10 melanoma. *Pigment. Cell Melanoma Res.* **2013**, *26*, 113–127. [CrossRef]
211. Maresca, V.; Flori, E.; Picardo, M. Skin phototype: A new perspective. *Pigment. Cell Melanoma Res.* **2015**, *28*, 378–389. [CrossRef] [PubMed]
212. Konger, R.L.; Derr-Yellin, E.; Travers, J.B.; Ocana, J.A.; Sahu, R.P. Epidermal PPARgamma influences subcutaneous tumor growth and acts through TNF-alpha to regulate contact hypersensitivity and the acute photoresponse. *Oncotarget* **2017**, *8*, 98184–98199. [CrossRef] [PubMed]
213. Meyer, S.; Vogt, T.; Landthaler, M.; Berand, A.; Reichle, A.; Bataille, F.; Marx, A.H.; Menz, A.; Hartmann, A.; Kunz-Schughart, L.A.; et al. Cyclooxygenase 2 (COX2) and Peroxisome Proliferator-Activated Receptor Gamma (PPARG) Are Stage-Dependent Prognostic Markers of Malignant Melanoma. *PPAR Res.* **2009**, *2009*, 848645. [CrossRef] [PubMed]
214. Ferrara, A.; Lewis, J.D.; Quesenberry, C.P.; Peng, T.; Strom, B.L.; Eeden, S.K.V.D.; Ehrlich, S.F.; Habel, L.A. Cohort study of pioglitazone and cancer incidence in patients with diabetes. *Diabetes Care* **2011**, *34*, 923–929. [CrossRef]
215. Pich, C.; Meylan, P.; Mastelic-Gavillet, B.; Nguyen, T.N.T.N.; Loyon, R.; Trang, B.K.B.K.; Moser, H.; Moret, C.; Goepfert, C.; Hafner, J.; et al. Induction of Paracrine Signaling in Metastatic Melanoma Cells by PPARγ Agonist Rosiglitazone Activates Stromal Cells and Enhances Tumor Growth. *Cancer Res.* **2018**, *78*, 6447–6461. [CrossRef]
216. Yousefnia, S.; Momenzadeh, S.; Forootan, F.S.; Ghaedi, K.; Esfahani, M.H.N. The influence of peroxisome proliferator-activated receptor γ (PPARγ) ligands on cancer cell tumorigenicity. *Gene* **2018**, *649*, 14–22. [CrossRef]

biomolecules

MDPI

Article

Activation of Peroxisome Proliferator-Activated Receptor-β/δ (PPARβ/δ) in Keratinocytes by Endogenous Fatty Acids

Bokai Zhu [1,†,‡], Xiaoyang Zhu [1,†], Michael G. Borland [1,2,§], Douglas H. Ralph [1], Christopher R. Chiaro [1,3], Kristopher W. Krausz [4], James M. Ntambi [5,6], Adam B. Glick [1], Andrew D. Patterson [1], Gary H. Perdew [1,2,3], Frank J. Gonzalez [4] and Jeffrey M. Peters [1,2,3,*]

1 Department of Veterinary and Biomedical Sciences, The Center for Molecular Toxicology and Carcinogenesis, The Pennsylvania State University, University Park, PA 16802, USA; bzhu@pitt.edu (B.Z.); xzz145@psu.edu (X.Z.); mborland@commonwealth.edu (M.G.B.); dougralph1@gmail.com (D.H.R.); cxc223@psu.edu (C.R.C.); abg11@psu.edu (A.B.G.); adp117@psu.edu (A.D.P.); ghp2@psu.edu (G.H.P.)
2 Department of Biochemistry, Microbiology and Molecular Biology, The Pennsylvania State University, University Park, PA 16802, USA
3 Department of Genetics, The Pennsylvania State University, University Park, PA 16802, USA
4 Laboratory of Metabolism, Center for Cancer Research, National Cancer Institute, Bethesda, MD 20892, USA; krauszk@intra.nci.nih.gov (K.W.K.); gonzalef@mail.nih.gov (F.J.G.)
5 Department of Nutritional Sciences, University of Wisconsin-Madison, Madison, WI 53706, USA; jmntambi@wisc.edu
6 Department of Biochemistry, University of Wisconsin-Madison, Madison, WI 53706, USA
* Correspondence: jmp21@psu.edu; Tel.: +1-814-863-1387
† These authors contributed equally to this work.
‡ Current address: Aging Institute of UPMC, School of Medicine, University of Pittsburgh, Pittsburgh, PA 15219, USA.
§ Current Address: Department of Biochemistry, Chemistry, Engineering, and Physics, Commonwealth University of PA, Bloomsburg Campus, Bloomsburg, PA 17815, USA.

Abstract: Nuclear hormone receptors exist in dynamic equilibrium between transcriptionally active and inactive complexes dependent on interactions with ligands, proteins, and chromatin. The present studies examined the hypothesis that endogenous ligands activate peroxisome proliferator-activated receptor-β/δ (PPARβ/δ) in keratinocytes. The phorbol ester treatment or HRAS infection of primary keratinocytes increased fatty acids that were associated with enhanced PPARβ/δ activity. Fatty acids caused PPARβ/δ-dependent increases in chromatin occupancy and the expression of angiopoietin-like protein 4 (*Angptl4*) mRNA. Analyses demonstrated that stearoyl Co-A desaturase 1 (*Scd1*) mediates an increase in intracellular monounsaturated fatty acids in keratinocytes that act as PPARβ/δ ligands. The activation of PPARβ/δ with palmitoleic or oleic acid causes arrest at the G2/M phase of the cell cycle of HRAS-expressing keratinocytes that is not found in similarly treated HRAS-expressing *Pparb/d*-null keratinocytes. HRAS-expressing *Scd1*-null mouse keratinocytes exhibit enhanced cell proliferation, an effect that is mitigated by treatment with palmitoleic or oleic acid. Consistent with these findings, the ligand activation of PPARβ/δ with GW0742 or oleic acid prevented UVB-induced non-melanoma skin carcinogenesis, an effect that required PPARβ/δ. The results from these studies demonstrate that PPARβ/δ has endogenous roles in keratinocytes and can be activated by lipids found in diet and cellular components.

Keywords: peroxisome proliferator-activated receptor; ligands; cell cycle; fatty acids; UVB-induced non-melanoma skin cancer

1. Introduction

Peroxisome proliferator-activated receptors (PPARs) are ligand-activated transcription factors. Three subtypes of PPARs exist, PPARα, PPARβ/δ, and PPARγ, each with some unique yet overlapping functions [1]. After ligand binding, PPARs heterodimerize with retinoid X receptor (RXR) causing the recruitment of other proteins used to remodel chromatin and alter the expression of PPAR target genes (e.g., histone acetyl transferases, RNA polymerase) [2]. Cells respond to specific ligands/signals and modulate cellular activities to maintain homeostasis by interacting with PPARs. Transcriptional regulation mediated by PPARs is dynamic. There are several dynamic mechanisms that modulate nuclear receptor activity including the following: (1) those that impact the availability of endogenous/exogenous ligands, (2) those that influence the relative expression of the receptor, and (3) those that modulate accessory proteins for chromatin remodeling [3–6]. For example, non-esterified fatty acids released from adipose and transported to the liver during periods of fasting are endogenous ligands for PPARα [7]. In this mechanism, the ligand activation of PPARα by endogenous fatty acids increases the expression of target genes encoding lipid transporters, lipid binding proteins, and fatty acid-catabolizing enzymes that facilitate the production of energy from lipid substrates to meet the needs of the cells during periods of fasting [7]. This PPARα-specific mechanism is particularly critical to meet the energy demands of cells during periods of fasting [8], a cellular state that predominates daily in most species.

The relative expression of PPARs has been used to help determine important roles of these transcription factors in different cells and tissues. For example, PPARα expression is relatively high in hepatocytes, and this is reflected by the requirement of this receptor for the oxidation of fatty acids to generate energy in the liver, a tissue that is central in the homeostatic regulation of lipid metabolism [7]. In contrast, PPARβ/δ is constitutively expressed at a relatively high level in epithelial cells including those in the intestine and skin, in particular in keratinocytes [9,10]. Based on studies using synthetic ligands and transgenic models, it is now recognized that PPARβ/δ has important functional roles in epithelial cells [11,12]. The ligand activation of PPARβ/δ inhibits the proliferation of both mouse primary keratinocytes and human keratinocytes by inducing terminal differentiation through a PPARβ/δ-dependent mechanism [12]. The topical application of a PPARβ/δ ligand is also chemopreventive against chemically induced non-melanoma skin cancer through a PPARβ/δ-dependent mechanism that targets premalignant tumors as a decrease in keratoacanthomas is only noted in ligand-treated wild-type mice but not in *Pparb/d*-null mice [13–15]. Interestingly, the inhibition of cell proliferation by PPARβ/δ in keratinocytes is greater in cells expressing high levels of HRAS [13] mediated by PPARβ/δ-dependent negative selection against HRAS-expressing cells by inducing a mitotic block at the G2/M phase of the cell cycle [16].

Synthetic ligands and genetic manipulations in cells, tissues, and animal models have also been instrumental for elucidating the role(s) of PPARs. However, the identification and functional characterization of endogenous PPAR ligands remain elusive [17]. There is evidence indicating that keratinocytes exhibit constitutive transcriptional activity of PPARβ/δ in the absence of any exogenous ligands [6]. The expression of more than four hundred genes is altered when PPARβ/δ is ablated in mouse primary keratinocytes [6]. Moreover, previous studies suggested that an endogenous PPARβ/δ ligand is produced in keratinocytes after treatment with phorbol ester [18]. For these reasons, the hypothesis that an endogenous PPARβ/δ ligand exists and can be identified and functionally characterized in keratinocytes was examined in the present study.

2. Materials and Methods

2.1. Keratinocyte Culture

Two-day-old neonates from either *Pparb/d* wild-type, *Pparb/d*-null [19], *Scd1* wild-type, or *Scd1*-null mice [20] were used to isolate primary keratinocytes and cultured as previously described [21]. As previous studies suggested that an endogenous PPARβ/δ ligand is produced in keratinocytes after treatment with phorbol ester (phorbol-12-myristate-13-acetate; TPA) [18], 80–90% confluent keratinocytes were cultured in medium containing 25 ng TPA/mL for eight hours as described previously [18]. Four 100 mm culture dishes were used per treatment. After this treatment, cells were trypsinized, and the cytosol was isolated after differential centrifugation (100,000× *g*) for one hour.

2.2. HPLC Fractionation

The cytosolic fraction was diluted with 1 volume cytosol/2 volumes solvent (hexane/ethyl acetate; 1:1), vortexed, and then centrifuged (3000 rpm). Two hundred milligrams of anhydrous sodium sulfite was added to each sample, then dried under nitrogen, and resuspended in 99% hexane, 1% isopropanol, 0.1% acetic acid. These lipid-enriched fractions were then used to separate compounds using high-performance liquid chromatography (HPLC). HPLC purifications were performed using a Waters system, consisting of a Waters 600E multi-solvent delivery unit and controller coupled with a Waters 996 photodiode array detector (Waters, Milford, MA, USA). The system was integrated and operated using Waters Millenium[32] software (version 3.20). Normal-phase HPLC purification was performed on a LiChrosphere 5 μm Silica-60 (4.6 × 250 mm) column (Supelco, Bellefonte, PA, USA) using a hexane/isopropanol/acetic acid solvent system [989:10:1 (*v*/*v*/*v*)] applied at 1 mL/min with a linear gradient increase in isopropanol concentration of 2.5%/min over 30 min. Fractions were collected every minute, dried under nitrogen, and resuspended in 100% ethanol before the examination of PPARβ/δ activity.

2.3. Quantitative Polymerase Chain Reaction (qPCR)

Primary mouse keratinocytes were isolated from 2-day-old neonates as described previously [21]. Keratinocytes were cultured in low-calcium (0.05 mM) Eagle's minimal essential medium with 8% chelexed fetal bovine serum at 37 °C and 7% CO_2 [21]. The keratinocytes were cultured until they were approximately 80% confluent and were then treated with either 0.2 μM GW0742 (positive control), 100 μM linoleic acid, 100 μM oleic acid, or 100 μM palmitoleic acid. After 8 h of treatment, total mRNA was isolated from the cells using TRIZOL and following the manufacturer's recommended protocol (Invitrogen, Carlsbad, CA, USA). The mRNA encoding *Angiopoietin-like 4* (*Angptl4*) was quantified using quantitative real-time polymerase chain reaction (qPCR) analysis. The cDNA was generated using 2.5 μg total RNA with the MultiScribe Reverse Transcriptase kit (Applied Biosystems, Foster City, CA, USA). Primers were designed for real-time PCR using the Primer Express software (Version 3.0, Applied Biosystems, Foster City, CA, USA). The sequence and GenBank accession number for the forward and reverse primers used to quantify mRNAs was *Angptl4* (NM_020581) forward, 5′-TTCTCGCCTACCAGAGAAGTTGGG-3′, and reverse, 5′-CATCCACAGCACCTACAACAGCAC-3′. The mRNA was normalized to the gene encoding glyceraldehyde 3-phosphate dehydrogenase (*Gapdh*; BC083149) using the following primers: forward, 5′-GGTGGAGCCAAAAGGGTCAT-3′, and reverse, 5′-GGTTCACACCCATCACAAACAT-3′. qPCR reactions were carried out using SYBR green PCR master mix (Finnzymes, Espoo, Finland) in the iCycler and detected using the MyiQ Real-Time PCR Detection System (Bio-Rad Laboratories, Hercules, CA, USA). The following conditions were used for PCR: 95 °C for 15 s, 94 °C for 10 s, 60 °C for 30 s, and 72 °C for 30 s and repeated for 45 cycles. The PCR included a no-template control reaction to control for contamination and/or genomic amplification. All reactions had >95% efficiency. The relative expression levels of mRNA were normalized to *Gapdh* and analyzed for statistical significance using a one-way analysis of variance (Prism 5.0, GraphPad Software Inc., La Jolla, CA, USA).

2.4. ChIP Assays

Pparb/d wild-type or *Pparb/d*-null primary keratinocytes were treated for 3 h with vehicle (DMSO), GW0742 (0.2 μM), palmitoleic acid (100 μM), or oleic acid (100 μM). Cross-linking was performed using 1% formaldehyde for 10 min with gentle agitation, followed by the addition of glycine to a final concentration of 125 mM for an additional 10 min of gentle agitation. Cells were washed twice with PBS before the addition of lysis buffer (50 mM Tris-HCl pH 8, 1% SDS, 10 mM EDTA, and protease inhibitor cocktail). The

DNA was sheared to obtain 500–1500 base pair fragments with the Diagenode Bioruptor™ (Diagenode, Sparta, NJ, USA). Protein A agarose (Santa Cruz Biotechnology, Dallas, TX, USA) beads were blocked with a solution of 1 μg/μL bovine serum albumin and 0.1 μg/μL salmon sperm DNA (Invitrogen) for 1 h prior to use. The sheared chromatin (2 units of 260 nm quantified DNA per immunoprecipitation) was precleared for 1 h with blocked protein A agarose before being immunoprecipitated with specific antibodies for rabbit IgG (Santa Cruz Biotechnology), acetylated histone H4 (Upstate Biotechnology, Charlottesville, VA, USA), or PPARβ/δ [9]. After 3 h, immune complexes were captured by the addition of blocked protein A agarose and were incubated overnight. The recovered beads were washed three times with a low-salt wash buffer (20 mM Tris-HCl pH 8, 2 mM EDTA, 0.1% sodium deoxycholate, 1% Triton-X, 150 mM NaCl, and protease inhibitor cocktail), once with a high-salt wash buffer (20 mM Tris-HCl pH 8, 2 mM EDTA, 0.1% sodium deoxycholate, 1% Triton-X, 500 mM NaCl, and protease inhibitor cocktail), and once with TE8 (10 mM Tris-HCl pH 8, 1 mM EDTA). The immune complexes were released by the addition of elution buffer (100 mM $NaHCO_3$, 1% SDS), and the cross-links were reversed by overnight incubation at 65 °C. Immunoprecipitated DNA was purified by phenol/chloroform/isoamyl alcohol (25:24:1) extraction and subjected to real-time PCR analysis for occupancy at a response element of the known PPARβ/δ target gene angiopoietin-like protein 4 (*Angptl4*). The primer set for *Angptl4* was designed based on a previous identification of PPREs in the intron 3 coding sequence [22,23]. The primers for *Angptl4* were 5′-CTAGCCAAGTAGAGGAAAGTTCAGAGC-3′ (forward) and 5′-CCAATCCCTCGGGCAGCTAGC-3′ (reverse). Real-time PCR reactions were carried out as previously described in the RNA analysis section. Each PCR reaction included a no-template control to detect contamination, and all PCR reactions had greater than 85% efficiency. The specific values were normalized to treatment inputs and verified to be greater than rabbit IgG controls. Accumulation was determined based on fold accumulation as normalized to the control treatment.

2.5. HRAS Infection

The *Hras* retrovirus was generated from ψ2 producer cells as described previously [24]. The virus titer was determined to be between 1 and 2 × 10^7 transforming units/mL using an NIH-3T3 focus-forming assay. Primary keratinocytes from newborn mice were prepared and cultured as previously described [21]. Keratinocytes were infected with the *Hras* retrovirus for two days at an estimated multiplicity of infection (M.O.I) of 1–12. Cells were subsequently cultured in control medium or medium containing palmitoleic acid, oleic acid, or the positive control GW0742.

2.6. Reporter Assays

Previously described plasmids containing the ligand-binding domain of mouse or human PPARβ/δ fused to the DNA binding domain of the yeast transcription factor Gal4 under the control of the SV40 promoter and the UAS-firefly luciferase reporter under the control of the Gal4 DNA response element were used to examine ligand activity [25]. Luciferase activity was measured using a luciferase reporter assay kit (Promega, Madison, WI, USA) and a Turner TD-20/20 Luminometer (Turner BioSystems, Sunnyvale, CA, USA) following the manufacturer's recommended procedures. Luciferase activity was normalized to the beta-galactosidase activity of each sample. The fold induction of normalized luciferase activity was calculated relative to vehicle-control cells (DMSO) and represents the mean of three independent samples per treatment group.

2.7. Quantitative Western Blot Analyses

Cell lysates and samples used for Western blots, and immunoprecipitations were prepared as previously described [9,26]. Western blot analysis using radioactive detection methods was performed as previously described [9]. The primary antibodies used were as follows: Acyl CoA carboxylase (ACC); fatty acid synthase (FAS); fatty acid desaturase1/2

(FADS1, FADS2); very-long-chain fatty acid elongase 6 (ELOVL6); cytochrome-b5 reductase1 (CYB5R1); and stearoyl CoA desaturase1 (SCD1), (Santa Cruz Biotechnology, Santa Cruz, CA, USA), or anti-β-ACTIN (Rockland, Gilbertsville, PA, USA). The anti-PPARβ/δ antibody was previously described [9]. Original figures can be found in Supplementary Materials.

2.8. Flow Cytometry

Control and HRAS-expressing *Scd1* wild-type or *Scd1*-null keratinocytes were cultured with or without GW0742, oleic acid, or palmitoleic acid for twenty-four hours and were stained with bromodeoxyuridine (BrdU) and/or propidium iodide and analyzed for cell cycle progression as previously described [16,27,28]. The percentage of cells at each phase of the cell cycle was determined with FCS Express software (version 4.00).

2.9. Coulter Counting and BrdU Labeling of Keratinocytes

Control and HRAS-expressing *Scd1* wild-type and *Scd1*-null keratinocytes were cultured with or without GW0742, oleic acid, or palmitoleic acid for six days as described above. The average cell number was determined using a Coulter counter as previously described [29]. After five days, cells were labeled with BrdU for twenty-four hours and immunohistochemical analysis was performed to determine the average number of BrdU-labeled cells as previously described [29].

2.10. GC/MS

Pparb/d wild-type keratinocytes, with or without HRAS infection, were cultured for seventy-two hours. Cells were trypsinized, homogenized, and fatty acids were quantified using GC/MS as previously described [30].

2.11. Topical Tetradecanoylphorbol-13-Acetate (TPA)Treatment in Wild-Type and Scd1-Null Mice

Scd1 wild-type and *Scd1*-null female mice aged 6–8 weeks, on an Sv/129 genetic background, were used for these studies [20]. Mice were shaved to remove back hair, and twenty-four hours later, either 25 μg of TPA (Sigma Chemical Co., St. Louis, MO, USA) dissolved in 200 μL of acetone or 200 μL of acetone was applied to the shaved area. Forty-eight hours post-TPA application, mice were euthanized, and the skin was removed and snap-frozen in liquid nitrogen. A section of skin was removed and placed in 10% phosphate buffered formalin for immunohistochemistry to assess the relative immunoreactivity for PCNA, keratin 5, or the nucleus as previously described. Skin samples were fixed with phosphate buffered formalin, embedded in paraffin, and sectioned at 5 μm thickness. Hematoxylin and eosin staining was performed using standard methods. Immunofluorescence staining was performed as previously described [29]. The following primary antibodies were used: anti-keratin 5 or anti-PCNA. Slides were kept in dark containers. Samples were washed three times and incubated with secondary antibodies at room temperature. The following secondary antibodies were used: anti-rabbit, anti-mouse, and anti-rat conjugated to AlexaFluor488 or AlexaFluor647. Stained slides were mounted in the mounting reagent containing DAPI for nuclei staining. The quantification of positive/negative cells was conducted manually using ImageJ (version 1.46).

2.12. UV-Induced Skin Cancer Using Pparb/d-Null Crossed with SKH Mice

Female SKH1 mice ($Ppar\beta/\delta^{+/+}$ or $Ppar\beta/\delta^{-/-}$) were UVB-irradiated (180 mJ/cm^2) 3 times per week for 25 weeks. The amount and frequency of UVB irradiation as well as the timeframe for this study was designed based on methods used in a previously published UVB-induced tumor study [28]. UV-emitting lights that produce wavelengths in the UVB and UVC range were used for this study. UVC light was specifically prevented from reaching the mouse by using a filter that reflects wavelengths higher than UVB. The time of irradiation exposure was calculated prior to each UVB- irradiation by using the intensity of the emitted UVB light which was measured with a UVB-specific radiometer. Irradiation was immediately followed by a topical dorsal application of 200 μL of vehicle control (acetone), the highly specific PPARβ/δ ligand (1.0 μM GW0742), or 25 μM oleic acid. The rationale for the concentration of GW0742 and oleic acid used was based on previous studies showing that these concentrations are in the range that can activate PPARβ/δ [31,32]. The topical dorsal application of control or ligand solution started in week 1. The lesion incidence, multiplicity, and volume were measured and recorded once per week. The lesion incidence for each mouse is defined as the week when the first measurable lesion appeared. The lesion multiplicity for each mouse is defined as the total number of lesions found weekly. Within 6 h after the last irradiation of week 25, mice were euthanized by overexposure to carbon dioxide and cervical dislocation. The sample sizes ranged from 7 to 11 mice per treatment group. Data were analyzed for statistical significance using ANOVA and the Bonferroni post hoc test (Prism 10.0, GraphPad Software, San Diego, CA, USA). Differences between treatment groups were considered significant when $p \leq 0.05$.

3. Results

3.1. PPARβ/δ Activity Is Increased in the Cytosol of Keratinocytes Post-TPA Treatment

Previous studies suggest that endogenous PPARβ/δ ligand(s) are produced in keratinocytes after treatment with TPA [18]. Since TPA activates phospholipases causing the release of fatty acids in keratinocytes [33], the organic fraction of cytosol obtained from either control mouse primary keratinocytes or mouse keratinocytes treated acutely with TPA were examined. Consistent with this past study [18], one fraction of organic keratinocyte cytosol caused an increase in PPARβ/δ reporter activity compared to controls (Figure 1A). Since TPA activates phospholipases causing the release of fatty acids in keratinocytes including linoleic and oleic acids [33], keratinocytes were cultured in linoleic, oleic, or palmitoleic acid, and the expression of the PPARβ/δ target gene, angiopoietin-like protein-4 (*Angplt4*), was measured. The expression of *Angptl4* mRNA was increased in wild-type primary keratinocytes in response to the synthetic ligand GW0742 and with linoleic acid, oleic acid, and palmitoleic acid compared to controls (Figure 1B). This effect was mitigated in *Pparb/d*-null keratinocytes (Figure 1B). Further, ChIP analyses demonstrated that palmitoleic acid increased the occupancy of PPARβ/δ on the *Angplt4* PPRE, and this effect was not observed in similarly treated *Pparb/d*-null keratinocytes (Figure 1C).

3.2. HRAS Activates PPARβ/δ

TPA activates PKCα and other proteins including HRAS [34,35]. Thus, the impact of HRAS in keratinocyte signaling was examined. Interestingly, wild-type keratinocytes expressing HRAS exhibit an increased expression of *Angptl4* mRNA compared to controls, and this effect was muted in *Pparb/d*-null keratinocytes (Figure 2A,B). Interestingly, the basal expression of *Angptl4* mRNA was higher in *Pparb/d*-null keratinocytes compared to wild-type keratinocytes (Figure 2A), consistent with past studies showing that the basal expression of this gene is constitutively negatively regulated by PPARβ/δ [36]. The expression of PPARβ/δ was not influenced by HRAS or GW0742 (Figure 2B). However, HRAS expression caused an increase in PPARβ/δ reporter activity for both mouse and human isoforms, comparable to that observed with the positive control GW0742 (Figure 2C).

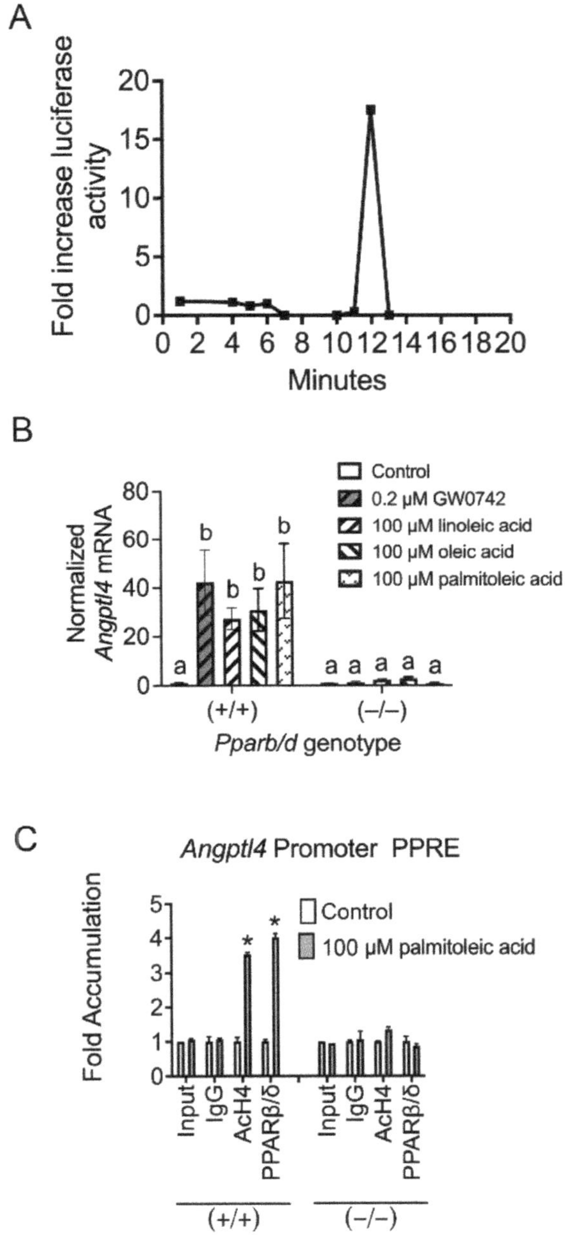

Figure 1. Phorbol ester (TPA) increases PPARβ/δ transcriptional activity in keratinocytes. (**A**) A reporter assay was used to detect PPARβ/δ activity in cytosolic isolates obtained from mouse primary keratinocytes after fractionation using HPLC. (**B**) Fatty acids known to be released by TPA treatment in wild-type keratinocytes increase the PPARβ/δ-dependent expression of *Angptl4* mRNA, and this effect is absent in similarly treated *Pparb/d*-null keratinocytes. $n = 5$ independent biological replicates. (**C**) A ChIP assay showing increased promoter occupancy on known *Angptl4* PPRE by palmitoleic acid in wild-type keratinocytes (+/+), an effect lacking in similarly treated *Pparb/d*-null keratinocytes (−/−), $n = 5$ independent biological replicates. Values represent the mean ± SD. Values with different letters are significantly different, $p \leq 0.05$. * Significantly different than control, $p \leq 0.05$.

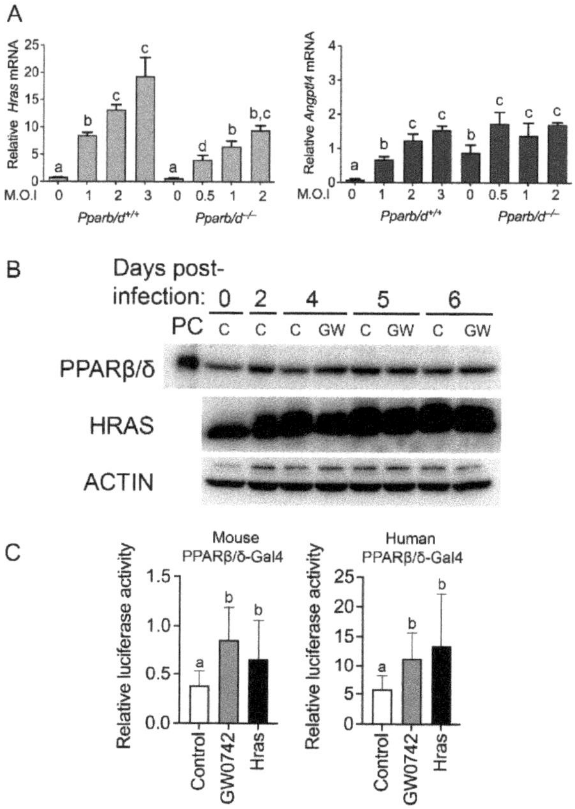

Figure 2. Activated HRAS increases PPARβ/δ transcriptional activity. (**A**) qPCR of *Hras* (**left panel**) or *Angptl4* (**right panel**) in mock-infected or HRAS-infected *Pparb/d* wild-type and *Pparb/d*-null keratinocytes with increasing M.O.I. $n = 6$ independent biological replicates. (**B**) Representative Western blot analysis of PPARβ/δ expression in HRAS-infected wild-type keratinocytes without GW0742 (**C**) or treated with GW0742 (GW). PC: positive control (COS cell lysate). (**C**) Luciferase assay in primary keratinocytes transiently transfected with either mouse (left panel) or human (right panel) PPARβ/δ-GAL4 fusion protein and UAS-luciferase plasmids. $n = 6$ independent biological replicates. Values represent mean ± SD. Values with different letters are significantly different, $p \leq 0.05$.

3.3. Bioinformatic Analyses Reveals Role of PPARβ/δ in Lipid Metabolism

Bioinformatic analyses were performed using previously published microarray data from *Pparb/d* wild-type or *Pparb/d*-null keratinocytes and HRAS-expressing *Pparb/d* wild-type or *Pparb/d*-null keratinocytes, with or without the ligand activation of PPARβ/δ with GW0742 (GEO accession number GSE32498 [6,16]). Of particular interest were the mRNAs that were co-modulated by the PPARβ/δ ligand GW0742 and HRAS in both wild-type and *Pparb/d*-null keratinocytes (Figure 3A). There was overlap in the expression of twelve genes (*Acsl3, Lss, Tmtc2, Bnip3, Syngr1, Angptl4, Fosl1, Gm5246, Insig1, Fgfbp1, Phgdh,* and *Acsbg1*) that occurred (Figure 3B), but the expression of these genes was generally higher in HRAS-expressing wild-type keratinocytes as compared to HRAS-expressing *Pparb/d*-null keratinocytes (Figure 3C). Six genes were regulated specifically by PPARβ/δ (Figure 3B, *Scd1, Gdpd1, Erc2, Vwa8, AU18091,* and *Sna1*). Since SCD1 is rate-limiting for monounsaturated fatty acid synthesis, qPCR analyses of control or HRAS-expressing wild-type or *Pparb/d*-null keratinocytes were performed to examine other enzymes that regulate lipogenesis (Figure 4A). The expression of *Fads1, Fads2, Fads3, Cyb5r1,* and *Scd1* mRNAs was increased by HRAS in wild-type keratinocytes compared to controls, and this effect was mirrored by the ligand activation of PPARβ/δ with GW0742 in HRAS-expressing wild-type keratinocytes (Figure 4A). This effect was mitigated in *Pparb/d*-null HRAS-expressing keratinocytes and in response to GW0742 (Figures 3C and 4A). Based on qPCR analyses, the expression of *Acc*, *Cyb5r2*, *Elovl6*, *Fasn*, and *Scd2* mRNAs was not differentially expressed in response to HRAS or GW0742 compared to controls. To determine whether the differential mRNA expression detected by qPCR analyses was similar for proteins, Western blot analyses were performed (Figure 4B). The ligand activation of PPARβ/δ with GW0742 did not influence the expression of any protein except CYB5R1 in wild-type keratinocytes, which was relatively higher, compared to the control (Figure 4B). Consistent with mRNA analyses, the relative expression of FADS1, FADS2, ELOLV6, CYB5R1, and SCD1 proteins was higher in HRAS-expressing wild-type keratinocytes, and this effect was not noted in HRAS-expressing *Pparb/d*-null keratinocytes (Figure 4B). These data suggest that the HRAS-dependent modulation of PPARβ/δ signaling could influence monounsaturated fatty acid synthesis and that these fatty acids may act as ligands for this receptor (Figure 4C).

A

	HRAS-induced in *Pparb/d*$^{+/+}$ cells	Not HRAS-induced in *Pparb/d*$^{+/+}$ cells
Induced by GW0742	18	40
Not induced by GW0742	1565	33137
	HRAS-induced in *Pparb/d*$^{-/-}$ cells	Not HRAS-induced in *Pparb/d*$^{-/-}$ cells
Induced by GW0742	13	45
Not induced by GW0742	1393	33309

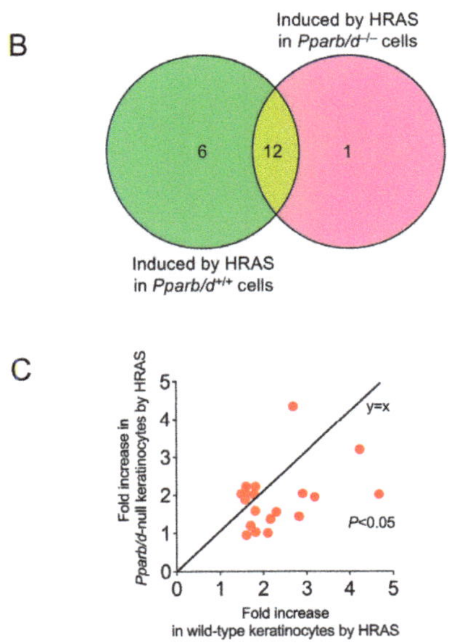

Figure 3. HRAS activation and GW0742 treatment induces common gene expression. (**A**) Overlap between the mRNAs that are induced by HRAS, and mRNAs are induced by the PPARβ/δ ligand GW0742 in primary keratinocytes. Only genes that are induced by ≥1.5 folds by either HRAS or GW0742 are included. n = 3 independent biological replicates. (**B**) A Venn diagram showing that 6 of 19 genes induced by GW0742 in wild-type keratinocytes are also induced by HRAS only in wild-type keratinocytes. In contrast, there was overlap in 12 genes between genotypes and treatments. (**C**) A plot of the fold change in genes in B induced by HRAS either in *Pparb/d* wild-type (x-axis) or *Pparb/d*-null keratinocytes (y-axis). Note that these genes are induced to a much higher fold by HRAS in wild-type keratinocytes compared to *Pparb/d*-null keratinocytes.

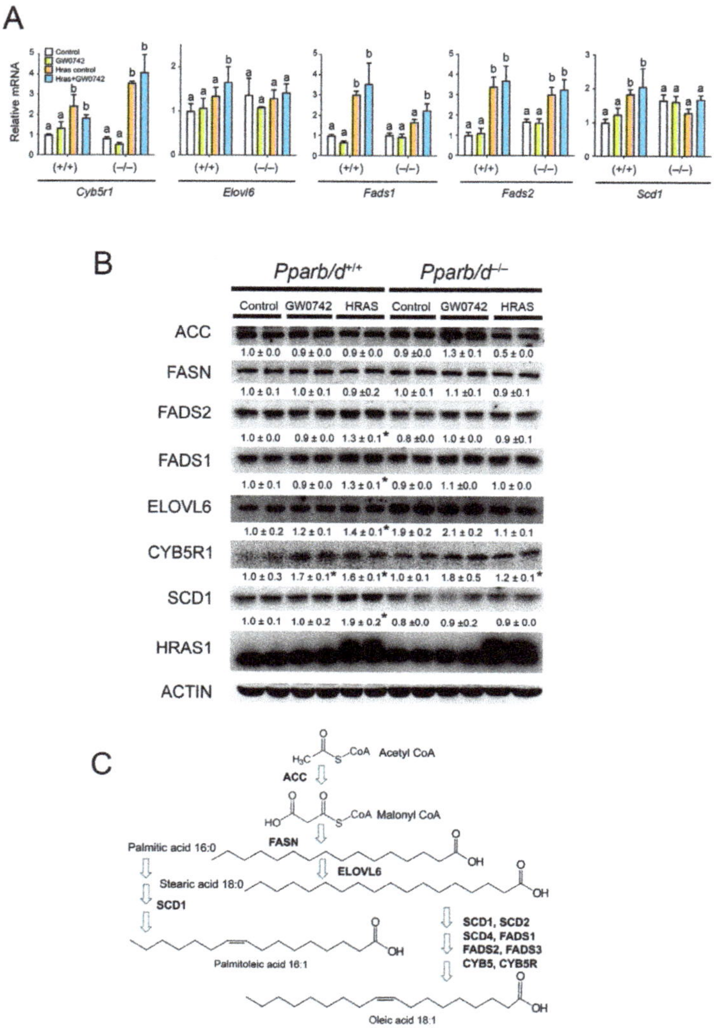

Figure 4. Activated HRAS increases expression of fatty acid synthesis genes. (**A**) qPCR of key fatty acid synthesis mRNAs in *Pparb/d* wild-type and *Pparb/d*-null keratinocytes treated with GW0742 or HRAS infection. n = 3–5 independent biological replicates. (**B**) Representative quantitative Western blot analysis of key fatty acid synthesis genes in *Pparb/d* wild-type and *Pparb/d*-null keratinocytes treated with GW0742 or HRAS infection. n = 4 independent biological replicates. (**C**) Diagram showing de novo fatty acid synthesis pathway influenced in these studies. Values represent mean ± SD. Values with different letters are significantly different, $p \leq 0.05$. * Significantly different than control, $p \leq 0.05$.

3.4. HRAS Increases Oleic and Palmitoleic Acid Levels in Primary Keratinocytes and Activates PPARβ/δ

To determine whether changes in the expression of lipogenic proteins found in HRAS-expressing keratinocytes influenced the synthesis of lipids, GC-MS analysis was performed. The expression of HRAS in keratinocytes caused an increase in the amount of palmitoleic acid, oleic acid, eladic acid, and linoleic acid compared to controls (Figure 5A). ChIP analyses showed that oleic acid increased the occupancy of PPARβ/δ on the *Angplt4* PPRE in wild-type keratinocytes similar to that observed with the synthetic PPARβ/δ ligand

GW0742, but this did not occur in *Pparb/d*-null keratinocytes (Figure 5B). As SCD1 is the rate-limiting enzyme catalyzing the synthesis of monounsaturated fatty acids (in particular for oleic and palmitoleic acids [37]), these observations collectively suggested that HRAS activation causes the synthesis of fatty acids that activate PPARβ/δ in keratinocytes. Since the ligand activation of PPARβ/δ results in the negative selection of cells expressing higher levels of the HRAS oncogene by inducing a mitotic block [16], cell cycle progression was examined. Consistent with previous studies, the ligand activation of PPARβ/δ in HRAS-expressing keratinocytes caused mitotic arrest at the G2/M phase of the cell cycle, and this effect was also observed in response to oleic acid or palmitoleic acid compared to controls (Figure 6). Importantly, this effect was not found in similarly treated HRAS-expressing *Pparb/d*-null keratinocytes (Figure 6).

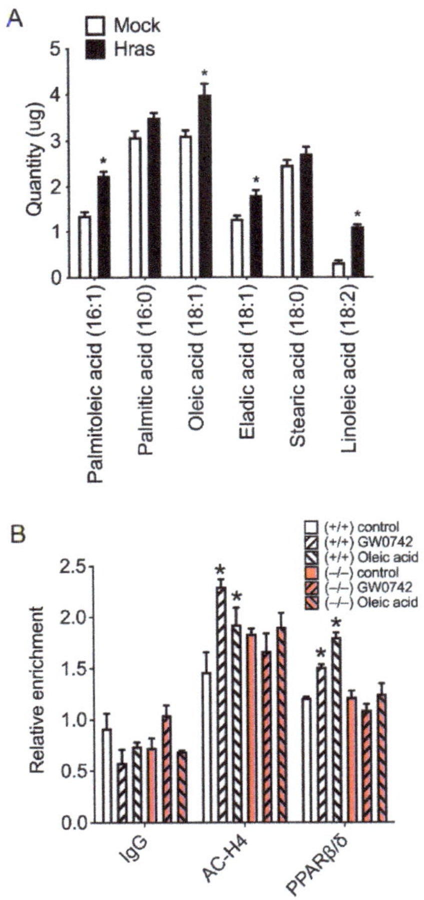

Figure 5. HRAS-expressing keratinocytes have higher palmitoleic and oleic acid levels, which are PPARβ/δ ligands. (**A**) Quantification of levels of key fatty acid in mock-infected or HRAS-infected *Pparb/d* wild-type keratinocytes by GC-MS. Data are compiled from three independent experiments. $n = 5$–6 independent biological replicates. (**B**) ChIP-qPCR analysis of Ac-H4 and PPARβ/δ binding to mouse *Angptl4* gene promoter in response to GW0742 or oleic acid treatment in *Pparb/d* wild-type or *Pparb/d*-null keratinocytes. $n = 3$ independent biological replicates. Values represent mean ± SD. * Significantly different than control, $p \leq 0.05$.

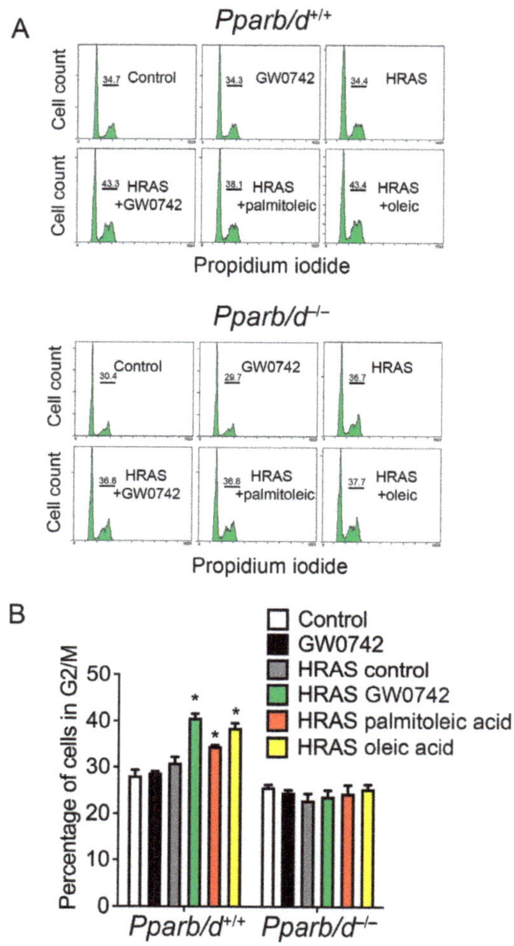

Figure 6. Palmitoleic and oleic acids induce G2/M arrest in HRAS-expressing keratinocytes through PPARβ/δ-dependent mechanism. Representative DNA histograms (**A**) and quantification of percentage of cells in G2/M phase, n = 3 independent biological replicates, (**B**) of mock-infected or HRAS-expressing *Pparb/d* wild-type or *Pparb/d*-null keratinocytes treated with GW0742, palmitoleic acid or oleic acid or 72 h. Representative peak of cells in G2/M phase shown above bar in each panel in (**A**). n = 3 independent biological replicates. Values in (**B**) represent mean ± SD. * Significantly different than control, $p \leq 0.05$.

3.5. Oleic Acid and Palmitoleic Acid Inhibit Hyperplasia in Keratinocytes

HRAS-expressing *Scd1*-null keratinocytes exhibited enhanced cell proliferation compared to the wild-type control (Figure 7), a phenotype remarkably similar to *Pparb/d*-null keratinocytes [16]. Interestingly, GW0742, oleic acid and palmitoleic acid all rescued this phenotype in *Scd1*-null keratinocytes compared to controls (Figure 7). Similarly, topical treatment with TPA caused epidermal hyperplasia in wild-type mice, and this effect was exacerbated in *Scd1*-null mice compared to the control (Figure 8). Similar to that observed with in vitro primary keratinocytes, this phenotype of the TPA-treated *Scd1*-null mouse skin was remarkably similar (Figure 8) to that observed in TPA-treated *Pparb/d*-null mouse skin that exhibited enhanced epidermal hyperplasia compared to controls [19]. More importantly GW0742, palmitoleic acid, and oleic acid inhibited TPA-induced hyperplasia in

Scd1-null mice (Figure 8). Collectively, these results suggest that the supplementation of exogenous monounsaturated fatty acids can rescue the hyperplasia phenotypes of *Scd1*-null mice, likely via the ligand activation of PPARβ/δ.

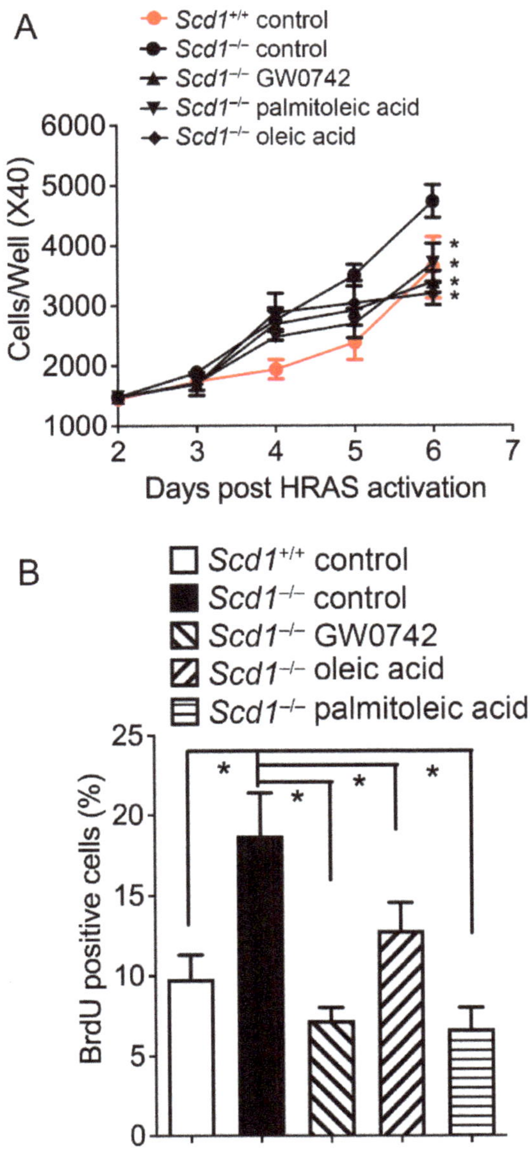

Figure 7. Palmitoleic and oleic acids rescue enhanced cell proliferation of HRAS-expressing keratinocytes resulting from *Scd1* ablation. (**A**) Cell counts of HRAS-expressing keratinocytes isolated from wild-type ($Scd1^{+/+}$), *Scd1* knock-out ($Scd1^{-/-}$), and $Scd1^{-/-}$ treated with GW0742, palmitoleic acid, or oleic acid for 6 days post-HRAS expression. n = 3–6 independent biological replicates. (**B**) BrdU-positive cells were identified by immunofluorescence after 6 days of HRAS expression. n = 4–6 independent biological replicates. Values represent mean ± SD. * Significantly different than control, $p \leq 0.05$.

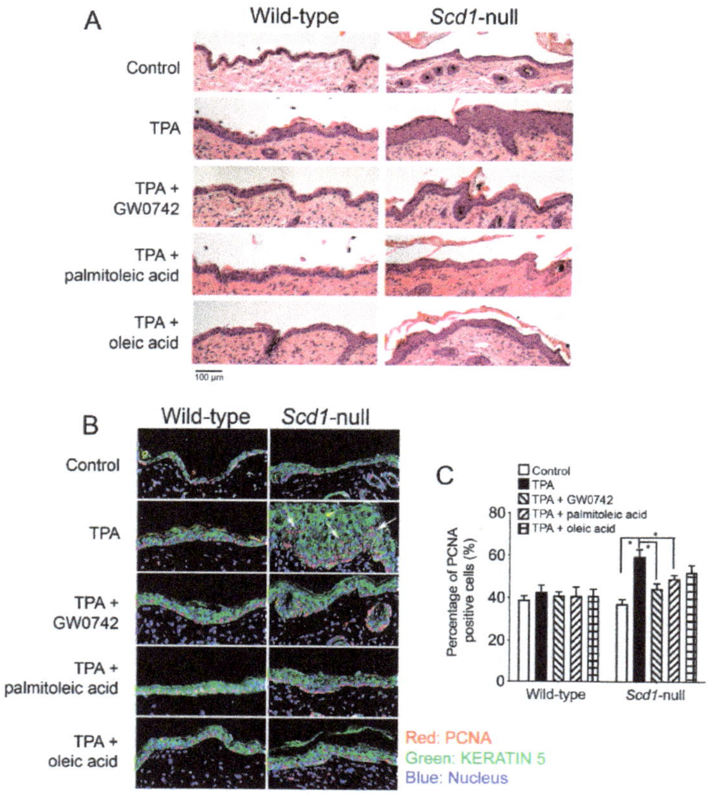

Figure 8. Palmitoleic and oleic acids rescue enhanced cell proliferation and hyperplasia resulting from *Scd1* ablation. (**A**) H&E staining of mouse skin from wild-type and $Scd1^{-/-}$ mice treated topically once with TPA and with or without GW0742, palmitoleic acid, or oleic acid for 48 h. n = 3–10 independent biological replicates. The 100 μm scale in (**A**) is the same for images in (**B**). (**B**) Representative immunofluorescence of PCNA (red), keratin 5 (green), and DAPI (blue), and (**C**) quantification of percentage of PCNA-positive cells in skin of wild-type and $Scd1^{-/-}$ mice treated with TPA and with or without GW0742, palmitoleic acid, or oleic acid for 48 h. Higher percentage of PCNA-positive cells shown with white arrows. n = 3–10 independent biological replicates. Values represent mean ± SD. * Significantly different than control, $p \leq 0.05$.

3.6. *PPARβ/δ-Dependent Inhibition of UVB-Induced Skin Cancer with Oleic Acid and Synthetic Ligand*

The ligand activation of PPARβ/δ with GW0742 is chemopreventive against chemically induced skin cancer [13–15]. This effect requires PPARβ/δ as the inhibition of malignant conversion was only noted in ligand-treated wild-type mice but not in ligand-treated *Pparb/d*-null mice [13–15]. The chemopreventive effect of the ligand activation of PPARβ/δ on the tumor incidence, tumor multiplicity, and tumor volume of UVB-induced skin tumors was examined in SKH-1 mice crossed with *Pparb/d*-wild-type or *Pparb/d*-null mice. The onset of UVB-induced tumor formation was delayed in SKH1 X $Pparb/d^{+/+}$ mice by either the topical application of oleic acid or GW0742 as compared to controls, and this effect was not observed in GW0742-treated or SKH1 X $Pparb/d^{-/-}$ mice (Figure 9A). The incidence of UVB-induced tumor formation was slightly delayed in SKH1 X $Pparb/d^{-/-}$ mice by the topical application of oleic acid as compared to controls, but this effect was not observed in GW0742-treated or control SKH1 X $Pparb/d^{-/-}$ mice (Figure 9A). The ligand activation of PPARβ/δ with either GW0742 or oleic acid inhibited tumor multiplicity in

SKH1 X *Pparb/d*$^{+/+}$ mice from weeks 17 to 25, and this effect was not observed in similarly treated or SKH1 X *Pparb/d*$^{-/-}$ mice (Figure 9B). Tumor multiplicity was higher during the last 5 weeks of treatment in all groups of SKH1 X *Pparb/d*$^{-/-}$ mice compared to in SKH1 X *Pparb/d*$^{+/+}$ mice (Figure 9B). The average tumor volume was not influenced by the ligand activation of PPARβ/δ with either GW0742 or oleic acid in SKH1 X *Pparb/d*$^{+/+}$ mice (Figure 9C). The average tumor volume was greater from weeks 23 to 25 in SKH1 X *Pparb/d*$^{-/-}$ mice compared to control SKH1 X *Pparb/d*$^{+/+}$ mice (Figure 9C). Interestingly, the average tumor volume was lower from weeks 23 to 25 in SKH1 X *Pparb/d*$^{-/-}$ mice in response to treatment with either GW0742 or oleic acid compared to control SKH1 X *Pparb/d*$^{-/-}$ mice (Figure 9C).

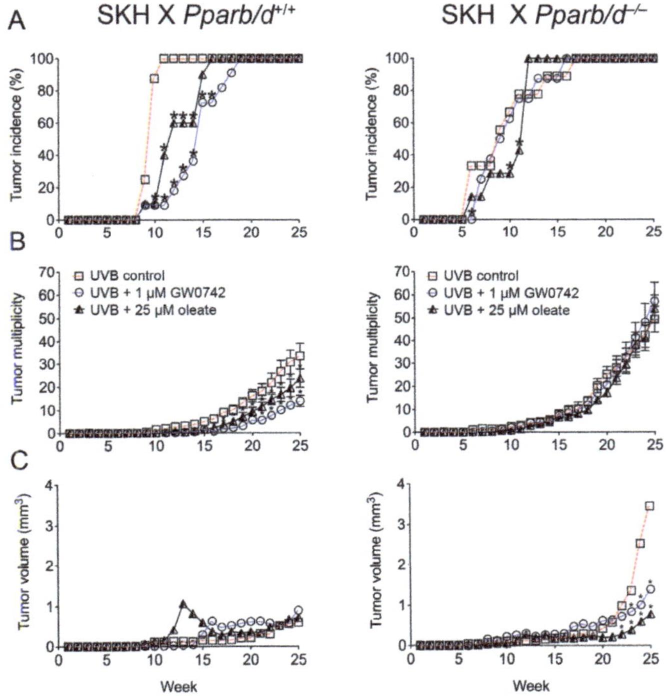

Figure 9. Ligand activation of PPARβ/δ with oleic acid or GW0742 inhibits UVB-induced skin cancer. SKH mice were crossed with *Pparb/d* wild-type or *Pparb/d*-null mice and irradiated with or without topical application of PPARβ/δ ligands after irradiation. Onset of tumor formation (**A**), number of tumors per mouse (**B**), and average tumor size (**C**) were measured weekly. n = 7–11 independent biological replicates. Each point represents mean ± S.D. * $p < 0.05$ compared with control.

4. Discussion

The premise for these studies is based on seminal work by the Wahli research group in 2001 suggesting that keratinocytes are a source of PPARβ/δ ligands [18]. The results from the present studies extend this original observation by showing that oleic acid and palmitoleic acid are increased in keratinocytes by HRAS, and this modulates PPARβ/δ activities in these cells. This is consistent with the observation that TPA causes the release of fatty acids in keratinocytes [33]. That endogenous ligands exist and dynamically modulate gene expression in keratinocytes is strongly supported by previous studies showing that the genetic silencing of *Pparb/d* in keratinocytes causes alterations in the constitutive expression of hundreds of PPARβ/δ target genes [6]. Additionally, increasing the basal expression of PPARβ/δ is known to inhibit the size and growth of tumors derived from human skin, neuroblastoma, breast cancer, melanoma, and testicular cancer cell lines implanted in mice, in the absence of exogenous ligands (reviewed in [12,38]). The constitutive, nuclear localization of PPARβ/δ is also relatively high in epithelial cells and PPARβ/δ co-immunoprecipitates with its heterodimerization partner, RXR [9]. These observations strongly support the hypothesis that PPARβ/δ is constitutively active and has critical roles in normal, basal keratinocyte physiology including the induction of terminal differentiation [39–41]. This appears particularly true for epithelial cells such as those found in the intestine or skin where the basal expression of PPARβ/δ is relatively high compared to other tissues in mice and humans [9,10].

SCD1 is the rate-limiting enzyme that catalyzes the synthesis of monounsaturated fatty acids including oleic and palmitoleic acids [37]. The results from the present studies show that high HRAS activity in keratinocytes caused a greater expression and activity of SCD1, and this effect requires PPARβ/δ. This is consistent with the higher levels of palmitoleic acid and oleic acid found in keratinocytes by increasing HRAS activity. As the ligand activation of PPARβ/δ causes the negative selection of cells expressing higher levels of HRAS by inducing a mitotic block at the G2/M phase of the cell cycle [16], it is critical to note that oleic acid, palmitoleic acid, and GW0742 prevented an increase in the percentage of cells remaining in the G2/M phase of the cell cycle induced by TPA in *Scd1*-null keratinocytes. Similarly, the epidermal hyperplastic response induced by TPA was exacerbated in *Scd1*-null mice, a phenotype that is remarkably similar to the phenotype noted in TPA-treated *Pparb/d*-null mice [15,19,40,42]. Importantly, palmitoleic acid, oleic acid, and GW0742 all reversed the TPA-induced hyperplasia in *Scd1*-null epidermis. These observations collectively support the hypothesis that HRAS-expressing cells synthesize PPARβ/δ ligands to inhibit the cell proliferation of mutant cells, likely the result of inducing terminal differentiation. Future studies should consider an examination of SCD1 inhibitors, as well as the influence of other monounsaturated fatty acids, to complement and extend the present studies.

The observations made in *Scd1*-null mouse skin support previous studies demonstrating that the ligand activation of PPARβ/δ with a synthetic ligand inhibits chemically induced skin carcinogenesis [13–15]. Interestingly, a higher expression of SCD1 is negatively associated with cancer cell growth and progression [43], and altered levels of monounsaturated fatty acids produced by SCD1 can both inhibit or promote cancer cell survival [43]. In contrast, UV exposure causes the inhibition of SCD1 expression and lower levels of fatty acids in human skin [44]. This is consistent with the present results showing that oleic and palmitoleic acid can act to inhibit UV-induced non-melanoma skin cancer. Given the growing incidence of non-melanoma skin cancer [45], further studies should examine whether the dietary administration of these lipids may prevent this disease. Further studies are also needed to delineate how SCD1-dependent regulation impacts PPARβ/δ function (and vice versa) in other cancers. These observations are also consistent with several two-stage chemical carcinogenesis studies showing that the ligand activation of PPARβ/δ inhibits skin carcinogenesis caused by the topical administration of dimethylbenzanthracene (DMBA) and TPA [13–15]. As the malignant conversion of tumors in the DMBA/TPA model requires the mutation of *HRAS*, the present studies suggest that the production of PPARβ/δ ligands during the time course of a two-stage chemical carcinogenesis bioassay may contribute to the observed inhibition noted in the previous studies [13–15]. Whether this mechanism reflects a negative feedback loop to prevent the proliferation of mutant keratinocytes requires additional evaluation. Since UVB-induced skin cancer is dependent on mutations in the TP53 gene, the sensitivity of TP53 mutant cells to the effects of the ligand activation of PPARβ/δ should be further examined. Another study found that UV-induced skin carcinogenesis was mitigated in SKH1 X *Pparb/d*-null compared to SKH1 X *Pparb/d* wild-type mice and that this effect was modulated by the regulation of SRC expression [46]. The results from the present studies cannot be compared to the former studies for several reasons. First, the UV light source used for the latter studies used both UVB and UVA [46], whereas the present studies used only UVB and eliminated the impact of UVA/UVC by using a filter. This is important to note because UVB is thought to be the most predominant source of solar radiation that causes cyclobutane pyrimidine dimers and pyrimidine (6-4) photoproducts [47–49] and is more commonly used for skin carcinogenesis bioassays. Second, the *Pparb/d*-null mice used for both studies used different approaches to silence PPARβ/δ expression. It remains possible that an off-target event(s) in the two transgenic lines underlie this difference. Lastly, important symbiotic roles of bacteria, viruses, and fungi are being revealed that may impact many routes of exposure. Thus, it is also possible that differences in the skin microbiome impacts results from these two studies. Further work is needed to clarify these issues.

Supplementary Materials: The following supporting information can be downloaded at: https://www.mdpi.com/article/10.3390/biom14060606/s1. Original Western Blots.

Author Contributions: Conceptualization, B.Z., G.H.P., A.B.G., A.D.P., F.J.G. and J.M.P.; methodology, B.Z., X.Z., M.G.B., D.H.R., C.R.C. and K.W.K.; formal analysis, B.Z., X.Z., M.G.B., D.H.R., G.H.P., A.D.P., F.J.G. and J.M.P.; investigation, B.Z., X.Z., M.G.B., D.H.R. and C.R.C.; writing—original draft preparation, B.Z., X.Z., M.G.B., D.H.R., C.R.C., K.W.K., J.M.N., A.B.G., A.D.P., G.H.P. and F.J.G.; writing—review and editing, B.Z., X.Z., M.G.B., D.H.R., C.R.C., K.W.K., J.M.N., A.B.G., A.D.P., G.H.P. and F.J.G.; funding acquisition, G.H.P., A.D.P., J.M.N., F.J.G. and J.M.P. All authors have read and agreed to the published version of the manuscript.

Funding: This work was supported by the USDA National Institute of Food and Agriculture and Hatch Appropriations under Project #PEN04954 and Accession #7006477 (J.M.P.); Project #PEN04772 and Accession #7000371 (G.H.P); #PEN4917 and Accession #7006412 (A.D.P.), and by the National Institutes of Health (CA124533, CA239256, J.M.P; ES028244, G.H.P.), and the National Cancer Institute Intramural Research Program (F.J.G.).

Institutional Review Board Statement: This study was approved by the Pennsylvania State University Institutional Animal Care and Use Committee (PRAMS200646080, 19 August 2019).

Informed Consent Statement: Not applicable.

Data Availability Statement: The data presented in this study are available on request from the corresponding author. The data are not publicly available due to the author's preference.

Acknowledgments: The authors gratefully acknowledge Walter Wahli for his leadership in this field and his indirect support/inspiration of all these studies. The authors also gratefully acknowledge Maggie Strable for technical assistance, the Penn State Cancer Institute Metabolomics Shared Resource (RRID:SCR_023864), and the Huck Institute of Life Sciences Flow Cytometry Core used for analyses in these studies.

Conflicts of Interest: The authors declare no conflicts of interest.

References

1. Green, S.; Wahli, W. Peroxisome proliferator-activated receptors: Finding the orphan a home. *Mol. Cell Endocrinol.* **1994**, *100*, 149–153. [CrossRef] [PubMed]
2. Berger, J.; Moller, D.E. The mechanisms of action of PPARs. *Annu. Rev. Med.* **2002**, *53*, 409–435. [CrossRef] [PubMed]
3. Biddie, S.C.; Hager, G.L. Glucocorticoid receptor dynamics and gene regulation. *Stress* **2009**, *12*, 193–205. [CrossRef]
4. Biddie, S.C.; John, S.; Hager, G.L. Genome-wide mechanisms of nuclear receptor action. *Trends Endocrinol. Metab.* **2010**, *21*, 3–9. [CrossRef] [PubMed]
5. Hager, G.L.; Varticovski, L. Chromatin in time and space. *Biochim. Biophys. Acta* **2012**, *1819*, 631. [CrossRef] [PubMed]
6. Khozoie, C.; Borland, M.G.; Zhu, B.; Baek, S.; John, S.; Hager, G.L.; Shah, Y.M.; Gonzalez, F.J.; Peters, J.M. Analysis of the peroxisome proliferator-activated receptor-β/δ (PPARβ/δ) cistrome reveals novel co-regulatory role of ATF4. *BMC Genom.* **2012**, *13*, 665. [CrossRef] [PubMed]
7. Kersten, S.; Stienstra, R. The role and regulation of the peroxisome proliferator activated receptor alpha in human liver. *Biochimie* **2017**, *136*, 75–84. [CrossRef] [PubMed]
8. Kersten, S.; Seydoux, J.; Peters, J.M.; Gonzalez, F.J.; Desvergne, B.; Wahli, W. Peroxisome proliferator-activated receptor α mediates the adaptive response to fasting. *J. Clin. Investig.* **1999**, *103*, 1489–1498. [CrossRef] [PubMed]
9. Girroir, E.E.; Hollingshead, H.E.; He, P.; Zhu, B.; Perdew, G.H.; Peters, J.M. Quantitative expression patterns of peroxisome proliferator-activated receptor-β/δ (PPARβ/δ) protein in mice. *Biochem. Biophys. Res. Commun.* **2008**, *371*, 456–461. [CrossRef]

10. Modica, S.; Gofflot, F.; Murzilli, S.; D'Orazio, A.; Salvatore, L.; Pellegrini, F.; Nicolucci, A.; Tognoni, G.; Copetti, M.; Valanzano, R.; et al. The intestinal nuclear receptor signature with epithelial localization patterns and expression modulation in tumors. *Gastroenterology* **2010**, *138*, 636–648. [CrossRef]
11. Montagner, A.; Wahli, W. Contributions of peroxisome proliferator-activated receptor β/δ to skin health and disease. *Biomol. Concepts* **2013**, *4*, 53–64. [CrossRef] [PubMed]
12. Peters, J.M.; Kim, D.J.; Bility, M.T.; Borland, M.G.; Zhu, B.; Gonzalez, F.J. Regulatory mechanisms mediated by peroxisome proliferator-activated receptor-β/δ in skin cancer. *Mol. Carcinog.* **2019**, *58*, 1612–1622. [CrossRef]
13. Bility, M.T.; Devlin-Durante, M.K.; Blazanin, N.; Glick, A.B.; Ward, J.M.; Kang, B.H.; Kennett, M.J.; Gonzalez, F.J.; Peters, J.M. Ligand activation of peroxisome proliferator-activated receptor-β/δ (PPARβ/δ) inhibits chemically-induced skin tumorigenesis. *Carcinogenesis* **2008**, *29*, 2406–2414. [CrossRef] [PubMed]
14. Bility, M.T.; Zhu, B.; Kang, B.H.; Gonzalez, F.J.; Peters, J.M. Ligand Activation of Peroxisome Proliferator-Activated Receptor-beta/delta and Inhibition of Cyclooxygenase-2 Enhances Inhibition of Skin Tumorigenesis. *Toxicol. Sci.* **2010**, *113*, 27–36. [CrossRef] [PubMed]
15. Kim, D.J.; Akiyama, T.E.; Harman, F.S.; Burns, A.M.; Shan, W.; Ward, J.M.; Kennett, M.J.; Gonzalez, F.J.; Peters, J.M. Peroxisome proliferator-activated receptor β (δ)-dependent regulation of ubiquitin C expression contributes to attenuation of skin carcinogenesis. *J. Biol. Chem.* **2004**, *279*, 23719–23727. [CrossRef] [PubMed]
16. Zhu, B.; Khozoie, C.; Bility, M.T.; Ferry, C.H.; Blazanin, N.; Glick, A.B.; Gonzalez, F.J.; Peters, J.M. Peroxisome proliferator-activated receptor β/δ cross talks with E2F and attenuates mitosis in HRAS-expressing cells. *Mol. Cell Biol.* **2012**, *32*, 2065–2082. [CrossRef] [PubMed]
17. Harmon, G.S.; Lam, M.T.; Glass, C.K. PPARs and lipid ligands in inflammation and metabolism. *Chem. Rev.* **2011**, *111*, 6321–6340. [CrossRef] [PubMed]
18. Tan, N.S.; Michalik, L.; Noy, N.; Yasmin, R.; Pacot, C.; Heim, M.; Fluhmann, B.; Desvergne, B.; Wahli, W. Critical roles of PPARβ/δ in keratinocyte response to inflammation. *Genes. Dev.* **2001**, *15*, 3263–3277. [CrossRef] [PubMed]
19. Peters, J.M.; Lee, S.S.; Li, W.; Ward, J.M.; Gavrilova, O.; Everett, C.; Reitman, M.L.; Hudson, L.D.; Gonzalez, F.J. Growth, adipose, brain and skin alterations resulting from targeted disruption of the mouse peroxisome proliferator-activated receptor β(δ). *Mol. Cell. Biol.* **2000**, *20*, 5119–5128. [CrossRef]
20. Ntambi, J.M.; Miyazaki, M.; Stoehr, J.P.; Lan, H.; Kendziorski, C.M.; Yandell, B.S.; Song, Y.; Cohen, P.; Friedman, J.M.; Attie, A.D. Loss of stearoyl-CoA desaturase-1 function protects mice against adiposity. *Proc. Natl. Acad. Sci. USA* **2002**, *99*, 11482–11486. [CrossRef]
21. Dlugosz, A.A.; Glick, A.B.; Tennenbaum, T.; Weinberg, W.C.; Yuspa, S.H. Isolation and utilization of epidermal keratinocytes for oncogene research. *Methods Enzymol.* **1995**, *254*, 3–20. [PubMed]
22. Heinaniemi, M.; Uski, J.O.; Degenhardt, T.; Carlberg, C. Meta-analysis of primary target genes of peroxisome proliferator-activated receptors. *Genome Biol.* **2007**, *8*, R147. [CrossRef] [PubMed]
23. Mandard, S.; Zandbergen, F.; Tan, N.S.; Escher, P.; Patsouris, D.; Koenig, W.; Kleemann, R.; Bakker, A.; Veenman, F.; Wahli, W.; et al. The direct peroxisome proliferator-activated receptor target fasting-induced adipose factor (FIAF/PGAR/ANGPTL4) is present in blood plasma as a truncated protein that is increased by fenofibrate treatment. *J. Biol. Chem.* **2004**, *279*, 34411–34420. [CrossRef]
24. Roop, D.R.; Lowy, D.R.; Tambourin, P.E.; Strickland, J.; Harper, J.R.; Balaschak, M.; Spangler, E.F.; Yuspa, S.H. An activated Harvey ras oncogene produces benign tumours on mouse epidermal tissue. *Nature* **1986**, *323*, 822–824. [CrossRef] [PubMed]
25. Bility, M.; Thompson, J.T.; McKee, R.H.; David, R.M.; Butala, J.H.; Vanden Heuvel, J.P.; Peters, J.M. Activation of mouse and human peroxisome proliferator-activated receptors (PPARs) by phthalate monoesters. *Toxicol. Sci.* **2004**, *82*, 170–182. [CrossRef]
26. Borland, M.G.; Khozoie, C.; Albrecht, P.P.; Zhu, B.; Lee, C.; Lahoti, T.S.; Gonzalez, F.J.; Peters, J.M. Stable over-expression of PPARβ/δ and PPARγ to examine receptor signaling in human HaCaT keratinocytes. *Cell Signal* **2011**, *23*, 2039–2050. [CrossRef]
27. Borland, M.G.; Foreman, J.E.; Girroir, E.E.; Zolfaghari, R.; Sharma, A.K.; Amin, S.M.; Gonzalez, F.J.; Ross, A.C.; Peters, J.M. Ligand Activation of Peroxisome Proliferator-Activated Receptor-β/δ (PPARβ/δ) Inhibits Cell Proliferation in Human HaCaT Keratinocytes. *Mol. Pharmacol.* **2008**, *74*, 1429–1442. [CrossRef]
28. He, P.; Borland, M.G.; Zhu, B.; Sharma, A.K.; Amin, S.; El-Bayoumy, K.; Gonzalez, F.J.; Peters, J.M. Effect of ligand activation of peroxisome proliferator-activated receptor-β/δ (PPARβ/δ) in human lung cancer cell lines. *Toxicology* **2008**, *254*, 112–117. [CrossRef]
29. Zhu, B.; Ferry, C.H.; Markell, L.K.; Blazanin, N.; Glick, A.B.; Gonzalez, F.J.; Peters, J.M. The nuclear receptor peroxisome proliferator-activated receptor-β/δ (PPARβ/δ) promotes oncogene-induced cellular senescence through repression of endoplasmic reticulum stress. *J. Biol. Chem.* **2014**, *289*, 20102–20119. [CrossRef]
30. Brocker, C.N.; Patel, D.P.; Velenosi, T.J.; Kim, D.; Yan, T.; Yue, J.; Li, G.; Krausz, K.W.; Gonzalez, F.J. Extrahepatic PPARα modulates fatty acid oxidation and attenuates fasting-induced hepatosteatosis in mice. *J. Lipid Res.* **2018**, *59*, 2140–2152. [CrossRef]
31. Forman, B.M.; Chen, J.; Evans, R.M. Hypolipidemic drugs, polyunsaturated fatty acids, and eicosanoids are ligands for peroxisome proliferator-activated receptors α and δ. *Proc. Natl. Acad. Sci. USA* **1997**, *94*, 4312–4317. [CrossRef]
32. Shearer, B.G.; Hoekstra, W.J. Recent advances in peroxisome proliferator-activated receptor science. *Curr. Med. Chem.* **2003**, *10*, 267–280. [CrossRef] [PubMed]

33. Li-Stiles, B.; Lo, H.H.; Fischer, S.M. Identification and characterization of several forms of phospholipase A2 in mouse epidermal keratinocytes. *J. Lipid Res.* **1998**, *39*, 569–582. [CrossRef] [PubMed]
34. Alexandropoulos, K.; Qureshi, S.A.; Foster, D.A. Ha-Ras functions downstream from protein kinase C in v-Fps-induced gene expression mediated by TPA response elements. *Oncogene* **1993**, *8*, 803–807. [PubMed]
35. Clark, J.A.; Black, A.R.; Leontieva, O.V.; Frey, M.R.; Pysz, M.A.; Kunneva, L.; Woloszynska-Read, A.; Roy, D.; Black, J.D. Involvement of the ERK signaling cascade in protein kinase C-mediated cell cycle arrest in intestinal epithelial cells. *J. Biol. Chem.* **2004**, *279*, 9233–9247. [CrossRef] [PubMed]
36. Kaddatz, K.; Adhikary, T.; Finkernagel, F.; Meissner, W.; Muller-Brusselbach, S.; Muller, R. Transcriptional profiling identifies functional interactions of TGFβ and PPARβ/δ signaling: Synergistic induction of ANGPTL4 transcription. *J. Biol. Chem.* **2010**, *285*, 29469–29479. [CrossRef]
37. Ntambi, J.M.; Miyazaki, M. Regulation of stearoyl-CoA desaturases and role in metabolism. *Prog. Lipid Res.* **2004**, *43*, 91–104. [CrossRef] [PubMed]
38. Peters, J.M.; Walter, V.; Patterson, A.D.; Gonzalez, F.J. Unraveling the role of peroxisome proliferator-activated receptor-β/δ (PPARβ/δ) expression in colon carcinogenesis. *NPJ Precis. Oncol.* **2019**, *3*, 26. [CrossRef] [PubMed]
39. Kim, D.J.; Bility, M.T.; Billin, A.N.; Willson, T.M.; Gonzalez, F.J.; Peters, J.M. PPARβ/δ selectively induces differentiation and inhibits cell proliferation. *Cell Death Differ.* **2006**, *13*, 53–60. [CrossRef]
40. Man, M.Q.; Barish, G.D.; Schmuth, M.; Crumrine, D.; Barak, Y.; Chang, S.; Jiang, Y.; Evans, R.M.; Elias, P.M.; Feingold, K.R. Deficiency of PPARβ/δ in the epidermis results in defective cutaneous permeability barrier homeostasis and increased inflammation. *J. Investig. Dermatol.* **2007**, *128*, 370–377. [CrossRef]
41. Schmuth, M.; Haqq, C.M.; Cairns, W.J.; Holder, J.C.; Dorsam, S.; Chang, S.; Lau, P.; Fowler, A.J.; Chuang, G.; Moser, A.H.; et al. Peroxisome proliferator-activated receptor (PPAR)-β/δ stimulates differentiation and lipid accumulation in keratinocytes. *J. Investig. Dermatol.* **2004**, *122*, 971–983. [CrossRef] [PubMed]
42. Michalik, L.; Desvergne, B.; Tan, N.S.; Basu-Modak, S.; Escher, P.; Rieusset, J.; Peters, J.M.; Kaya, G.; Gonzalez, F.J.; Zakany, J.; et al. Impaired skin wound healing in peroxisome proliferator-activated receptor (PPAR)α and PPARβ mutant mice. *J. Cell Biol.* **2001**, *154*, 799–814. [CrossRef] [PubMed]
43. Sen, U.; Coleman, C.; Sen, T. Stearoyl coenzyme A desaturase-1: Multitasker in cancer, metabolism, and ferroptosis. *Trends Cancer* **2023**, *9*, 480–489. [CrossRef] [PubMed]
44. Kim, E.J.; Jin, X.J.; Kim, Y.K.; Oh, I.K.; Kim, J.E.; Park, C.H.; Chung, J.H. UV decreases the synthesis of free fatty acids and triglycerides in the epidermis of human skin in vivo, contributing to development of skin photoaging. *J. Dermatol. Sci.* **2010**, *57*, 19–26. [CrossRef] [PubMed]
45. Hu, W.; Fang, L.; Ni, R.; Zhang, H.; Pan, G. Changing trends in the disease burden of non-melanoma skin cancer globally from 1990 to 2019 and its predicted level in 25 years. *BMC Cancer* **2022**, *22*, 836. [CrossRef] [PubMed]
46. Montagner, A.; Delgado, M.B.; Tallichet-Blanc, C.; Chan, J.S.; Sng, M.K.; Mottaz, H.; Degueurce, G.; Lippi, Y.; Moret, C.; Baruchet, M.; et al. Src is activated by the nuclear receptor peroxisome proliferator-activated receptor β/δ in ultraviolet radiation-induced skin cancer. *EMBO Mol. Med.* **2014**, *6*, 80–98. [CrossRef] [PubMed]
47. Hung, K.F.; Sidorova, J.M.; Nghiem, P.; Kawasumi, M. The 6-4 photoproduct is the trigger of UV-induced replication blockage and ATR activation. *Proc. Natl. Acad. Sci. USA* **2020**, *117*, 12806–12816. [CrossRef] [PubMed]
48. Osakabe, A.; Tachiwana, H.; Kagawa, W.; Horikoshi, N.; Matsumoto, S.; Hasegawa, M.; Matsumoto, N.; Toga, T.; Yamamoto, J.; Hanaoka, F.; et al. Structural basis of pyrimidine-pyrimidone (6-4) photoproduct recognition by UV-DDB in the nucleosome. *Sci. Rep.* **2015**, *5*, 16330. [CrossRef]
49. Yokoyama, H.; Mizutani, R. Structural biology of DNA (6-4) photoproducts formed by ultraviolet radiation and interactions with their binding proteins. *Int. J. Mol. Sci.* **2014**, *15*, 20321–20338. [CrossRef]

biomolecules

MDPI

Article

Central Actions of Leptin Induce an Atrophic Pattern and Improves Heart Function in Lean Normoleptinemic Rats via PPARβ/δ Activation

Blanca Rubio [1,2], Cristina Pintado [3,4], Lorena Mazuecos [1,4], Marina Benito [5], Antonio Andrés [1,4,*] and Nilda Gallardo [1,4,*]

[1] Biochemistry Section, Faculty of Sciences and Chemical Technologies, University of Castilla-La Mancha, Avda. Camilo José Cela 10, 13071 Ciudad Real, Spain; blancamaria.rubio@cnic.es (B.R.); lorena.mazuecos@uclm.es (L.M.)

[2] Molecular Regulation of Heart Failure Research Group, National Cardiovascular Research Center Carlos III (CNIC), Melchor Fernández Almagro 3, 28029 Madrid, Spain

[3] Biochemistry Section, Faculty of Environmental Sciences and Biochemistry, University of Castilla-La Mancha, Avda. Carlos III s/n, 45071 Toledo, Spain; cristina.pintado@uclm.es

[4] DOE Research Group, Institute of Biomedicine, University of Castilla-La Mancha (IB-UCLM), 13071 Ciudad Real, Spain

[5] ICTS Bioimagen Complutense (BioImaC), Universidad Complutense de Madrid, P°. de Juan XXIII 1, 28040 Madrid, Spain; marben14@ucm.es

* Correspondence: antonio.andres@uclm.es (A.A.); nilda.gallardo@uclm.es (N.G.)

Abstract: Leptin, acting centrally or peripherally, has complex effects on cardiac remodeling and heart function. We previously reported that central leptin exerts an anti-hypertrophic effect in the heart via cardiac PPARβ/δ activation. Here, we assessed the impact of central leptin administration and PPARβ/δ inhibition on cardiac function. Various cardiac properties, including QRS duration, R wave amplitude, heart rate (HR), ejection fraction (EF), end-diastolic left ventricular mass (EDLVM), end-diastolic volume (EDV), and cardiac output (CO) were analyzed. Central leptin infusion increased cardiac PPARβ/δ protein content and decreased HR, QRS duration, and R wave amplitude. These changes induced by central leptin suggested a decrease in the ventricular wall growth, which was confirmed by MRI. In fact, the EDLVM was reduced by central leptin while increased in rats co-treated with leptin and GSK0660, a selective antagonist of PPARβ/δ activity. In summary, central leptin plays a dual role in cardiac health, potentially leading to ventricular atrophy and improving heart function when PPARβ/δ signaling is intact. The protective effects of leptin are lost by PPARβ/δ inhibition, underscoring the importance of this pathway. These findings highlight the therapeutic potential of targeting leptin and PPARβ/δ pathways to combat cardiac alterations and heart failure, particularly in the context of obesity.

Keywords: PPARβ/δ; leptin; MRI; cardiac remodeling; Wistar rat

1. Introduction

Accumulating evidence indicates that an excess of nutrient intake is harmful to the heart as a result of its role in cardiac remodeling [1,2]. In fact, cardiovascular diseases (CVD) represent a significant global health burden, especially among individuals with obesity and its associated metabolic disturbances, including metabolic syndrome and diabetes. Obesity and insulin resistance are recognized inducers of cardiac hypertrophy. While cardiac hypertrophy, which involves the enlargement of heart muscle cells, is the more commonly discussed form of remodeling, cardiac atrophy is also an important aspect.

Interestingly, multiple studies have shown that an atrophic pattern caused by reduced fuel supply to the heart leads to a lower cardiovascular risk profile. This is characterized by a decreased cardiac workload due to enhanced metabolic efficiency and alterations in signaling pathways associated with cardiac remodeling, potentially improving cardiac function and efficiency [1,3,4]. Although the mechanisms behind cardiac atrophy resulting from metabolic unloading are still not well understood, some researchers have suggested that signals from the intermediary metabolism of energy-providing substrates might play significant roles [1].

Leptin, a hormone predominantly secreted by adipose tissue, is well-known for its role in regulating energy balance by inhibiting hunger, thereby reducing food intake and promoting energy expenditure. Moreover, the circulating levels of leptin correlate positively with body mass index (BMI), reflecting the body's fat stores. Leptin, acting through the long isoform of the leptin receptor (Ob-Rb) in the hypothalamus, stimulates both thyrotropin-releasing hormone (TRH) and corticotropin-releasing hormone (CRH) secretion by the paraventricular nucleus (PVH), mediating leptin's stimulation of the sympathetic nervous system (SNS) [5]. Nevertheless, the neural circuitry that integrates leptin and other metabolic signals to control energy homeostasis and body weight remains poorly defined.

Beyond its metabolic functions, leptin has been shown to exert complex effects on the cardiovascular system. Notably, leptin influences cardiac structure and function, supported by studies demonstrating its involvement in both hypertrophic and protective mechanisms within the heart [6,7].

For instance, studies in obese and leptin-deficient mouse models (*ob/ob* and *db/db*) [8, 9] suggest leptin might protect against LV hypertrophy during the progression of obesity. Importantly, it has been demonstrated that chronic intracerebroventricular (ICV) leptin infusion protects the heart from adverse remodeling and cardiac contractile dysfunction after myocardial infarction (MI) [10].

Despite these insights, the overall effects of leptin on cardiac remodeling in humans remain largely unknown. Recently, a cross-sectional study by Kamimura et al. [11] examined the relationship between plasma leptin concentrations and LV structure and function indices in a community-based cohort of 1172 African Americans with preserved EF. The study found sex- and BMI-specific associations between leptin levels and LV mass, suggesting that leptin's protective effects against LV hypertrophy might be more pronounced in certain subpopulations.

Similarly, PPARβ/δ plays a significant role as a nuclear receptor acting as a nutrient sensor. Activation of PPARβ/δ regulates various inflammatory processes, lipid metabolism, and energy utilization pathways, which collectively enhance cardiac efficiency and protect against metabolic stress [10,11]. Moreover, enhancing PPARβ/δ activity, either through natural means or pharmacological intervention, represents a potential therapeutic approach to mitigate the metabolic complications associated with obesity. Therefore, we propose that PPARβ/δ may interact with leptin signaling pathways, potentially influencing cardiac remodeling and function in response to metabolic stress and nutrient availability.

In fact, recent data from our laboratory support a potential protective role of central leptin signaling against cardiac hypertrophy and provide new insights into the mechanisms by which leptin, acting at a central level, activates PPARβ/δ, harmonizing anti-hypertrophic responses, redox state, proteasome-dependent protein degradation, and autophagy in the heart [12]. Therefore, balancing their activities could help manage obesity and its comorbidities, which often include insulin resistance and cardiovascular diseases. However, the *in vivo* mechanisms underlying the effects of both leptin and PPARβ/δ on cardiac structure and function are not fully understood.

In this study, we focused on investigating the influence of central leptin on heart rate and cardiac output in normoleptinemic rats with normal leptin sensitivity. Specifically, we aimed to establish in vivo the role of PPARβ/δ on cardiac remodeling when the stimulus applied to induce it was the activation of central leptin signaling. To achieve this, we administered the selective PPARβ/δ antagonist GSK0660.

Our findings support the concept that leptin, acting centrally, induces an atrophic pattern in the heart and contribute to a healthier myocardial profile [10,12]. More importantly, our data indicate that administration of the PPARβ/δ antagonist led to a condition resembling left ventricular hypertrophy (LVH) in obesity, characterized by an abnormal increase in left ventricular myocardial volume while still maintaining a preserved ejection fraction. This supports the notion that pharmacological inhibition of PPARβ/δ throughout the body abolishes the central actions of leptin in cardiac remodeling.

2. Materials and Methods

2.1. Experimental Animals

Experiments were performed on 3-month-old male Wistar rats. All the animals were fed with a standard chow diet and water and maintained in ventilated-controlled quarters (20–25 °C temperature, 50–55% humidity, daily 12-h light cycle 7 a.m.–7 p.m.). Animals were randomly housed individually to control their food intake and thereby avoid differences in adipose tissue weight and serum levels of hormones and metabolites that depend on the amount of feed eaten by animals. Body weight and food intake were monitored daily during all the treatments. Animals were handled according to the European Union's laws (2010/63/EU) and following Spanish regulations (RD 53/2013) for laboratory animals' use. The experimental protocols were approved by the Institutional Scientific Ethics Committee under project license es PR-31-2019. All efforts were made to reduce the number of animals used and minimize animal suffering.

2.2. Intracerebroventricular Leptin Administration

Leptin infusion was performed as previously described [12,13]. Rats were anesthetized in an induction chamber with 4% isoflurane (0.8 L/min oxygen flow) (Pharmacia-Upjohn, Barcelona, Spain) and then were placed in a stereotaxic frame (David Kopf, Los Angeles, CA, USA) with a thermal blanket below. Leptin or saline (PBS) was administered for 7 days in the lateral ventricle through a cannula connected to an osmotic minipump (Alzet, Palo Alto, CA, USA), with a releasing rate of 1 μL/h and filled with 0.0082 μg/μL (0.2 μg/day) rat leptin (Sigma), or its vehicle (saline). To analyse the effect of central infusion of leptin, rats were treated with either saline (SS, n = 5) or leptin (Lep, n = 5) for 7 days. A third group of rats (PF, n = 5) was treated with saline and *pair-fed* to the amount of food consumed by the leptin-treated group. Body weight and food intake were measured daily during the experiment.

2.3. Pharmacological Administration of the PPARβ/δ Antagonist GSK0660

To analyse the contribution of PPARβ/δ, we administer in vivo GSK0660, a selective antagonist inhibitor of PPARβ/δ, to a group of rats at a dose that does not induce toxic side effects but abolished the effects on body weight mediated either by central leptin infusion or by caloric restriction in *pair-fed* rats, as previously described [11,12]. Briefly, GSK0660 was diluted first in DMSO (less than 1%) and later in 0.9% NaCl. Then, it was infused daily intraperitoneally (i.p; 1 mg/kg per day) for 7 days in the PF+GSK0660 (n = 5) and Lep+GSK0660 (n = 5) group of rats. Control groups PF+DMSO (n = 5) and Lep+DMSO (n = 5) received an intraperitoneal injection of the vehicle at 2 mL/kg (0.062% DMSO) in parallel. Seven days after minipump implantation, the animals were fasted overnight. Rats were anesthetized by CO_2 inhalation and killed by decapitation. Blood was removed and centrifuged (2000× *g*, 15 min), and serum was recovered and frozen in liquid nitrogen at −70 °C until use. Hearts were rapidly excised and washed twice in Henseleit buffer at 37 °C. After removing the major blood vessels and connective tissue, hearts were dried and weighed. Epididymal (eWAT), perirenal (PrWAT), and brown adipose tissue (BAT) were rapidly excised and weighted. The tibia from each animal was dissected, and after removing the muscle, the length of the tibia was measured using a digital micrometer caliper, and the heart weight/tibial length ratio was calculated. Hereafter, atria were removed, and both ventricles were used in all analyses after being flash-frozen in liquid nitrogen and stored at −70 °C until use. The hypothalamic regions were carefully dissected as previously described [14]. After that, hypothalamic regions were frozen in liquid nitrogen and stored at −70 °C until further processing.

2.4. Biochemical Assessment

Serum hormone and metabolite levels were measured as described [12,13,15] following the manufacturer's instructions. Glucose was measured using an Accutrend Glucose Analyser, (#05050472223, Roche, Basel, Switzerland), leptin was quantified using a Rat Elisa Kit (#RD291001200R, Biovendor, Heidelberg, Germany), and insulin was determined using a Rat Elisa Kit (#10-1251-01, Mercodia, Uppsala, Sweden).

2.5. Quantitative Transcription Analysis with Real-Time Polymerase Chain Reaction (qRT-PCR)

RNA from the hypothalamus was obtained using All Prep DNA/RNA/Protein Mini kit (Cat. No. 80004, Qiagen, Venlo, The Netherlands) following the manufacturer's instructions. The cDNA was synthesised from 1.5 µg of DNase-treated RNA. Relative quantification of the long isoform of the leptin receptor (*Ob-Rb*), corticotropin-releasing hormone (*Crh*), and thyrotropin-releasing hormone (*Trh*) mRNA levels was performed by real-time PCR according to the manufacturer's protocol on an ABI PRISM 7500 FAST Sequence Detection System instrument and software (PE Applied Biosystem, Foster City, CA, USA). To standardize the amount of cDNA added to the reaction, amplification of endogenous control 18S rRNA was included in separate wells using VIC (TaqMan Assay) or primers as a real-time reporter. The ΔΔCT method was used to calculate the relative differences between experimental conditions and control groups as a fold of change in gene expression. Details about the genes used in this study are provided in Supplementary Data Table S1.

2.6. Western Blot Analysis

100 mg of frozen ventricles were ground under liquid N_2 before homogenization (2 mL buffer/g tissue) in Henseleit buffer (1 mM PMSF, 100 mM EDTA, 2 mM Na_3VO_4, 10 µg/mL leupeptin, 10 µg/mL aprotinin and 1 µg/mL pepstatin), using a manual Dounce homogenizer, followed centrifugation at 800× *g* for 5 min at 4 °C to produce a total extract. Protein lysates (equal amounts of 50 µg) were separated under reducing conditions (10% polyacrylamide concentration gels) SDS-PAGE. Samples were previously mixed with SDS sample buffer and boiled at 95 °C for 10 min. Proteins were transferred to nitrocellulose sheets (0.2 µm, Bio-Rad, Hercules, CA, USA) and incubated overnight (12–16 h) at 4 °C with the appropriate primary antibodies, followed by incubation at room temperature for 2 h with corresponding secondary antibody conjugated with horseradish peroxidase. Primary polyclonal antibody was anti-PPARβ/δ (1:1000, ab23676) and anti-β-actin (1:1000, ab8226) from Abcam, Cambridge, UK. The secondary antibody used was goat anti-rabbit conjugated with horseradish peroxidase (1:4000, 172-1019) from Bio-Rad, Madrid, Spain. Blots were repeated three times to ensure the reproducibility of the results. The immunocomplexes formed were visualized using the ECL Western-blotting detection kit (Amersham Biosciences, Inc., Piscataway, NJ, USA), and the images were subjected to a densitometric analysis with a G-Box Densitometer. Bands were quantified by scanning densitometry with the exposure in the linear range using Gene Tools software (Syngene, Cambridge, UK). Samples from rats infused with vehicle, leptin, or GSK0660 in all experimental conditions were run on the same gel to allow a direct comparison. β-actin was used as a control for protein loading.

2.7. Cardiac Magnetic Resonance Imaging and Analysis

Magnetic Resonance Imaging (MRI) studies were performed in a Bruker Biospec 70/30 scanner using a combination of a linear coil (for transmission) with a cardiac phase array coil (for reception). Animals were anesthetized with isoflurane (3% for induction and 1% for maintenance) and placed in an MRI-adapted stereotaxic holder. Respiration rate and ECG were continuously monitored during the scans (SA Instruments). MRI acquisition protocol included an initial flash sequence to center the Field of View (FOV) and plan the short axis (Table 1). Each short-axis slice (total n = 20) was acquired with an integrated sequence.

Table 1. MRI parameters for LVEF acquisition.

Repetition time	8 ms
Echo time	2.5 ms
Slices thickness	1.5 mm
FOV	40 × 40 mm
Matrix	256 × 256
Resolution	0.156 × 0.156

FOV: Field of View.

Cardiac function and morphology were assessed from the cinema sequences using the freely available software Segment v4.0 R11044b (http://segment.heiberg.se, accessed on 29 Jaunary 2024) [16]. The primary outcome measure was the LVEF (left ventricular ejection fraction), determined by cardiac MRI. Segmentation of the left ventricle was performed manually. Papillary muscles were included when defining endo- and epicardial borders. Definitions of end-diastole and end-systole were calculated by the software. Following manual LV segmentation, the software calculated end-diastolic volume (EDV), end-diastolic left ventricular mass (EDLVM), ejection fraction (EF), and cardiac output (CO) automatically [17,18]. Regional wall analyses were performed on a slice-by-slice basis, according to the user manual. In brief, the left ventricle was covered by five short-axis slices (one slice in the base/three slices in the middle/one slice in the apex; Figure 1).

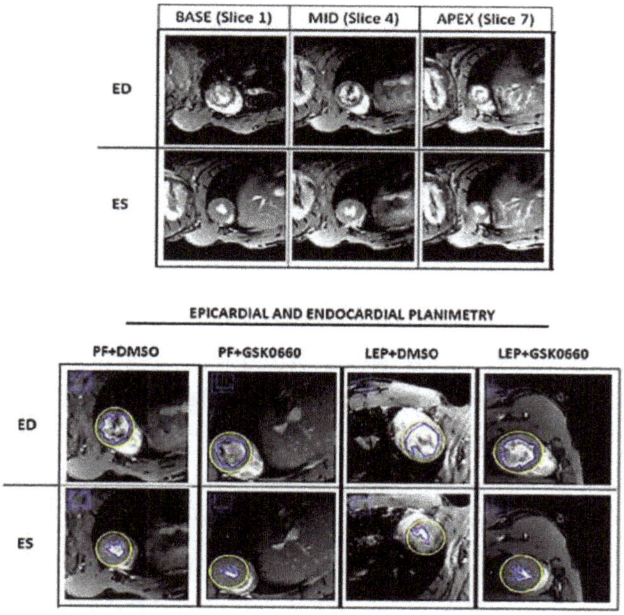

Figure 1. Multiplane assessment of the left ventricular wall by MRI. The figure shows a representative image base to apex (base [base], mid zone [mid] and apex [ape]) short axis end of diastole (End-Diastole, ED) and the end of systole (End-Systole, ES; **upper panel**). Additionally, a representative image of the planimetry of the epicardium (yellow line) and endocardium (blue line) used for the analysis of the four treated groups is shown (**lower panel**).

2.8. Statistical Analysis

Data are expressed as mean ± SEM. Statistical analysis was performed using the GraphPad Prism version 8.0.2 for Windows (GraphPad Software). Differences between the two groups were assessed using the unpaired Student's *t*-test. Significant differences between more than two groups were assessed by one-way ANOVA followed by the Tukey test as a post-hoc analysis (different letters indicate significant differences). Specific analysis and symbols used have been specified in each figure legend. A *p* value of ≤0.05 was considered statistically significant. The number of rats used per experiment is stated in each figure legend.

3. Results and Discussion

3.1. Validation of Central Leptin Infusion and Pharmacological Inhibition of PPARβ/δ Activity

In this work, we infused leptin ICV in 3-month-old Wistar rats for 7 days to stimulate PPARβ/δ activity in the heart and co-administered GSK0660 (i.p) to abolish the effects of this transcription factor in leptin-treated rats. Initially, we wanted to confirm the ability of exogenous leptin to activate hypothalamic leptin signaling through its own receptor by measuring the hypothalamic mRNA levels of the long-form *Ob-Rb* leptin receptor and those of *Crh* and *Trh*, two target genes for the action of leptin in saline, *pair-fed* and leptin-treated rats. As expected, leptin-induced hypothalamic *Ob-Rb*, *Crh*, and *Trh* gene expression compared to the saline and *pair-feeding*, indicating that central leptin sensitivity was maintained during the 7 days of leptin treatment and supporting its role in regulating energy balance and metabolic processes (Table 2).

Table 2. Effects of central leptin on hypothalamic and cardiac gene expression levels.

Hypothalamic Gene	SS	PF	LEP
Ob-Rb	1.0 ± 0.3 [a]	2.7 ± 0.2 [b]	3.7 ± 0.4 [c]
Crh	0.9 ± 0.1 [a]	0.6 ± 0.1 [b]	1.96 ± 0.1 [c]
Trh	1.1 ± 0.1 [a]	0.9 ± 0.1 [a]	2.6 ± 0.3 [b]

Hypothalamic mRNA levels of *Ob-Rb*, *Crh*, and *Trh* in leptin (Lep) or saline-infused rats (SS and PF). The values are expressed relative to the SS group. Statistical analysis was performed using one-way ANOVA followed by Tukey's test. Data are the mean ± SEM (n = 5 per treatment). Different letters indicate significant differences ($p \leq 0.05$). Groups: SS: saline-infused rats fed *ad libitum*; PF: saline-infused *pair-fed* rats; Lep: leptin-infused rats.

We next confirmed that chronic central infusion of leptin increased PPARβ/δ protein content in the heart, compared to the saline and *pair-feeding* (Figure 2), highlighting its role in leptin's cardiac effects. Moreover, the pharmacological inhibition of PPARβ/δ, via the in vivo administration of the selective antagonist GSK0660, blunted the induction of PPARβ/δ in the heart mediated by central leptin (Figure 2).

3.2. Effect of Leptin-GSK0660 Co-Treatment on Biological Characteristics of the Rats

Confirming previous data [13,15], rats infused with central leptin (Lep+DMSO) did not change significantly their body weight (Figure 3A), average food consumption (Figure 3B), and adiposity (Table 3) compared to *pair-fed* groups infused with vehicle (PF+DMSO). The intraperitoneal (i.p) administration of GSK0660 during 7-day (Lep+GSK0660 and PF+GSK0660) abolished the effects on body weight and food intake (Figure 3A,B) mediated either by central leptin infusion (Lep+GSK0660) or by caloric restriction (PF+GSK0660) in *pair-fed* rats.

Central leptin infusion did not induce changes in serum glucose and leptin levels compared to the pair-fed group infused with vehicle (PF+DMSO) (Table 3), as previously reported [12]. Nevertheless, GSK0660 treatment notably increased serum levels of insulin (Table 3), indicating that GSK0660 induced insulin resistance in 3-month-old Wistar rats.

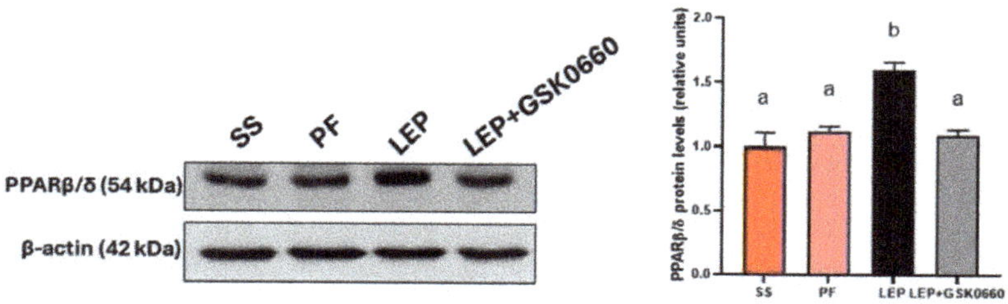

Figure 2. Effect of central leptin on PPARβ/δ protein levels in the heart. Western-Blot analysis of PPARβ/δ in 50 µg of total extracts from cardiac ventricles after central saline or leptin infusion. The values are expressed relative to the SS group. Data are the mean ± SEM (n = 5) per group of animals. Statistical analysis was performed using one-way ANOVA followed by Tukey's test. Different letters indicate significant differences ($p \leq 0.05$). Groups: SS: saline-infused rats fed *ad libitum*; PF: saline-infused *pair-fed* rats; Lep: leptin-infused rats; Lep+GSK0660: leptin-infused rats and treated with GSK0660. A representative Red Ponceau staining of the nitrocellulose membrane for total protein normalization prior to immunodetection and the quantitative densitometry readings of PPARβ/δ protein from the Western-Blot analysis are depicted in Figure S1 and Table S2, respectively, in the Supplemental Data.

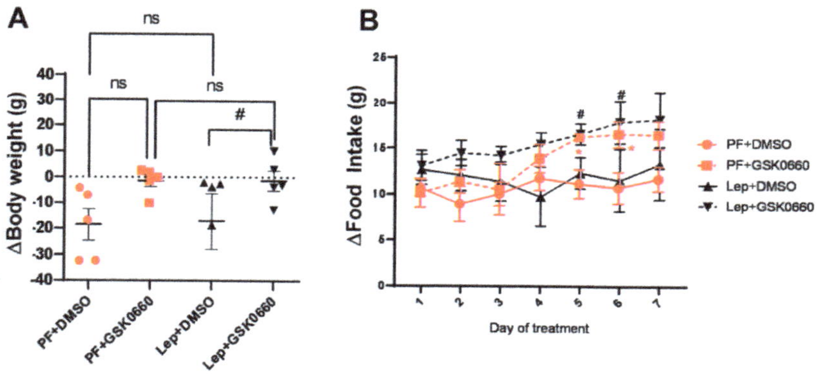

Figure 3. Variation in body weight (**A**) and food intake (**B**) after 7 days of central administration of leptin plus DMSO, leptin plus GSK0660, or vehicle-infused *pair-fed* rats plus DMSO and vehicle-infused *pair-fed* rats plus GSK0660. Data are the mean ± SEM (n = 5) per group of animals. Differences between two groups were assessed using the unpaired Student's *t*-test (# $p \leq 0.05$ Lep+GSK0660 vs. Lep+DMSO; * $p \leq 0.05$ PF+GSK0660 vs. PF+DMSO; Figure 2A,B). Groups: PF+DMSO: vehicle-infused *pair-fed* rats plus DMSO; PF+GSK0660: vehicle-infused pair-fed rats plus GSK0660; Lep+DMSO: leptin plus DMSO; Lep+GSK0660: leptin plus GSK0660.

In small animals with rapid changes in body weight, as in rats, the normalization of heart size relative to body weight is not a good indicator of adverse cardiac remodeling development [19]. Instead, as the heart weight/tibia length index remains constant after maturity [19], we used this index to analyze the effects of leptin and GSK0660 on cardiac remodeling. As observed in Figure 4, this index did not show significant differences between rats treated for 7 days with leptin or vehicle, indicating that leptin treatment did not seem to induce a significant alteration of the ventricular wall. However, in the group of animals treated with GSK0660, the ratio was slightly higher compared to its control group treated with DMSO (Figure 4).

Table 3. Effects of central leptin on adipose tissue weights (g) and serum profiles.

	PF + DMSO	PF + GSK0660	LEP + DMSO	LEP + GSK0660
PrWAT (g)	4.62 ± 0.86 [a]	4.86 ± 1 [a]	3.04 ± 0.85 [a]	3.71 ± 0.68 [a]
eWAT (g)	4.21 ± 0.36 [a]	4.80 ± 0.86 [a]	3.10 ± 0.54 [a]	3.53 ± 0.40 [a]
BAT (g)	0.42 ± 0.05 [a]	0.40 ± 0.07 [a]	0.29 ± 0.02 [a]	0.39 ± 0.04 [a]
Serum glucose (mg/dL)	99 ± 19 [a]	105 ± 12 [a]	112 ± 12 [a]	118 ± 10 [a]
Serum leptin (ng/mL)	1.07 ± 0.1 [a]	1.60 ± 0.5 [a]	1.10 ± 0.3 [a]	1.14 ± 0.3 [a]
Serum insulin (ng/mL)	2.49 ± 0.8 [a]	2.67 ± 0.6 [a]	2.46 ± 0.6 [a]	5.19 ± 2.1 [b]

Variations in adipose tissue mass (eWAT, PrWAT and BAT), glucose (mg/dL), leptin (ng/mL) and insulin (ng/mL) values in the serum of animals after 7 days of central administration of leptin plus DMSO, leptin plus GSK0660 or vehicle-infused *pair-fed* rats plus DMSO and vehicle-infused *pair-fed* rats plus GSK0660. Data are the mean ± SEM (n = 5) per group of animals. One-way ANOVA followed by Tukey test (different letters indicate significant differences, $p \leq 0.05$). Groups: PF+DMSO: vehicle-infused *pair-fed* rats plus DMSO; PF+GSK0660: vehicle-infused *pair-fed* rats plus GSK0660; Lep+DMSO: leptin plus DMSO; Lep+GSK0660: leptin plus GSK0660; PrWAT: peritoneal white adipose tissue; eWAT: epididymal white adipose tissue; BAT: brown adipose tissue.

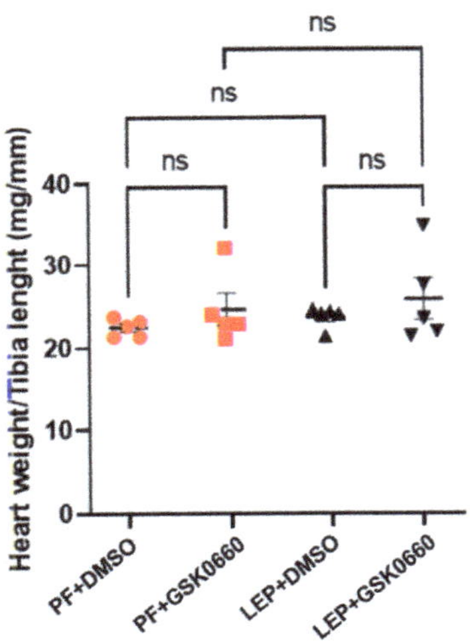

Figure 4. Graphic representation of heart weight/tibia length ratio (mg/mm). Data are the mean ± SEM (n = 5) per group of animals. Differences between the two groups were assessed using the unpaired Student's *t*-test ($p \leq 0.05$). Groups: PF+DMSO: vehicle-infused *pair-fed* rats plus DMSO; PF+GSK0660: vehicle-infused *pair-fed* rats plus GSK0660; Lep+DMSO: leptin plus DMSO; Lep+GSK0660: leptin plus GSK0660.

3.3. Effect of Leptin-GSK0660 Co-Treatment on Cardiac Electrical Properties

To gain a deeper insight into the effects of central leptin on cardiac electrical properties and remodeling, we studied the influence of central leptin treatment in the absence or presence of GSK0660 on cardiac function by analyzing both electrical and physiological patterns.

Left ventricular hypertrophy (LVH) is defined as an increase in left ventricular mass (LVM) associated with structural changes in the myocardium. This occurs in response to mechanical and/or neurohormonal stimuli, leading to an increased workload on the myocyte, which, under biomechanical stress, can induce hypertrophy development [20].

It has previously been reported that ventricular wall thickening is associated with changes in cardiac depolarization and repolarization states, manifested as altered patterns in the duration and amplitude of the QRS complex, as well as QT interval prolongation [21,22]. Accordingly, we used this approach to assess cardiac function and electrical activity in our experiments.

Indeed, as presented in Table 4, we compare the mean values of QRS duration, R wave amplitude as a representation of the QRS complex voltage, and corrected QT duration (QTc) at baseline (t0) and final time points of the treatments (t7) to underscore physiological changes or other alterations in cardiac structure.

Table 4. ECG signal at (t0) and final time (t7).

	QRS Duration (ms)		ΔQRS (ms)	R Amplitude (mV)		ΔR (mV)
	t0	t7		t0	t7	
PF+DMSO	22.7 ± 5.1	19.6 ± 5.1	−3.1 ± 3.2 [a]	147 ± 19	149 ± 24	2.0 ± 15 [a]
PF+GSK0660	28.8 ± 5.7	26.3 ± 5.0	−2.5 ± 3.6 [a]	103 ± 26	95 ± 7	−8 ± 14 [a]
LEP+DMSO	29.2 ± 7.1	20.4 ± 3.8 *	−8.8 ± 3.6 [a]	137 ± 23	93 ± 12 *	−44 ± 13 [b]
LEP+GSK0660	21.4 ± 3.0	17.1 ± 8	−4.3 ± 3.8 [a]	126 ± 12	132 ± 27	6 ± 13 [a]
	QT Duration (ms)		**ΔQT (ms)**	**QTc Duration (ms)**		**ΔQTc (ms)**
	t0	t7		t0	t7	
PF+DMSO	71.9 ± 13.4	70.9 ± 11.1	−1.0 ± 7.8 [a]	6.2 ± 1.2	5.6 ± 0.8	−0.6 ± 0.6 [a]
PF+GSK0660	65.2 ± 7.3	55.1 ± 7.3	−10.1 ± 4.9 [a,b]	5.4 ± 0.6	4.2 ± 0.5 *	−1.2 ± 0.4 [a]
LEP+DMSO	79.9 ± 10.5	65.5 ± 12.1	−14.4 ± 7.2 [b]	6.9 ± 1.0	5.6 ± 1.1	−1.3 ± 0.7 [a]
LEP+GSK0660	55.3 ± 7.2	60.3 ± 3.6	5.0 ± 3.6 [a,c]	4.6 ± 0.6	5.1 ± 0.3	0.5 ± 0.3 [b]

Duration of the QRS wave (ms), the amplitude of the R wave (mV), the duration of the QT interval (ms), and the duration of QTc (QT corrected by Bazett's formula) of the rats belonging to the four treatment groups at baseline and end time. Values are expressed relative to control rats (PF+DMSO). Results are the mean ± SEM (n = 5) per group of animals. Differences between the two times were assessed using the unpaired Student's *t*-test (* $p \leq 0.05$ t7 vs. t0). Differences between treatments were assessed using one-way ANOVA followed by the Tukey test (different letters indicate significant differences, $p \leq 0.05$). Groups: PF+DMSO: vehicle-infused *pair-fed* rats plus DMSO; PF+GSK0660: vehicle-infused *pair-fed* rats plus GSK0660; Lep+DMSO: leptin plus DMSO; Lep+GSK0660: leptin plus GSK0660.

The QRS duration and R wave amplitude in lead II voltages, at different times, showed a reduction in the group of animals treated with central leptin, being significant only for the R wave amplitude measurement. These changes suggest a decrease in abnormal ventricular wall growth, which could be indicative of a more atrophic pattern. This is in consonance with previous studies performed in rodents [10,12]. Interestingly, these differences were not observed in the group of animals treated with the PPARβ/δ antagonist. By comparison, QTc duration only increased in the group of animals treated with the PPARβ/δ antagonist when comparing baseline (t0) and final time (t7), indicating a potential alteration in ventricular repolarization, which can have implications for arrhythmogenesis and electrical instability.

In fact, the longer QRS duration and higher R wave voltage are used as measures of relative sensitivity and specificity of abnormal ventricular wall growth, which, in combination with other physiological parameters, facilitate the detection of left ventricular hypertrophy. The results presented in this work indicate that rats treated with the PPARβ/δ antagonist present an altered ECG, possibly due to increased ventricular wall thickness, contrasting with the characteristic atrophic pattern observed in rats treated centrally with leptin, suggesting that the effects observed with central leptin treatment are mediated by leptin-specific pathways rather than PPARβ/δ signaling.

Given that PPARβ/δ has been implicated in the regulation of ion channels and cellular processes involved in cardiac repolarization [23–25], antagonism of PPARβ/δ may disrupt these processes, leading to QT prolongation.

Moreover, previous studies have demonstrated the role of disturbed PPARβ/δ activity in the development of metabolic syndrome and its components, including insulin resistance [26], which are associated with QT interval prolongation and increased cardiovascular risk [25,27–29].

On the other hand, it is known that leptin signaling improves insulin sensitivity, which is important for changing QTc duration and potentially reducing arrhythmogenic risk [30,31]. Improved insulin sensitivity through effective leptin signaling can lead to better regulation of glucose and lipid metabolism, thereby stabilizing cardiac electrophysiology and reducing the likelihood of QTc prolongation and related arrhythmias.

In addition to these measures facilitating screening of early-stage pathology development (QRS duration, R amplitude, and long QTc), an increase in heart rate (HR) can be used as an indicator of poor prognosis. Various studies associate increased mortality with an elevated heart rate, finding a positive correlation between the two. Similarly, some authors suggest that reducing HR could be used as a potential target for treating patients with heart failure [32].

As seen in Table 5, animals treated with central leptin were able to reduce heart rate, which was not observed in animals additionally treated with the PPARβ/δ antagonist. In the GSK0660-treated groups (PF+GSK0660 and Lep+GSK0660), an increase in heart rate was induced compared to their controls (PF+DMSO and Lep+DMSO). Surprisingly, a significant increase in R-R interval duration (ms) occurred in *pair-fed* control animals treated with the vehicle (PF+DMSO).

Table 5. Sinus rhythm (R-R interval, R-R) and mean heart rate (HR) for treated groups.

	R-R INTERVAL (ms)		**ΔR-R (ms)**	**Heart Rate (bpm)**		**ΔHR (bpm)**
	t0	t7		t0	t7	
PF+DMSO	136 ± 6	159 ± 7 *	23 ± 4.1 [a]	210 ± 22	274 ± 30 *	64 ± 16.64 [a]
PF+GSK0660	145 ± 4	196 ± 31 *	51 ± 14.0 [b]	205 ± 12	275 ± 34 *	70 ± 16.12 [a]
LEP+DMSO	135 ± 5	139 ± 4	4 ± 2.9 [c]	231 ± 30	219 ± 20	−12 ± 16.12 [b]
LEP+GSK0660	141 ± 4	142 ± 6	1 ± 3.2 [c]	199 ± 4	224 ± 14 *	25 ± 6.51 [c]

Sinus rhythm (R-R), measured in milliseconds (ms), and heart rate (HR), measured in beats per minute (bpm) of the rats belonging to the four treatment groups at baseline and final time. Values are expressed relative to control rats (PF+DMSO). Results are the mean ± SEM (n = 5) per group of animals. One-way ANOVA followed by Tukey test (different letters indicate significant differences, $p \leq 0.05$). Differences between the two times were assessed using the unpaired Student's *t*-test (* $p \leq 0.05$ t7 vs. t0). Differences between treatments were assessed using one-way ANOVA followed by the Tukey test (different letters indicate significant differences, $p \leq 0.05$). Groups: PF+DMSO: vehicle-infused *pair-fed* rats plus DMSO; PF+GSK0660: vehicle-infused *pair-fed* rats plus GSK0660; Lep+DMSO: leptin plus DMSO; Lep+GSK0660: leptin plus GSK0660.

This result indicates that central leptin signaling may act as a regulator against the development of cardiac arrhythmias, probably due to its ability to impact cardiomyocyte size [33,34], improve insulin sensitivity, and modulate the autonomic nervous system activity and the expression of ion channels involved in cardiac repolarization [30,31]. Interestingly, Lin et al. [35] have suggested that circulating leptin directly modulates cardiac electrical properties, including heart rate and QT interval, via its receptors expressed in cardiac tissue. These findings suggest a mechanism by which leptin may influence cardiovascular function independently of the central pathways, indicating the potential implications of dysregulated leptin signaling in cardiovascular disorders, including the development of bradycardia, QT interval prolongation, and ventricular arrhythmias.

Therefore, while the precise role of leptin in protecting against cardiac arrhythmias is still being elucidated, our data suggests that central leptin actions may have beneficial effects on cardiac electrophysiology and rhythm regulation

3.4. *Effect of Leptin-GSK0660 Co-Treatment on Cardiac Remodeling and Function*

Next, we evaluated alterations in cardiac function using Magnetic Resonance Imaging (MRI) to monitor the response to the progression of central leptin or vehicle treatment. As expected, at baseline (t0), there were no differences in ejection fraction (EF, %) between groups (Figure 5). However, a decrease in EF was induced at the final time (t7) in the Lep+GSK0660 group, which was statistically significant compared to its control group

(Lep+DMSO). This decrease only fell below the 50% threshold when the PPARβ/δ activity was inhibited in the central leptin-treated group (Lep+GSK0660). Although EF% values are considered borderline between 41 and 49% [36–38], this should not be considered an indicator of heart failure development; instead, it indicates a loss of contractile capacity.

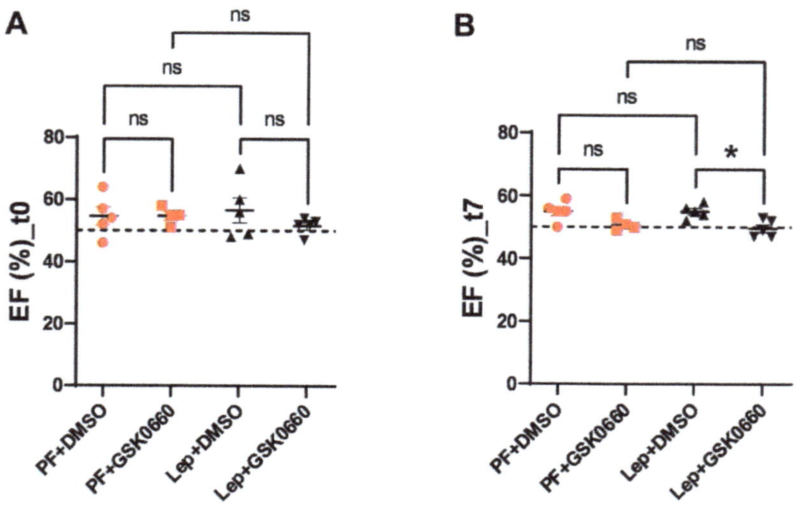

Figure 5. EF values (%) obtained after analysis of the MRI images with the Segment program at t0 (**A**) and t7 (**B**) times. Values are expressed relative to control rats (PF+DMSO). Results are the mean ± SEM (n = 5) per group of animals. One-way ANOVA followed by Tukey test ($p \leq 0.05$). Differences between treatments were assessed using the unpaired Student's *t*-test (* $p \leq 0.05$) Lep+GSK0660 vs. Lep+DMSO. Groups: PF+DMSO: vehicle-infused pair-fed rats plus DMSO; PF+GSK0660: vehicle-infused pair-fed rats plus GSK0660; Lep+DMSO: leptin plus DMSO; Lep+GSK0660: leptin plus GSK0660.

Studies in animal models and some human studies have shown that high-fat diets can lead to cardiac dysfunction, including reduced ejection fraction, as evidenced [39,40]. These effects appear to be mediated by various mechanisms, including inflammation, oxidative stress, mitochondrial dysfunction, and insulin resistance, all of which can impair cardiac contractility and function. While these studies primarily focus on insulin resistance and mitochondrial dynamics in heart disease, there could be implications for the role of leptin signaling in cardiac function. Importantly, vehicle-treated and centrally leptin-treated groups showed preserved ejection fraction after 7 days of treatment (Figure 5).

As already mentioned, analysis of MRI allows for a more effective estimation of variations in ventricular mass. We observed that animals treated with the vehicle did not present differences in end-diastolic left ventricular mass (EDLVM) (Figure 6) at baseline and post-treatment times. However, in animals treated centrally with leptin, EDLVM was significantly greater in the group where PPARβ/δ action was blocked using GSK0660. Surprisingly, these animals already showed differences in ventricular mass at baseline. These differences increased at the final time (t7), with EDLVM being significantly greater in this group (Lep+GSK0660).

Thus, this study supports that central leptin treatment at a low dose (0.2 μg/day) promotes an atrophic pattern in the myocardium and that PPARβ/δ is required for leptin signaling to remodel cardiac tissue in rats with normal leptin sensitivity.

To measure the potential impact of increased ventricular mass on ventricular filling volume, end-diastolic volume (EDV) was measured at the final time (t7). Although differences were not significant, a slight decrease in final ventricular filling volume was observed for the group of animals with dual treatment (ICV-Leptin and IP-GSK0660) (Figure 7).

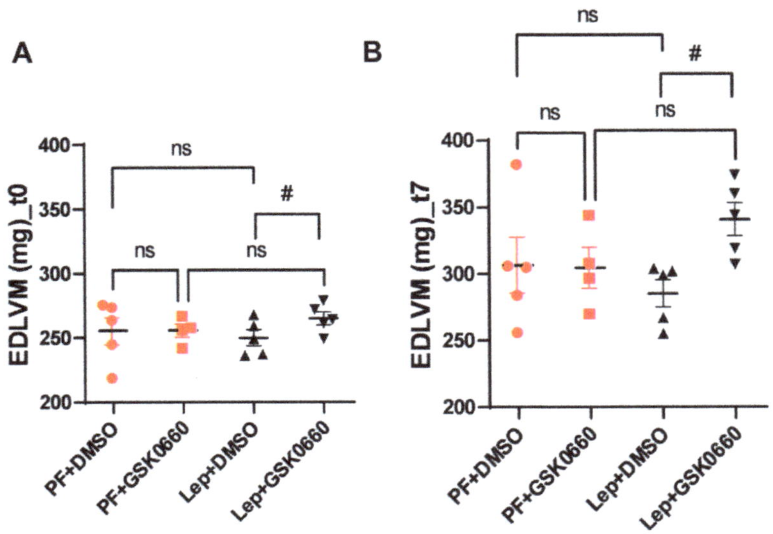

Figure 6. EDLVM values, expressed in gr, obtained after analysis of the MRI images with the Segment program at t0 (**A**) and t7 (**B**) times. Values are expressed relative to the control group (PF+DMSO). Results are the mean ± SEM (n = 5) per group of animals. One-way ANOVA followed by Tukey test ($p \leq 0.05$). Differences between treatments were assessed using the unpaired Student's *t*-test (# $p \leq 0.05$) Lep+GSK0660 vs. Lep+DMSO. Groups: PF+DMSO: vehicle-infused *pair-fed* rats plus DMSO; PF+GSK0660: vehicle-infused *pair-fed* rats plus GSK0660; Lep+DMSO: leptin plus DMSO; Lep+GSK0660: leptin plus GSK0660.

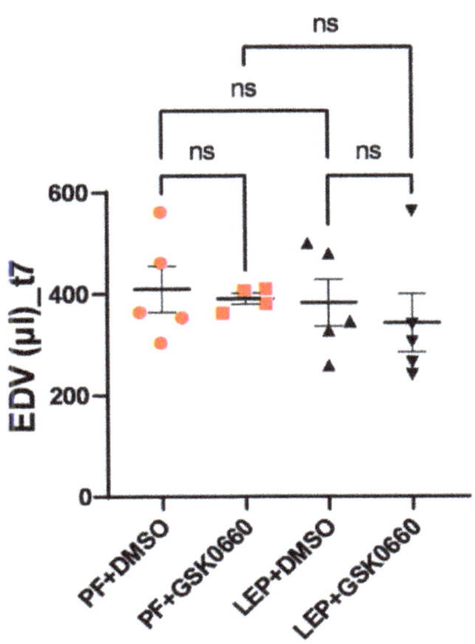

Figure 7. EDV values at t7, expressed in µL, obtained after analysis of the MRI images with the Segment program. Values are expressed relative to the control group (PF+DMSO). Results are the mean ± SEM (n = 5) per group of animals. One-way ANOVA followed by Tukey test ($p \leq 0.05$). Groups: PF+DMSO: vehicle-infused *pair-fed* rats plus DMSO; PF+GSK0660: vehicle-infused *pair-fed* rats plus GSK0660; Lep+DMSO: leptin plus DMSO; Lep+GSK0660: leptin plus GSK0660.

Finally, possible changes in cardiac output (CO) were studied. As seen in Figure 8, the differences in ventricular mass did not induce changes in CO pre- and post-treatment, despite an increase in the Lep+GSK0660 group (t7) being observed, which was not significant (p = 0.078). The lack of significant changes in CO may be attributed, at least in part, to compensatory mechanisms that were initially induced during the development of hypertrophy to prevent an increase in CO [41]. These compensatory mechanisms could involve various physiological adaptations aimed at maintaining cardiac output despite changes in ventricular mass. One such mechanism could involve an increase in heart rate (HR), which was observed in the animal treated with leptin and the PPARβ/δ antagonist (Lep+GSK0660), as depicted in Figure 8.

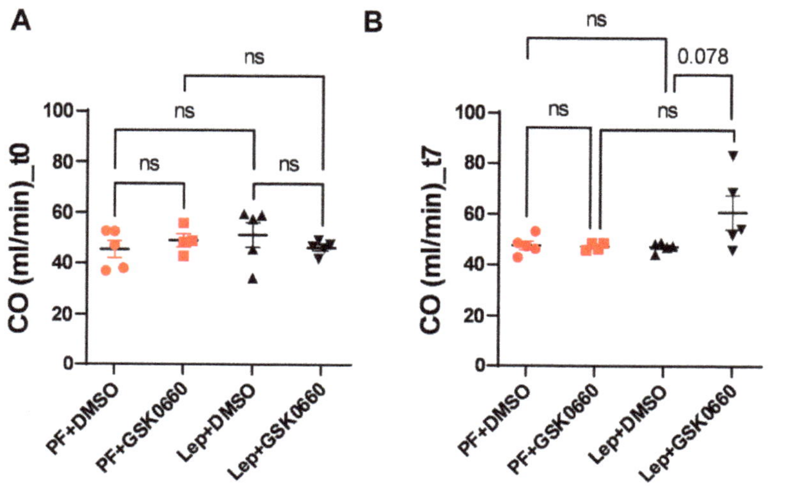

Figure 8. CO values (mL/min) obtained after analysis of the MRI images with the Segment program at t0 (**A**) and t7 (**B**) times. Values are expressed relative to the control group (PF+DMSO). Results are the mean ± SEM (n = 5) per group of animals. One-way ANOVA followed by Tukey test ($p \leq 0.05$). Differences between treatments were assessed using the unpaired Student's t-test ($p \leq 0.05$). Groups: PF+DMSO: vehicle-infused *pair-fed* rats plus DMSO; PF+GSK0660: vehicle-infused *pair-fed* rats plus GSK0660; Lep+DMSO: leptin plus DMSO; Lep+GSK0660: leptin plus GSK0660.

Finally, supporting the cardioprotective effects of central leptin reported herein in normoleptinemic adult rats with normal leptin sensitivity, we have previously reported that leptin, acting in the central nervous system, regulates cardiac lipid metabolism increasing the expression of genes involved in the myocardial intracellular lipolysis (ATGL and HSL) and mitochondrial/peroxisomal fatty acid utilization (PDK4, UCP3, Acox1), whereas decreased those involved in TAG synthesis (SCD-1 and DGAT1), leading these changes to a significant reduction in cardiac TAG and TBARS content [15]. In addition, central leptin markedly increased heart glucose transport independently of insulin. Moreover, the pharmacological inhibition of PPARβ/δ decreased the effects induced by central leptin on the expression of genes involved in lipid metabolism and increased cardiac TAG content [15]. These results support that central leptin plays a critical role in protecting the heart from excess lipid accumulation that, when uncontrolled, can drive cardiac lipotoxicity [7] and align with the suggestion of Baskin and Taegtmeyer [1] who proposed that signals from intermediary metabolism of energy-providing substrates are likely mediators of atrophic remodeling.

Beyond the regulation of cardiac metabolism and protection against cardiac lipotoxicity, we have recently demonstrated that, in adult rats with normal leptin sensitivity, central leptin-induced an atrophy-related gene program in cardiac tissue activating FoxO1/3 [12], master regulators of the atrophy program [42,43], increasing the content of Atrogin-1 and MURF-1, proteins involved in limiting cardiac hypertrophy. Besides, central leptin activated the anti-hypertrophic kinase GSK3β, while the phenotypic markers of cardiac hypertrophy atrial natriuretic peptide (ANP) and β-myosin heavy chain, gene product of Myh7 were decreased in response to central leptin infusion. Finally, the pharmacological inhibition of PPARβ/δ, via in vivo administration of the selective antagonist GSK0660, blunted the induction of FOXO1/3, Atrogin-1, MuRF1, and GSK3β in the heart mediated by central leptin infusion [12].

On the other hand, central leptin inhibited mTORC1 pathway by decreasing protein and basal phosphorylation levels of mTORC1, RAPTOR protein abundance, and basal phosphorylation levels of 4E-BP1 and S6K1. These data may suggest that central leptin modulates the mTORC1 pathway to suppress anabolic processes in the heart [12]. In this sense, it is known that the inhibition of the mTORC-S6K1 pathway regresses established cardiac hypertrophy in mice with transverse aortic constriction [44].

Taken together, the results presented in this study and our prior findings [12,15] suggest that central leptin may exert cardioprotective effects in adult rats with normal leptin sensitivity by facilitating atrophic remodeling, potentially linked with metabolic unloading. This process could imply the induction of FoxO1/3, the inhibition of mTORC1, and the activation of PPARβ/δ pathways.

In addition, this study examines the impact of central leptin infusion and the pharmacological inhibition of PPARβ/δ on cardiac remodeling and cardiac function in Wistar rats. We confirm that leptin may have protective effects against abnormal ventricular growth, as confirmed by MRI, while PPARβ/δ antagonism might lead to changes in cardiac electrical activity indicative of left ventricular hypertrophy (LVH). Our results provide new insights into the physiological mechanisms by which the leptin-PPARβ/δ crosstalk impacts cardiac remodeling and cardiac function in rats.

In contrast, treatment with the PPARβ/δ antagonist GSK0660 resulted in increased QTc duration, indicating potential alterations in ventricular repolarization and an increased risk of arrhythmias. Hence, antagonists like GSK0660 could potentially be used to study metabolic pathways but might also have implications for disrupting energy balance if not used appropriately.

These results underscore the importance of leptin in mediating cardiac remodeling through metabolic pathways. Future research should focus on identifying specific metabolic intermediates and signaling pathways involved in leptin's cardioprotective effects. Understanding these mechanisms could provide new insights into potential therapeutic strategies for managing cardiac diseases linked with metabolic disorders.

4. Conclusions and Future Directions

This study explores the specific interactions between leptin and PPARβ/δ in cardiac function. It was conducted using animal models with normal leptin sensitivity. However, the findings support that inhibition of PPARβ/δ-associated pathways may interfere with central leptin's beneficial effects in the heart, which may be relevant in patients with conditions like generalized lipodystrophy where leptin therapy is used to mitigate cardiac hypertrophy [45]. Further research is needed to explore the long-term effects of leptin and PPARβ/δ modulation on cardiac health.

Supplementary Materials: The following supporting information can be downloaded at https://www.mdpi.com/article/10.3390/biom14081028/s1; Table S1: Primers sequences of genes used for quantification of mRNAs by qRT-PCR; Figure S1: Representative Red Ponceau staining of the nitrocellulose membrane for total protein normalization prior to immunodetection; Table S2: Quantitative densitometry analysis of PPARβ/δ protein bands from the Western-Blot.

Author Contributions: Conceptualization, A.A. and N.G.; Funding acquisition, A.A. and N.G.; Investigation, B.R., C.P. and L.M.; Methodology, B.R., C.P., L.M. and M.B.; Project administration, A.A. and N.G.; Resources, A.A. and N.G.; Software, B.R. and M.B.; Supervision, A.A. and N.G.;

Validation, C.P., A.A. and N.G.; Visualization, B.R., C.P. and L.M.; Writing—original draft, B.R. and N.G.; Writing—review & editing, A.A. and N.G. All authors have read and agreed to the published version of the manuscript.

Funding: This research was funded by Ministerio de Ciencia, Innovación y Universidades, Spain, under project grants RTI2018-098643-B-I00. Ministerio de Ciencia e Innovación, Spain, under project grants PID2021-128243OB-I00. University of Castilla-La Mancha, Spain and the European Regional Development Fund (FEDER) under project grants 2021 GRIN-30987.

Institutional Review Board Statement: Not applicable.

Informed Consent Statement: Not applicable.

Data Availability Statement: Data will be made available on request.

Acknowledgments: The authors acknowledge their gratitude to Sergio Moreno for the excellent technical assistance.

Conflicts of Interest: The authors declare no conflicts of interest.

References

1. Baskin, K.K.; Taegtmeyer, H. Taking Pressure off the Heart: The Ins and Outs of Atrophic Remodelling. *Cardiovasc. Res.* **2011**, *90*, 243–250. [CrossRef] [PubMed]
2. Gibb, A.A.; Hill, B.G. Metabolic Coordination of Physiological and Pathological Cardiac Remodeling. *Circ. Res.* **2018**, *123*, 107–128. [CrossRef] [PubMed]
3. Sudi, S.B.; Tanaka, T.; Oda, S.; Nishiyama, K.; Nishimura, A.; Sunggip, C.; Mangmool, S.; Numaga-Tomita, T.; Nishida, M. TRPC3-Nox2 Axis Mediates Nutritional Deficiency-Induced Cardiomyocyte Atrophy. *Sci. Rep.* **2019**, *9*, 9785. [CrossRef] [PubMed]
4. Bacova, B.S.; Andelova, K.; Sykora, M.; Egan Benova, T.; Barancik, M.; Kurahara, L.H.; Tribulova, N. Does Myocardial Atrophy Represent Anti-Arrhythmic Phenotype? *Biomedicines* **2022**, *10*, 2819. [CrossRef] [PubMed]
5. Morris, D.L.; Rui, L. Recent advances in understanding leptin signaling and leptin resistance. *Am. J. Physiol. Endocrinol. Metab.* **2009**, *297*, E1247–E1259. [CrossRef] [PubMed]
6. Karmazyn, M.; Purdham, D.M.; Rajapurohitam, V.; Zeidan, A. Leptin as a Cardiac Hypertrophic Factor: A Potential Target for Therapeutics. *Trends Cardiovasc. Med.* **2007**, *17*, 206–211. [CrossRef]
7. Hall, M.E.; Harmancey, R.; Stec, D.E. Lean Heart: Role of Leptin in Cardiac Hypertrophy and Metabolism. *World J. Cardiol.* **2015**, *7*, 511–524. [CrossRef] [PubMed]
8. Me, H.; Mw, M.; Je, H.; De, S. Rescue of Cardiac Leptin Receptors in Db/Db Mice Prevents Myocardial Triglyceride Accumulation. *Am. J. Physiol. Endocrinol. Metab.* **2014**, *307*, E316–E325. [CrossRef]
9. Barouch, L.A.; Berkowitz, D.E.; Harrison, R.W.; O'Donnell, C.P.; Hare, J.M. Disruption of Leptin Signaling Contributes to Cardiac Hypertrophy Independently of Body Weight in Mice. *Circulation* **2003**, *108*, 754–759. [CrossRef]
10. Gava, F.N.; da Silva, A.A.; Dai, X.; Harmancey, R.; Ashraf, S.; Omoto, A.C.M.; Salgado, M.C.; Moak, S.P.; Li, X.; Hall, J.E.; et al. Restoration of Cardiac Function After Myocardial Infarction by Long-Term Activation of the CNS Leptin-Melanocortin System. *JACC Basic Transl. Sci.* **2021**, *6*, 55–70. [CrossRef] [PubMed]
11. Kamimura, D.; Suzuki, T.; Wang, W.; de Shazo, M.; Hall, J.E.; Winniford, M.D.; Kullo, I.J.; Mosley, T.H.; Butler, K.R.; Hall, M.E. Higher Plasma Leptin Levels Are Associated with Reduced Left Ventricular Mass and Left Ventricular Diastolic Stiffness in Black Women: Insights from the Genetic Epidemiology Network of Arteriopathy (GENOA) Study. *Hypertens. Res.* **2018**, *41*, 629–638. [CrossRef] [PubMed]
12. Rubio, B.; Mora, C.; Pintado, C.; Mazuecos, L.; Fernández, A.; López, V.; Andrés, A.; Gallardo, N. The Nutrient Sensing Pathways FoxO1/3 and mTOR in the Heart Are Coordinately Regulated by Central Leptin through PPARβ/δ. Implications in Cardiac Remodeling. *Metabolism* **2020**, *115*, 154453. [CrossRef] [PubMed]
13. Mazuecos, L.; Pintado, C.; Rubio, B.; Guisantes-Batán, E.; Andrés, A.; Gallardo, N. Leptin, Acting at Central Level, Increases FGF21 Expression in White Adipose Tissue via PPARβ/δ. *Int. J. Mol. Sci.* **2021**, *22*, 4624. [CrossRef]
14. Rodríguez, M.; Pintado, C.; Torrillas-de la Cal, R.; Moltó, E.; Gallardo, N.; Andrés, A.; Arribas, C. Ageing Alters the Lipid Sensing Process in the Hypothalamus of Wistar Rat. Effect of Food Restriction. *Nutr. Neurosci.* **2022**, *25*, 1509–1523. [CrossRef] [PubMed]
15. Mora, C.; Pintado, C.; Rubio, B.; Mazuecos, L.; López, V.; Fernández, A.; Salamanca, A.; Bárcena, B.; Fernández-Agulló, T.; Arribas, C.; et al. Central Leptin Regulates Heart Lipid Content by Selectively Increasing PPAR β/δ Expression. *J. Endocrinol.* **2018**, *236*, 43–56. [CrossRef] [PubMed]
16. Heiberg, E.; Sjögren, J.; Ugander, M.; Carlsson, M.; Engblom, H.; Arheden, H. Design and Validation of Segment--Freely Available Software for Cardiovascular Image Analysis. *BMC Med. Imaging* **2010**, *10*, 1. [CrossRef] [PubMed]
17. Schweins, M.; Gäbel, R.; Raitza, M.; Vasudevan, P.; Lemcke, H.; Joksch, M.; Schildt, A.; Kurth, J.; Lindner, T.; Meinel, F.G.; et al. Multi-Modal Assessment of a Cardiac Stem Cell Therapy Reveals Distinct Modulation of Regional Scar Properties. *J. Transl. Med.* **2024**, *22*, 187. [CrossRef]

18. Lang, C.I.; Vasudevan, P.; Döring, P.; Gäbel, R.; Lemcke, H.; Lindner, T.; Steinhoff, G.; Krause, B.J.; Vollmar, B.; Meinel, F.G.; et al. Expedient Assessment of Post-Infarct Remodeling by Native Cardiac Magnetic Resonance Imaging in Mice. *Sci. Rep.* **2021**, *11*, 11625. [CrossRef] [PubMed]
19. Yin, F.C.; Spurgeon, H.A.; Rakusan, K.; Weisfeldt, M.L.; Lakatta, E.G. Use of Tibial Length to Quantify Cardiac Hypertrophy: Application in the Aging Rat. *Am. J. Physiol.* **1982**, *243*, H941–H947. [CrossRef] [PubMed]
20. Samak, M.; Fatullayev, J.; Sabashnikov, A.; Zeriouh, M.; Schmack, B.; Farag, M.; Popov, A.-F.; Dohmen, P.M.; Choi, Y.-H.; Wahlers, T.; et al. Cardiac Hypertrophy: An Introduction to Molecular and Cellular Basis. *Med. Sci. Monit. Basic Res.* **2016**, *22*, 75–79. [CrossRef] [PubMed]
21. Domain, G.; Chouquet, C.; Réant, P.; Bongard, V.; Vedis, T.; Rollin, A.; Mandel, F.; Delasnerie, H.; Voglimacci-Stephanopoli, Q.; Mondoly, P.; et al. Relationships between Left Ventricular Mass and QRS Duration in Diverse Types of Left Ventricular Hypertrophy. *Eur. Heart J. Cardiovasc. Imaging* **2022**, *23*, 560–568. [CrossRef] [PubMed]
22. Okin, P.M.; Roman, M.J.; Devereux, R.B.; Kligfield, P. Gender Differences and the Electrocardiogram in Left Ventricular Hypertrophy. *Hypertension* **1995**, *25*, 242–249. [CrossRef] [PubMed]
23. Papatheodorou, I.; Galatou, E.; Panagiotidis, G.-D.; Ravingerová, T.; Lazou, A. Cardioprotective Effects of PPARβ/δ Activation against Ischemia/Reperfusion Injury in Rat Heart Are Associated with ALDH2 Upregulation, Amelioration of Oxidative Stress and Preservation of Mitochondrial Energy Production. *Int. J. Mol. Sci.* **2021**, *22*, 6399. [CrossRef] [PubMed]
24. Planavila, A.; Rodríguez-Calvo, R.; Jové, M.; Michalik, L.; Wahli, W.; Laguna, J.C.; Vázquez-Carrera, M. Peroxisome Proliferator-Activated Receptor Beta/Delta Activation Inhibits Hypertrophy in Neonatal Rat Cardiomyocytes. *Cardiovasc. Res.* **2005**, *65*, 832–841. [CrossRef] [PubMed]
25. Barish, G.D.; Narkar, V.A.; Evans, R.M. PPAR Delta: A Dagger in the Heart of the Metabolic Syndrome. *J. Clin. Investig.* **2006**, *116*, 590–597. [CrossRef] [PubMed]
26. Manickam, R.; Wahli, W. Roles of Peroxisome Proliferator-Activated Receptor β/δ in Skeletal Muscle Physiology. *Biochimie* **2017**, *136*, 42–48. [CrossRef] [PubMed]
27. DeBoer, M.D.; Gurka, M.J.; Woo, J.G.; Morrison, J.A. Severity of Metabolic Syndrome as a Predictor of Cardiovascular Disease Between Childhood and Adulthood: The Princeton Lipid Research Cohort Study. *J. Am. Coll. Cardiol.* **2015**, *66*, 755–757. [CrossRef] [PubMed]
28. Lee, C.-H.; Olson, P.; Evans, R.M. Minireview: Lipid Metabolism, Metabolic Diseases, and Peroxisome Proliferator-Activated Receptors. *Endocrinology* **2003**, *144*, 2201–2207. [CrossRef]
29. Wang, Y.-X. PPARs: Diverse Regulators in Energy Metabolism and Metabolic Diseases. *Cell Res.* **2010**, *20*, 124–137. [CrossRef] [PubMed]
30. Morrison, C.D.; Huypens, P.; Stewart, L.K.; Gettys, T.W. Implications of Crosstalk between Leptin and Insulin Signaling during the Development of Diet-Induced Obesity. *Biochim. Biophys. Acta* **2009**, *1792*, 409–416. [CrossRef]
31. Sierra-Johnson, J.; Romero-Corral, A.; Lopez-Jimenez, F.; Gami, A.S.; Sert Kuniyoshi, F.H.; Wolk, R.; Somers, V.K. Relation of Increased Leptin Concentrations to History of Myocardial Infarction and Stroke in the United States Population. *Am. J. Cardiol.* **2007**, *100*, 234–239. [CrossRef]
32. Hori, M.; Okamoto, H. Heart Rate as a Target of Treatment of Chronic Heart Failure. *J. Cardiol.* **2012**, *60*, 86–90. [CrossRef] [PubMed]
33. Fraley, M.A.; Birchem, J.A.; Senkottaiyan, N.; Alpert, M.A. Obesity and the Electrocardiogram. *Obes. Rev. Off. J. Int. Assoc. Study Obes.* **2005**, *6*, 275–281. [CrossRef] [PubMed]
34. Wiegerinck, R.F.; Verkerk, A.O.; Belterman, C.N.; van Veen, T.A.B.; Baartscheer, A.; Opthof, T.; Wilders, R.; de Bakker, J.M.T.; Coronel, R. Larger Cell Size in Rabbits with Heart Failure Increases Myocardial Conduction Velocity and QRS Duration. *Circulation* **2006**, *113*, 806–813. [CrossRef] [PubMed]
35. Lin, Y.-C.; Huang, J.; Hileman, S.; Martin, K.H.; Hull, R.; Davis, M.; Yu, H.-G. Leptin Decreases Heart Rate Associated with Increased Ventricular Repolarization via Its Receptor. *Am. J. Physiol. Heart Circ. Physiol.* **2015**, *309*, H1731–H1739. [CrossRef] [PubMed]
36. Lang, R.M.; Badano, L.P.; Mor-Avi, V.; Afilalo, J.; Armstrong, A.; Ernande, L.; Flachskampf, F.A.; Foster, E.; Goldstein, S.A.; Kuznetsova, T.; et al. Recommendations for Cardiac Chamber Quantification by Echocardiography in Adults: An Update from the American Society of Echocardiography and the European Association of Cardiovascular Imaging. *J. Am. Soc. Echocardiogr. Off. Publ. Am. Soc. Echocardiogr.* **2015**, *28*, 1–39.e14. [CrossRef]
37. Savarese, G.; Stolfo, D.; Sinagra, G.; Lund, L.H. Heart Failure with Mid-Range or Mildly Reduced Ejection Fraction. *Nat. Rev. Cardiol.* **2022**, *19*, 100–116. [CrossRef]
38. Golla, M.S.G.; Hajouli, S.; Ludhwani, D. Heart Failure and Ejection Fraction. In *StatPearls*; StatPearls Publishing: Treasure Island, FL, USA, 2024.
39. Abel, E.D.; O'Shea, K.M.; Ramasamy, R. Insulin Resistance: Metabolic Mechanisms and Consequences in the Heart. *Arterioscler. Thromb. Vasc. Biol.* **2012**, *32*, 2068–2076. [CrossRef] [PubMed]
40. Dorn, G.W. Mitochondrial Dynamics in Heart Disease. *Biochim. Biophys. Acta* **2013**, *1833*, 233–241. [CrossRef] [PubMed]
41. Saheera, S.; Krishnamurthy, P. Cardiovascular Changes Associated with Hypertensive Heart Disease and Aging. *Cell Transplant.* **2020**, *29*, 963689720920830. [CrossRef] [PubMed]

42. Sandri, M.; Sandri, C.; Gilbert, A.; Skurk, C.; Calabria, E.; Picard, A.; Walsh, K.; Schiaffino, S.; Lecker, S.H.; Goldberg, A.L. FoxO transcription factors induce the atrophy-related ubiquitin ligase atrogin-1 and cause skeletal muscle atrophy. *Cell* **2004**, *117*, 399–412. [CrossRef] [PubMed]
43. Skurk, C.; Izumiya, Y.; Maatz, H.; Razeghi, P.; Shiojima, I.; Sandri, M.; Sato, K.; Zeng, L.; Schiekofer, S.; Pimentel, D.; et al. The FOXO3a transcription factor regulates cardiac myocyte size downstream of AKT signaling. *J. Biol. Chem.* **2005**, *280*, 20814–20823. [CrossRef] [PubMed]
44. McMullen, J.R.; Sherwood, M.C.; Tarnavski, O.; Zhang, L.; Dorfman, A.L.; Shioi, T.; Izumo, S. Inhibition of mTOR signaling with rapamycin regresses established cardiac hypertrophy induced by pressure overload. *Circulation* **2004**, *109*, 3050–3055. [CrossRef] [PubMed]
45. Nguyen, M.-L.; Sachdev, V.; Burklow, T.R.; Li, W.; Startzell, M.; Auh, S.; Brown, R.J. Leptin Attenuates Cardiac Hypertrophy in Patients With Generalized Lipodystrophy. *J. Clin. Endocrinol. Metab.* **2021**, *106*, e4327–e4339. [CrossRef] [PubMed]

biomolecules

MDPI

Article

Docosahexaenoic Acid Supplementation in Postnatal Growth Restricted Rats Does Not Normalize Lung Function or PPARγ Activity

Adrienne J. Cohen [1,†], Wesley R. Chidester [1,†], Daniel T. Wray [1], Nicolette Jessen [1], Aimee Jones [1], Cheylah Bitsui [1], James Zhao [1], J. Alan Maschek [2,3], James E. Cox [2,3], Camilia R. Martin [4] and Lisa A. Joss-Moore [1,*]

1 Department of Pediatrics, University of Utah, Salt Lake City, UT 84108, USA; wesley.r.chidester@utah.edu (W.R.C.)
2 Health Science Center Cores, University of Utah Health Sciences Center, Salt Lake City, UT 84108, USA
3 Department of Biochemistry, University of Utah, Salt Lake City, UT 84108, USA
4 Division of Neonatology, Weill Cornell Medicine, New York, NY 10065, USA
* Correspondence: lisa.joss-moore@hsc.utah.edu
† These authors contributed equally to this work.

Abstract: The development of BPD in preterm neonates is increased by poor growth and nutritional deficits. The involvement of the fatty acid DHA in the development of BPD has been a focus for over a decade. However, recent clinical trials show that isolated DHA supplementation may increase BPD in subgroups of preterm neonates. One explanation for poor lung outcomes in DHA-supplemented neonates is a disruption of global fatty acid profiles and increased expression of a dominant-negative splice variant of a key driver of lung development, PPARγ. We previously developed a rat model of postnatal growth restriction (PGR) in which pups have impaired lung function and altered PPARγ activity. Here, we use our PGR rat model to assess the effects of DHA supplementation on lung outcomes. We hypothesize that the PPARγ splice variant, PPARγΔ5, will be expressed in the rat lung, and that DHA supplementation of PGR rat pups will alter circulating lipid profiles, lung mechanics, and PPARγ variant expression. Our findings demonstrate that PPARγΔ5 is expressed in the developing rat lung and that DHA supplementation of PGR rat pups alters global circulating fatty-acid profiles and does not normalize PGR-induced impaired lung mechanics or PPARγ activity.

Keywords: lung development; growth restriction; bronchopulmonary dysplasia; DHA; PPARγ; alternative splicing

1. Introduction

Preterm birth and poor growth increase the risk of developing the chronic lung disease bronchopulmonary dysplasia (BPD). While the development of BPD in preterm neonates is multifactorial, poor growth during the pre- and postnatal periods, along with nutritional deficits contribute to the BPD lung phenotype of impaired alveolar formation. Important nutrients in the context of BPD and lung development include long-chain polyunsaturated fatty acids (LCPUFAs), particularly the long-chain omega 3 fatty acid docosahexaenoic acid (DHA) [1].

The involvement of DHA deficiency in the development of BPD, and potential protective effects of DHA supplementation in resolving lung outcomes in preterm neonates and in animal studies has been a focus for over a decade. In preterm infants born <30 weeks gestation, DHA decreases rapidly over the first week of life, and decreased DHA is associated with increased BPD [2]. Other studies conducted to assess the effects of DHA supplementation on neurodevelopment, with BPD as a secondary outcome, showed a positive impact of enteral DHA supplementation on BPD in a subset of preterm neonates [3]. The concept that DHA is beneficial to the developing lung, particularly in the context of pre- and postnatal growth restriction and neonatal lung injury, is also supported by animal studies [4–7]. As a result of the body of research supporting a role for DHA in the development of BPD, several clinical trials examining the effects of DHA supplementation of preterm infants on BPD rates and severity have been conducted. However, these trials show that supplemental DHA administration to preterm neonates may actually increase the incidence of BPD in subsets of neonates [8,9]. Given the need for, and potential consequences of, lipid supplementation of preterm neonates, a comprehensive understanding of the effects of lipid supplementation on the developing lung is essential.

One key lipid responsive molecular mediator of lung development is the transcriptional activator PPARγ. PPARγ is well established to have important functions in the lung and is necessary for the epithelial–mesenchymal interactions required for lung development and lung vascular integrity [10–12]. PPARγ is also essential in the lung response to injury, and injury effects can often be mitigated by PPARγ activation, which can be accomplished by DHA [4,13–16]. Like many transcription factors, PPARγ is subject to extensive alternative splicing. One splice variant of PPARγ, the PPARγ delta 5 (PPARγΔ5) variant, has been reported in adipose and kidney tissue, and functions as a dominant-negative protein isoform due to lack of a transactivation domain [17,18]. However, whether PPARγΔ5 is expressed in the lung, and whether its expression is altered by PGR and/or DHA supplementation is unknown.

In order to elucidate mechanistic effects of growth and DHA on lung outcomes, we use a rat model of postnatal growth restriction (PGR) following normal in utero growth [19]. We recently showed that prolonged PGR, without additional lung injury, impairs lung structure and function in the rat. In this model, the lung characteristics are similar to those of neonates with BPD, including increased alveolar wall thickness, decreased lung compliance, and increased tissue damping [19]. Here, we aim to determine the effects of DHA supplementation on lung outcomes in our previously described PGR rat model, as well as expression of PPARγ and the PPARγΔ5 splice variant. We hypothesize that the PPARγ splice variant, PPARγΔ5, will be expressed in the rat lung, and that DHA supplementation of PGR rat pups will alter circulating lipid profiles, lung mechanics, and PPARγ variant expression.

2. Materials and Methods

2.1. Rat Model of PGR

All animal procedures were reviewed and approved by the University of Utah Animal Care and Use Committee, following the NIH Guidelines for the Care and Use of Laboratory Animals. Timed pregnant Sprague Dawley rats (Charles River, Wilmington, MA, USA) were housed in a controlled environment with a 12 h light/dark cycle and unrestricted access to food and water. We employed the postnatal growth restriction (PGR) model as previously described by our group [19]. Briefly, at birth, pups were assigned to foster dams to establish litter sizes of either eight (control) or sixteen (PGR). Pups remained within their respective litters for the study duration. On postnatal day 21, animals were either euthanized for tissue and serum collection (decapitated under anesthesia with ketamine/xylazine at 80/12 mg/kg) or assigned to lung mechanics assessments. Each litter contributed one male and one female pup for experimental analyses.

2.2. DHA Supplementation

At delivery, rat dams from the PGR groups were randomized to receive either standard rat chow (6% fat from soybean oil, EnVIGO Teklad, Indianapolis, IN, USA) or rat chow supplemented with DHA at 0.01% of total fats (LoDHA) or DHA at 0.1% of total fats (HiDHA) (Nu-Chek-Prep, Inc., Elysian, MN, USA). The DHA chow was formulated by substituting the desired percentage of soybean oil with pure DHA. All rat chow formulations were isocaloric and equally tolerated by the dams. All dams had ad libitum access to chow and water during lactation.

2.3. Serum Fatty Acids

Serum fatty acid composition was analyzed via gas chromatography following transesterification, as previously described by our group [19,20]. Fatty acid methyl esters (FAMEs) were generated via transesterification using heptadecanoic acid as an internal standard for quantification. Analyses were performed on an Agilent 7890A gas chromatograph (Agilent Technologies, Santa Clara, CA, USA) equipped with a capillary column, enabling the detection of fatty acids with chain lengths ranging from 10 to 24 carbons. FAMEs were identified and quantified using the Supelco 37 FAME mix (Sigma-Aldrich, St. Louis, MO, USA), and data processing was conducted with OpenLAB Chromatography software (Agilent Technologies).

2.4. Lung Mechanics

Lung mechanics were assessed using the FlexiVent FX 2 system (SCIREQ, Montreal, QC, Canada) and analyzed with FlexiWare 7.2 software (Service Pack 2, Build 728; SCIREQ, Montreal, QC, Canada). On postnatal day 21, rat pups were anesthetized via intraperitoneal injection of ketamine (50 mg/kg) and xylazine (8 mg/kg). Following tracheostomy, a 16- or 18-gauge cannula was inserted, and animals were connected to the FlexiVent system. Ventilation was maintained at a rate of 150 breaths per minute with tidal volumes of 10 mL/kg and a positive end-expiratory pressure of 3 cm H_2O. After a stabilization period of 3 min to establish consistent breathing patterns and confirm the absence of air leaks, vecuronium bromide (1 mg/kg, IP) was administered as a paralytic agent.

Lung function was evaluated using automated maneuvers, as previously described [21]. Inspiratory capacity (IC) and static compliance (Cst) were derived from quasistatic pressure–volume (PV) loops recorded over a pressure range of 3 to 30 cm H_2O. To differentiate airway mechanics from parenchymal properties, the forced oscillation technique (FOT) was applied [21]. Respiratory impedance data were modeled using the constant phase equation to obtain Newtonian resistance (Rn), an index of airway narrowing, as well as tissue elastance (H) and tissue damping (G), which reflect alveolar stiffness and resistance, respectively. Tissue hysteresivity (η, G/H) was also calculated to assess viscoelastic properties of the lung.

2.5. Identification of PPAR$\gamma\Delta5$ in the Lung

Lung RNA was extracted from lung tissue from control rats at birth, postnatal day 12, and postnatal day 21 and reverse transcribed as previously described [19]. Gene specific primers were used to amplify the region of the Pparγ transcript between exon 4 and exon 6. Primer sequences were forward (within exon 4), CGAGAAGGAGAAGCTGTTGG, and reverse (within exon 6), GCACGTGCTCTGTGACAATC. The PCR product was subject to gel electrophoresis using 2.2% agarose FlashGel® Recovery Cassettes and measured (57022, Lonza Bioscience, Rockland, ME, USA). Resulting bands were excised and sequenced by the University of Utah Sequencing Core.

2.6. mRNA Transcript Levels

Real-time reverse transcriptase PCR was used as described by our group [19,22] to measure the abundance of lung Pparγ and of the PparγΔ5 splice variant. We used an assay-on-demand primer/probe set for Pparγ, which covers the exon boundary between exon 5 and 6 (Rn00440945_m1, ThermoFisher Scientific, Waltham, MA, USA). To measure levels of PparγΔ5, we used a custom primer/probe set spanning the exon 4–6 junction (forward, CGAGAAGGAGAAGCTGTTGG; reverse, GCGGTTGATTTGTCTGTTGT; probe, CCCTGGCAAAGCATTTGTAT). mRNA levels of PPARγ target gene, Perilipin 2 (Plin2) was also measured using the following Assay on Demand: Rn01399516_m1. For all measures, the comparative CT method was used, with GADPH as an internal control.

2.7. Protein Abundance

We utilized immunoblot to measure protein levels of PPARγ and PPARγΔ5. Proteins were separated NuPAGE™ Bis-Tris Midi Protein Gels, 4 to 12% (WG1403BOX, Invitrogen, Waltham, MA, USA). Following electrophoresis, protein was transferred using PVDF Transfer Stacks (IB34001, Invitrogen, Waltham, MA, USA) and dry-transfer with the iBlot™ 3 Western blot Transfer Device (IB31001, Invitrogen, Waltham, MA, USA). Total protein normalization was performed using No-Stain™ Protein Labeling Reagent according to manufacturer instructions (A44449, Invitrogen, Waltham, MA, USA). Membranes were probed using anti-PPARγ polyclonal antibody (16643-1-AP, Proteintech, Rosemont, IL, USA) in SuperSignal™ Western blot Enhancer. Membranes were incubated in SuperSignal™ West Pico PLUS Chemiluminescent Substrate according to manufacturer instructions (34580, Thermo Scientific, Waltham, MA, USA), and imaged using Universal settings of the iBright™ CL1500 Imaging System (A44114, Invitrogen, Waltham, MA, USA). Unedited and uncropped blots are available in Supplemental Data.

2.8. Statistics

Groups were the control receiving a regular diet (Control), PGR with a regular diet (PGR), PGR with 0.01% DHA diet (PGR + LoDHA), and PGR with 0.1% DHA diet (PGR + HiDHA). Males and females were treated as separate groups. A one-way analysis of variance (ANOVA) with Fisher's post hoc protected least-significance difference was used to determine differences between groups. Statistical significance was defined as $p < 0.05$.

3. Results

3.1. Rat Model of PGR and DHA Supplementation

As we previously reported, our model of PGR resulted in a significant decrease in weight at day 21 in both males and females compared to control (Table 1). PGR with a regular diet resulted in a 33% decrease in body weight in males. The addition of DHA supplementation in male rat pups did not improve weight gain in PGR, with PGR + LoDHA and PGR + HiDHA rat pups weighing 38% and 27% less than male controls, respectively. Similarly, in females, PGR resulted in a 34% reduction in body weight in the regular diet group, and a 41% and 28% reduction in body weight with PGR + LoDHA and PGR + HiDHA, respectively.

Table 1. Rat body weight at study time, postnatal day 21.

	Control	PGR	PGR + LoDHA	PGR + HiDHA
Weight, g ± SD				
Male	53.4 ± 3.3	35.7 ± 4.5 *	33.0 ± 2.7 *	39.2 ± 3.5 *
Female	52.2 ± 3.8	34.6 ± 4.6 *	31.2 ± 3.6 *	37.6 ± 2.8 *

* $p < 0.05$ compared to control.

3.2. Serum Fatty Acids

Both PGR and DHA supplementation significantly affected serum fatty acid levels (Table 2). PGR with a regular diet altered serum levels of several fatty acids in male rat pups. In male rat pups, PGR on a regular diet decreased serum levels of stearic acid (18:0) by 40%, oleic acid (18:1) by 43%, linoleic acid (18:2n6) by 40%, and DHA (22:6n3) by 33% compared to controls. PGR also increased the ARA/DHA ratio by 40% in male rat pups. In contrast, in female rat pups, PGR on a regular diet did not alter levels of any serum fatty acid.

Table 2. Serum Fatty acids.

	Male				Female			
Fatty Acid (μg/mL ± SD)	Control	PGR	PGR +LoDHA	PGR +HiDHA	Control	PGR	PGR +LoDHA	PGR +HiDHA
Saturated								
Palmitic Acid (16:0)	537 ± 36	485 ± 142	775 ± 192 *†	699 ± 91 †	633 ± 117	674 ± 133	676 ± 51	700 ± 81
Stearic Acid (18:0)	458 ± 162	268 ± 106 *	436 ± 113	352 ± 28	402 ± 158	477 ± 139	376 ± 60	380 ± 65
Monounsaturated								
Palmitoleic Acid (16:1)	34 ± 8	28 ± 7	14 ± 5 *†	18 ± 3 *†	44 ± 19	45 ± 20	11 ± 2 *†	21 ± 2 *†
Oleic Acid (18:1)	421 ± 171	240 ± 89 *	356 ± 94	392 ± 89	475 ± 182	561 ± 198	321 ± 62 †	386 ± 46
ω-6								
Linoleic Acid (18:2n6)	1068 ± 320	642 ± 228 *	1261 ± 192 †	1346 ± 261 †	1047 ± 350	1258 ± 224	1135 ± 181	1191 ± 226
Arachidonic Acid (20:4n6)	1335 ± 439	1187 ± 972	2153 ± 445 †	2367 ± 226 *†	1337 ± 498	1631± 658	1835 ± 542	2239 ± 508
ω-3								
Docosahexaenoic Acid (22:6n3)	142 ± 31	95 ± 44 *	228 ± 70 †	589 ± 123 *†‡	145 ± 50	143 ± 62.0	175 ± 46	524 ± 96 *†‡
LA/DHA	8 ± 3	10 ± 6	6 ± 2	2 ± 0.1 *†‡	8 ± 2	10 ± 4	7 ± 2	2 ± 0.2 *†‡
ARA/DHA	9 ± 1	12 ± 2 *	10 ± 1 †	4 ± 1 *†‡	9 ± 1	12 ± 4	10 ± 1	4 ± 0.3 *†‡

* $p \leq 0.05$ compared to sex-matched control, † $p \leq 0.05$ compared to sex-matched PGR with regular diet, ‡ $p \leq 0.05$ compared to sex-matched PGR with low-dose DHA diet.

Consistent with the experimental design, serum DHA was increased by maternal DHA supplementation in PGR rat pups. In male PGR rat pups, the LoDHA diet increased serum DHA by 140% compared to PGR on a regular diet, while the HiDHA diet increased serum DHA by 520% compared to PGR on a regular diet and 315% compared to controls. In female PGR rat pups, the HiDHA diet increased serum DHA by 370% compared to PGR on a regular diet and 360% compared to controls.

DHA supplementation of PGR rat pups also significantly altered serum levels of other fatty acids. In male PGR rat pups, the Lodha diet normalized levels of oleic acid (18:1), linoleic acid (18:2n6), and the ARA/DHA ratio, while serum levels of palmitoleic acid (16:1) decreased by 60% compared to controls, and palmitic acid (16:0) increased by 45% compared to controls. In female PGR rat pups, the Lodha diet also decreased levels of palmitoleic acid (16:1) by 75% compared to controls. In male PGR rat pups, the HiDHA diet increased serum levels of ARA by 77% compared to controls while also decreasing the LA/DHA ratio by 75% and the ARA/DHA ratio by 56% compared to controls. Similarly, in female PGR rat pups, the HiDHA diet decreased the LA/DHA ratio by 75% and the ARA/DHA ratio by 55%.

3.3. Lung Mechanics

Using closed chest pressure–volume loops to determine quasi-static lung compliance and the forced oscillation technique to differentiate between central and peripheral lung tissue, we measured lung mechanics in all four groups (control, PGR, PGR + LoDHA, PGR + HiDHA).

Similar to our previous report [19], PGR altered lung mechanics in male and female rat pups. In male rat pups, PGR on a regular diet decreased inspiratory capacity by 12%, decreased static compliance by 46%, and increased tissue elastance by 85% and tissue damping by 66% compared to controls (Figure 1A). In female rat pups, PGR increased inspiratory capacity by 11%, decreased static compliance by 39%, and increased tissue damping by 52% compared to controls (Figure 1B).

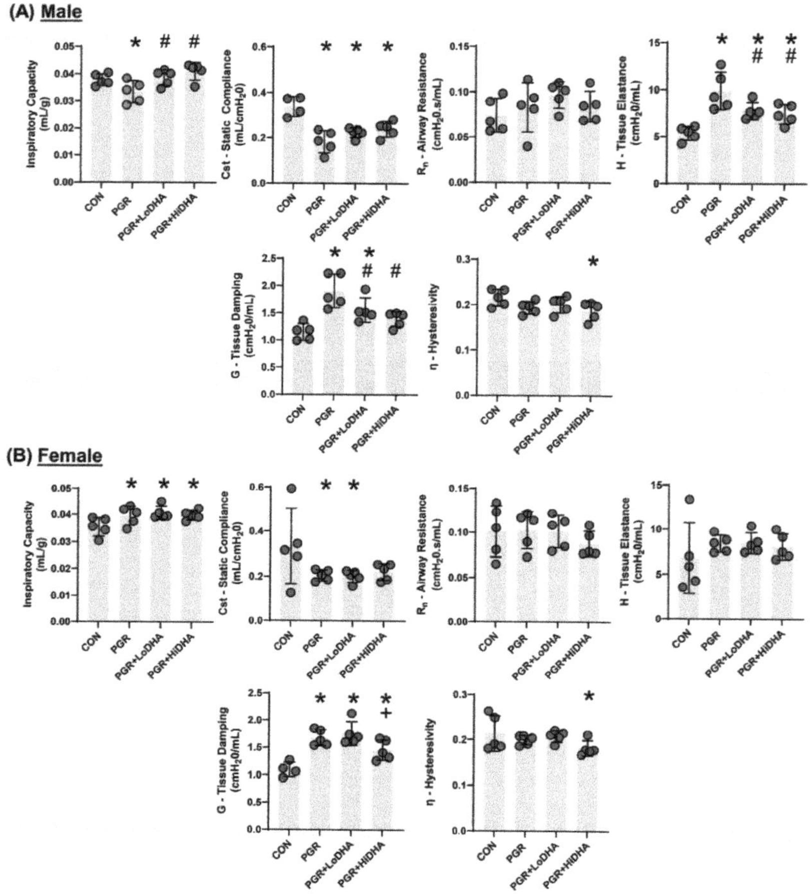

Figure 1. PGR with and without DHA supplementation alters rat lung mechanics. (**A**) Lung function measurements in male and (**B**) female rat pups. Data points are individual rat pups, bars are mean with SD. * $p \leq 0.05$ compared to sex-matched control, # $p \leq 0.05$ compared to PGR, + $p \leq 0.05$ compared to PGR + LoDHA.

DHA supplementation of PGR rat pups did not restore lung function to that of controls in either sex. In male PGR rat pups, while inspiratory capacity was normalized, static lung compliance remained reduced (34% and 30% decrease compared to controls with LoDHA and Hi DHA diets, respectively), and tissue elastance remained increased (45% and 38% increase compared to controls with LoDHA and HiDHA diets, respectively). Tissue damping remained increased male PGR rat pups with LoDHA diet (35% increase relative to control), but was normalized with the HiDHA diet. Additionally, the HiDHA diet decreased lung hysterestivity by 14% compared to the control (Figure 1A). In female PGR rat pups, inspiratory capacity remained increased (15% and 12% increase compared to controls with LoDHA and Hi DHA diets, respectively) and static lung compliance remained reduced with the LoDHA diet (40% compared to controls) but was normalized with the HiDHA diet. Tissue damping remained increased female PGR rat pups on both diets (59% and 32% increase compared to controls with LoDHA and HiDHA diets, respectively). Similarly to male PGR rat pups, in female PGR rat pups, the HiDHA diet decreased lung hysterestivity by 15% compared to the control (Figure 1B).

Overall, the effects of PGR on a regular diet were consistent with our previously published data [19]. The addition of DHA to PGR rat pups resulted in some sex-divergent improvements in lung function. Improvements included, in male rat pups, normalized IC, improved tissue elastance, and tissue damping, with normalization at high DHA levels. In female rat pups, static compliance was normalized at the high DHA dose. However, significant lung function deficits remain in both male and female PGR rat pups on a DHA diet.

3.4. *PPAR$\gamma\Delta$5 in the Lung*

To determine whether *Ppar$\gamma\Delta$5* is expressed in the lung, we performed PCR on cDNA isolated from rat lungs, using primers contained within exon 4 (forward) and exon 6 (reverse). Gel electrophoresis of our PCR product produced two bands, one measuring approximately 850 bp and the other measuring approximately 450 bp, consistent with the estimated molecular weights of the targeted amplicons of full-length *Pparγ* and the *Ppar$\gamma\Delta$5* splice variant, respectively (Figure 2B). Sequencing confirmed that the 850 bp band amplified *Pparγ* containing exon 5, and the 450 bp band amplified the *Ppar$\gamma\Delta$5* variant with exon four spliced directly to exon 6 (Figure 2C).

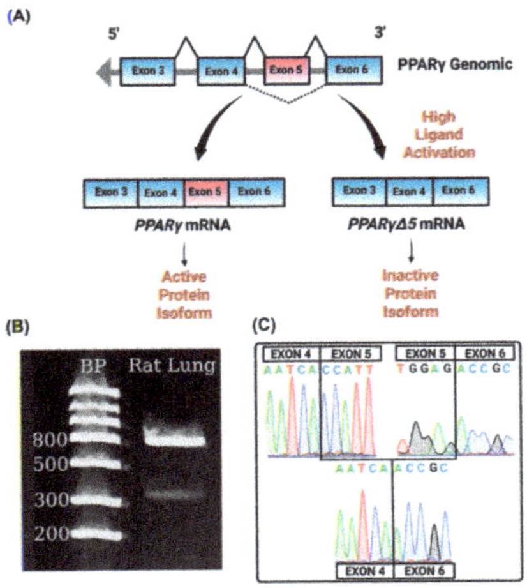

Figure 2. *Ppar$\gamma\Delta$5* splice variant is expressed in the lung. (**A**) Schematic of *Pparγ* alternative splicing, (**B**) gel electrophoresis of PCR across exon 4 to 6. Products are full-length *Pparγ* (850 bp) and *Ppar$\gamma\Delta$5* splice variant in lung tissue (450 bp), (**C**) corresponding sequencing results of the upper and lower bands.

3.5. *PPARγ mRNA and Protein Quantification*

We quantified levels of *Pparγ* and *Ppar$\gamma\Delta$5* mRNA and PPARγ and PPARγΔ5 protein in all study groups. In male rat lung, PGR with a regular diet decreased *Pparγ* mRNA by 39% compared to control but did not alter *Ppar$\gamma\Delta$5* mRNA (Figure 3A). The *Pparγ/Ppar$\gamma\Delta$5* ratio was decreased in male the PGR rat lung on a regular diet by 35% relative to control (Figure 3A). In male rat lung, PGR with a regular diet decreased PPARγ protein levels by 30% compared to control, without affecting protein levels of PPARγΔ5. Similarly to mRNA findings, in the male rat lung, PGR with a regular diet decreased the PPARγ/PPARγΔ5 ratio by 26% compared to control (Figure 3B,C). In the female rat lung, PGR with a regular diet did not affect mRNA levels of *Pparγ* or *PparγΔ5* (Figure 3D). However, the resulting *Pparγ/PparγΔ5* ratio was increased in the female PGR rat lung on a regular diet by 65% compared to control (Figure 3D). In the female rat lung, PGR with a regular diet decreased PPARγ protein levels by 24% compared to control, without affecting protein levels of PPARγΔ5. In a female rat lung, PGR with a regular diet decreased the PPARγ/PPARγΔ5 ratio by 27% compared to control (Figure 3E,F).

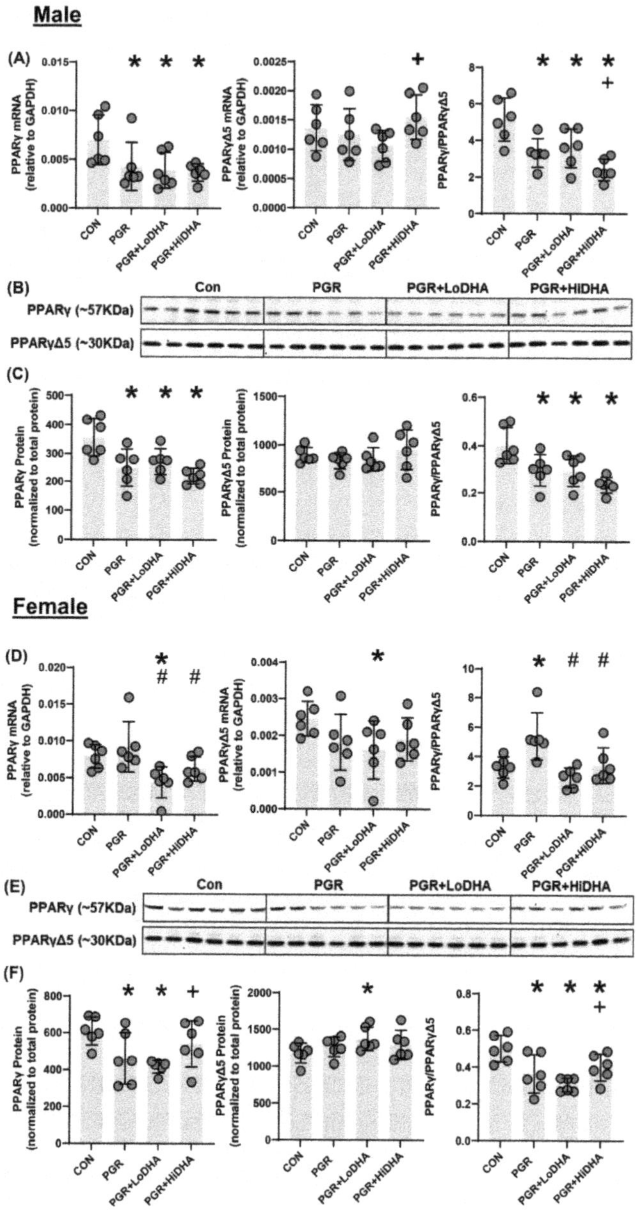

Figure 3. PGR with and without DHA supplementation alters expression of PPARγ mRNA variants and protein isoforms in the lung. (**A**) Male rat lung Pparγ, mRNA PparγΔ5 mRNA, and Pparγ/PparγΔ5

mRNA ratio. (**B**) Western blot images of male rat lung PPARγ and PPARγΔ5, (**C**) Male rat lung protein levels of PPARγ and PPARγΔ5 and the resulting ratio (normalized to total protein). (**D**) Female rat lung Pparγ mRNA, PparγΔ5 mRNA, and Pparγ/PparγΔ5 mRNA ratio. (**E**) Western blot images of female rat lung PPARγ and PPARγΔ5. (**F**) Female rat lung protein levels of PPARγ and PPARγΔ5 and the resulting ratio (normalized to total protein). Data points are individual rat pups, bars are mean with SD. * $p \leq 0.05$ compared to sex-matched control, # $p \leq 0.05$ compared to PGR, + $p \leq 0.05$ compared to PGR + LoDHA.

In male PGR rat lung, *Pparγ* mRNA levels remained decreased with both DHA diets (44% and 46% compared to control for LoDHA and HiDHA, respectively), and *PparγΔ5* mRNA remained unchanged (Figure 3A). Similarly, in male PGR rat lung, PPARγ protein levels remained decreased with both DHA diets (24% and 38% compared to control for LoDHA and HiDHA, respectively), and PPARγΔ5 protein levels were unaffected (Figure 3B,C). The resulting PPARγ/PPARγΔ5 protein ratio also remained reduced in male PGR rat lung with both DHA diets (27% and 41% compared to control for LoDHA and HiDHA, respectively).

In female PGR rat lung, *Pparγ* and *PparγΔ5* mRNA levels decreased with the LoDHA diet (44% and 34%, respectively, compared to control) (Figure 3D). In the female PGR rat lung, PPARγ protein levels decreased with the LoDHA diet by 32% compared to control, while PPARγΔ5 protein levels increased by 17% relative to the control (Figure 3E,F). The resulting PPARγ/PPARγΔ5 protein ratio, however, remained reduced in female PGR rat lung with both DHA diets (39% and 20% compared to control for LoDHA and HiDHA, respectively).

Overall, DHA supplementation did not normalize any PPARγ measure in the male PGR rat lung but did normalize PPARγ mRNA and PPARγ protein at a high dose in female rat pups. However, the PPARγ/PPARγΔ5 ratio was not normalized in the female rat lung.

3.6. *mRNA of Downstream PPARγ Target Gene Plin2*

To assess PPARγ activity, we measured mRNA levels of the PPARγ target gene Perilipin-2 (*Plin2*). In the male rat lung, PGR with a regular diet decreased *Plin2* mRNA levels by 32% compared to controls (Figure 4A). In the female rat lung, PGR with a regular diet decreased *Plin2* mRNA levels by 27% compared to controls (Figure 4B). In the male rat PGR lung, *Plin2* mRNA remained decreased with both DHA diets (40% and 56% compared to control for LoDHA and HiDHA, respectively) (Figure 4A). In the female rat PGR lung, *Plin2* mRNA also remained decreased with both DHA diets (66% and 37% compared to control for LoDHA and HiDHA, respectively) (Figure 4B).

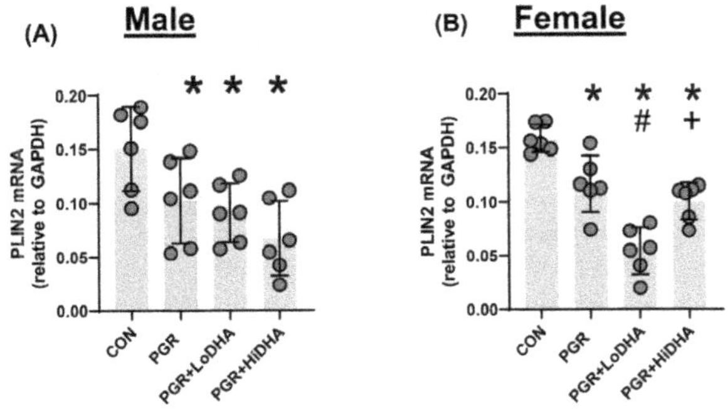

Figure 4. PGR and DHA supplementation decreases the expression of PPARγ target gene Plin2. (**A**) Male rat lung Plin2 mRNA levels. (**B**) Female rat lung Plin2 mRNA levels. * $p \leq 0.05$ compared to sex-matched control, # $p \leq 0.05$ compared to PGR, + $p \leq 0.05$ compared to PGR + LoDHA.

Overall, the addition of DHA to PGR rat pups did not normalize lung mRNA levels of the PPARγ target gene Plin2 in either male or female rat lung.

4. Discussion

The development of bronchopulmonary dysplasia is multifactorial, and the molecular drivers and prevention of this disease remain poorly understood. Given the confounding data demonstrating the importance of DHA in lung development and prevention of BPD, combined with the recent clinical trials demonstrating that DHA supplementation of preterm neonates may increase BPD, a comprehensive understanding of the molecular effects of DHA supplementation in the developing lung is required. Using a combined model of postnatal growth restriction and DHA supplementation, we show that postnatal growth restriction in the absence of in utero growth restriction, or additional lung injury, alters circulating fatty acids and impairs lung mechanics in rats. We also show that these effects are not normalized by DHA supplementation. We also demonstrate, for the first time, that a dominant negative splice variant of PPARγ, a key regulator of lung development, is expressed in the developing lung. We show that PPARγ activity is decreased in PGR and not normalized with DHA supplementation. Collectively, our data support the view that postnatal DHA supplementation does not normalize lung outcomes in PGR.

The negative impact of nutritional deficits in the development of BPD is well accepted [1,23–25]. In utero growth restriction and lower birth weight are associated with higher rates of BPD and worse lung outcomes in human neonates [26–28]. Even when prenatal growth is not impaired, postnatal growth deficits have been linked to elevated BPD risk and adverse pulmonary outcomes [29–31]. Postnatally, poor nutrition in human preterm infants occurs secondary to feeding intolerance and clinically indicated feeding volume restriction. In this study, we focused on PGR in the context of normal in utero growth and the absence of additional lung injury, such as hyperoxia. We found that PGR on a regular diet significantly impaired lung function parameters in both male and female rat pups. The addition of DHA to PGR pups resulted in improvements in some lung function parameters, and no change in others. In addition, these effects were sex-divergent. For example, in male PGR rats, DHA supplementation normalized tissue damping without improving static compliance, suggesting that DHA had beneficial effects on dynamic aspects of lung function without changing the overall lung structure or elasticity at total lung capacity. One possible mechanism for improved dynamic lung function by DHA is via effects on surfactant production, as improved surfactant production following PPARγ activation has been demonstrated in other models of lung injury [32,33]. In contrast, the lack of improvement in static lung compliance by DHA in PGR rats may reflect a structural alteration in the lung caused by PGR, which is not improved by DHA. We previously demonstrated that PGR on a regular diet impairs alveolar formation with a concomitant increase in static compliance, parameters consistent with failed alveolar formation in BPD [19,34]. Examination of DHA administration to PGR rat pups at an earlier postnatal time point, i.e., during the saccular-to-alveolar transition, is an important next step in determining whether the timing of DHA administration increases favorable outcomes. Sex-divergent outcomes in lung parameters in human neonates are also well accepted, with males generally at a risk of higher incidence and greater severity of BPD. A complete understanding of sex-divergent outcomes in human neonates remains elusive [35]. However, studies suggest that chromosomal sex, as opposed to gonadal sex, is likely a driving force [36].

Recent clinical trials examining the effects of DHA supplementation of preterm neonates used either direct enteral supplementation of DHA to preterm neonates (N-3 fatty acids for improvement in respiratory outcomes (N3RO) trial), or DHA administration via maternal DHA supplementation and the provision of breastmilk to preterm neonates

(The Maternal Omega-3 Supplementation to Reduce Bronchopulmonary Dysplasia in Very Preterm Infants (MOBYDIck) trial) [8,9]. The N3RO trial ultimately concluded that neonatal DHA supplementation did not improve the overall incidence of BPD and may increase the incidence of BPD in neonates born at less than 27 weeks. Considering these results from the N3RO trial, an early interim analysis of the MOBYDIck trial was performed. The analysis favored the placebo over DHA in BPD-free survival, and the study was terminated due to concern for potential harm to future participants. These outcomes were surprising given the dearth of information demonstrating the importance of DHA in lung development and response to injury, and several rationales for these outcomes and currently being explored. One potential contributor to the negative effects of isolated DHA supplementation is the effect of increased DHA on other circulating fatty acids, including the omega-6 fatty acid ARA.

In the N3RO trial, blood levels of ARA were decreased in the DHA group, as was the ARA/DHA ratio [8]. In the MOBYDIck trial, blood lipids were not assessed. However, the average lipid composition of the maternal milk used in the trial was analyzed at two weeks [37]. Findings from this analysis showed that while maternal DHA supplementation increased milk DHA levels, supplementation decreased the ARA/DHA ratio [37]. The potential importance of including ARA in supplementation is highlighted by other studies that used supplementation with a combination of ARA and DHA. Collectively, the studies that utilized a combination of ARA and DHA (at a 2:1 ratio) did not report negative effects on BPD outcomes [38,39]. The changes to the ω3-to-ω6 ratio and to arachidonic acid levels are especially notable, given the role ARA plays in modulating the inflammatory cascade. In our study, we examined two doses of DHA administered via the maternal diet. Our low-dose DHA diet contained 0.01% DHA, which increased serum DHA levels by approximately 1% of total lipids in male and female PGR pups compared to those on a regular diet. Our low DHA diet increased serum DHA to a similar degree as the 60 mg/kg dose of DHA used in the N3RO trial [8]. Our high dose DHA diet contained 0.1% DHA, which increased serum DHA by approximately 6.5% of total fatty acids. In our study, DHA supplementation also significantly reduced the ARA/DHA ratio in both male and female rat pups, again with the low dose DHA diet similar to that of the N3RO trial.

An important consideration in the context of DHA supplementation, and disturbed circulating fatty acid profiles during lung development, is the transcriptional activator, PPARγ. PPARγ is required for appropriate lung development, particularly the formation of alveoli. Conditional knockout of PPARγ in mouse airway epithelial cells leads to airspace enlargement and disruption of epithelial–mesenchymal interactions [10]. PPARγ activation also promotes myofibroblast differentiation to a lipid-laden lipofibroblast phenotype in the developing lung, characterized by high triglyceride levels, which has a significant role in lung maturation, including alveolar maturation [40–42]. We and others have previously shown that neonatal rat pups exposed to fetal growth restriction or postnatal hyperoxia have decreased PPARγ expression and impaired alveolar formation. In both cases, lung phenotypes are reversed by PPARγ activation [4,16]. An important consideration, however, is that to achieve greater regulation, transcriptional activators also undergo alternative splicing and express dominant negative variants [43].

PPARγ produces a dominant negative isoform, PPARγΔ5, via the alternative splicing of exon 5 [17]. Studies in adipose tissue have shown that expression of PPARγΔ5 is increased by ligand activation of PPARγ. The protein isoform produced by alternative splicing of PPARγ exon 5 results in a premature stop codon and truncated protein isoform lacking the ligand binding domain and one transactivation domain [17]. The truncation also results in the removal of a ubiquitin binding domain, thus potentially making the truncated protein more stable than the native isoform [44]. In adipose tissue, a decrease in the cellular

PPARγ/PPARγΔ5 ratio reduces transcription of downstream target genes. In this study, we demonstrate, for the first time, that the PPARγΔ5 variant is expressed in the rat lung under control conditions. Given that DHA can act as a ligand for PPARγ, we considered whether supplemental DHA would increase the expression of PPARγΔ5 in the lung [45]. Our data show that supplemental DHA has a limited effect on the levels of PPARγΔ5 mRNA or protein in the rat lung. However, given that the PPARγ/PPARγΔ5 ratio conveys overall functional potential, this may be a more appropriate metric. We demonstrated that PGR alone decreases the PPARγ/PPARγΔ5 ratio in the lung in both male and female rat pups, and that this decrease is associated with a decrease in the mRNA levels of the target gene Plin2 (also known as adipose differentiation-related protein), another critical player in lung development [46]. Reduced PPARγ activity and expression of Plin2 are both associated with impaired lung development and the structural changes consistent with the altered lung function measures we detected. The addition of DHA to PGR rats in this study did not completely normalize PGR-induced impaired lung function, nor did the diets significantly alter the PGR-induced reduction in PPARγ activity. An important caveat for our study is the timing of rat lung development. At our study time point, postnatal day 21, the rat lung has completed bulk alveolar formation. It will be important in subsequent studies to assess the effects of DHA on the rat lung during the transition from the saccular to alveolar stages. Also important will be studies aimed at determining the precise mechanisms by which PGR decreases the PPARγ/PPARγΔ5 ratio, and the cause-and-effect consequences of this change on lung development.

Our study is not without limitations. In this study, we only examined circulating fatty acid profiles. An understanding of how PGR and DHA supplementation affect intrapulmonary fatty acid profiles is needed. An advantage of understanding intrapulmonary fatty acid profiles and circulating fatty acid profiles in the same system is the potential for the identification of fatty acid biomarkers that may be useful in identifying neonates with detrimental fatty acid complements. It will also be important in future studies to determine the expression of PPARγΔ5 in human lung cells. Other studies have identified the expression of PPARγΔ5 in human tissue [17], but to date, not in the human lung. We also have not explored the origins of the sex differences observed in our study. Several potential interactions may contribute to the differences in sex outcomes in the lung, a topic that should be addressed in future studies. Lastly, an analysis assessing correlations between PPARγ activity and lung function parameters would be informative to understand the overall effects of DHA supplementation of PGR rat pups. However, as a correlation analysis requires both variables to be assed in the same experimental subject, we were not able to perform this analysis. Because the protocol for lung function experiments requires administration of vecuronium bromide to the rats, we opted not to collect lung tissue for molecular experiments following the function studies. Therefore, our molecular measurements and our lung function experiments were performed in different rats and, thus, not appropriate for correlation analysis.

5. Conclusions

In conclusion, we demonstrate that the novel splice variant of PPARγ, PPARγΔ5, is expressed in the lung, and that PGR reduces the PPARγ/PPARγΔ5 ratio in association with impaired lung mechanics. DHA supplementation at and beyond clinically relevant doses improved some lung functional and molecular parameters. However, neither dose of DHA normalized all functional or molecular profiles of PGR rat pups. DHA supplementation did, however, disturb circulating fatty acid profiles for other LCPUFAs. Ongoing examination of the timing of DHA supplementation in the developing rat lung in the context of PGR is an important next step.

Supplementary Materials: The following supporting information can be downloaded at: https://www.mdpi.com/article/10.3390/biom15040551/s1, Western Blot uncropped images.

Author Contributions: Conceptualization, L.A.J.-M., A.J.C. and C.R.M.; methodology, L.A.J.-M., J.A.M. and J.E.C.; validation, L.A.J.-M., J.A.M. and A.J.C.; formal analysis, L.A.J.-M., A.J.C. and J.A.M.; investigation, A.J.C., W.R.C., D.T.W., N.J., A.J., C.B., J.Z. and J.A.M.; writing—original draft preparation, L.A.J.-M.; writing—review and editing, L.A.J.-M., C.R.M., J.A.M. and A.J.C.; funding acquisition, L.A.J.-M. All authors have read and agreed to the published version of the manuscript.

Funding: This study was supported in part by National Institutes of Health (NIH) DK084036 (LJM) and 1-S10-OD021505-01 (JEC), and the Division of Neonatology, Department of Pediatrics at the University of Utah. AJ and CB were supported by the University of Utah Native American Internship (R25HL108828).

Institutional Review Board Statement: The animal study protocol was approved by the Institutional Review Board (or Ethics Committee) of the University of Utah (protocol number 21-09019/1827, approval 30 November 2023).

Informed Consent statement: Not applicable.

Data Availability Statement: The raw data supporting the conclusions of this article will be made available by the authors on request.

Acknowledgments: We are grateful to the Division of Neonatology and the Department of Pediatrics at the University of Utah for support, and to Haimei Wang for managing the animal work used in this study.

Conflicts of Interest: The authors declare no conflicts of interest. The funders had no role in the design of the study; in the collection, analyses, or interpretation of data; in the writing of the manuscript; or in the decision to publish the results.

Abbreviations

The following abbreviations are used in this manuscript:

BPD	Bronchopulmonary Dysplasia
DHA	Docosahexaenoic Acid
PPARγ	Peroxisome Proliferator Activated Receptor gamma
PGR	Postnatal Growth Restriction
ARA	Arachidonic Acid
PLIN2	Perilipin 2

References

1. Poindexter, B.B.; Martin, C.R. Impact of Nutrition on Bronchopulmonary Dysplasia. *Clin. Perinatol.* **2015**, *42*, 797–806. [CrossRef] [PubMed]
2. Martin, C.R.; DaSilva, D.A.; Cluette-Brown, J.E.; DiMonda, C.; Hamill, A.; Bhutta, A.Q.; Coronel, E.; Wilschanski, M.; Stephens, A.J.; Driscoll, D.F.; et al. Decreased postnatal docosahexaenoic and arachidonic acid blood levels in premature infants are associated with neonatal morbidities. *J. Pediatr.* **2011**, *159*, e1–e2. [CrossRef] [PubMed]
3. Makrides, M.; Gibson, R.A.; McPhee, A.J.; Collins, C.T.; Davis, P.G.; Doyle, L.W.; Simmer, K.; Colditz, P.B.; Morris, S.; Smithers, L.G.; et al. Neurodevelopmental Outcomes of Preterm Infants Fed High-Dose Docosahexaenoic Acid. *JAMA* **2009**, *301*, 175–182. [CrossRef] [PubMed]
4. Joss-Moore, L.A.; Wang, Y.; Baack, M.L.; Yao, J.; Norris, A.W.; Yu, X.; Callaway, C.W.; McKnight, R.A.; Albertine, K.H.; Lane, R.H. IUGR decreases PPARγ and SETD8 Expression in neonatal rat lung and these effects are ameliorated by maternal DHA supplementation. *Early Hum. Dev.* **2010**, *86*, 785–791. [CrossRef]
5. Rogers, L.K.; Valentine, C.J.; Pennell, M.; Velten, M.; Britt, R.D.; Dingess, K.; Zhao, X.; Welty, S.E.; Tipple, T.E. Maternal docosahexaenoic acid supplementation decreases lung inflammation in hyperoxia-exposed newborn mice. *J. Nutr.* **2011**, *141*, 214–222. [CrossRef]
6. Ali, M.; Heyob, K.M.; Velten, M.; Tipple, T.E.; Rogers, L.K. DHA suppresses chronic apoptosis in the lung caused by perinatal inflammation. *Am. J. Physiol. Cell. Mol. Physiol.* **2015**, *309*, L441–L448. [CrossRef]

7. Velten, M.; Britt, R.D.; Heyob, K.M.; Tipple, T.E.; Rogers, L.K. Maternal dietary docosahexaenoic acid supplementation attenuates fetal growth restriction and enhances pulmonary function in a newborn mouse model of perinatal inflammation. *J. Nutr.* **2014**, *144*, 258–266. [CrossRef]
8. Collins, C.T.; Makrides, M.; McPhee, A.J.; Sullivan, T.R.; Davis, P.G.; Thio, M.; Simmer, K.; Rajadurai, V.S.; Travadi, J.; Berry, M.J.; et al. Docosahexaenoic Acid and Bronchopulmonary Dysplasia in Preterm Infants. *N. Engl. J. Med.* **2017**, *376*, 1245–1255. [CrossRef]
9. Marc, I.; Piedboeuf, B.; Lacaze-Masmonteil, T.; Fraser, W.; Mâsse, B.; Mohamed, I.; Qureshi, M.; Afifi, J.; Lemyre, B.; Caouette, G.; et al. Effect of Maternal Docosahexaenoic Acid Supplementation on Bronchopulmonary Dysplasia-Free Survival in Breastfed Preterm Infants: A Randomized Clinical Trial. *JAMA* **2020**, *324*, 157–167. [CrossRef]
10. Simon, D.M.; Arikan, M.C.; Srisuma, S.; Bhattacharya, S.; Tsai, L.W.; Ingenito, E.P.; Gonzalez, F.; Shapiro, S.D.; Mariani, T.J. Epithelial cell PPARγ contributes to normal lung maturation. *FASEB J.* **2006**, *20*, 1507–1509. [CrossRef]
11. Cerny, L.; Torday, J.S.; Rehan, V.K. Prevention and treatment of bronchopulmonary dysplasia: Contemporary status and future outlook. *Lung* **2008**, *186*, 75–89. [CrossRef] [PubMed]
12. Gien, J.; Tseng, N.; Seedorf, G.; Roe, G.; Abman, S.H. Abman Peroxisome proliferator activated receptor-gamma-Rho-kinase interactions contribute to vascular remodeling after chronic intrauterine pulmonary hypertension. *Am. J. Physiol. Lung Cell Mol. Physiol.* **2014**, *306*, L299–L308. [CrossRef] [PubMed]
13. Kulkarni, A.A.; Woeller, C.F.; Thatcher, T.H.; Ramon, S.; Phipps, R.P.; Sime, P.J. Sime Emerging PPARgamma-Independent Role of PPARgamma Ligands in Lung Diseases. *PPAR Res.* **2012**, *2012*, 705352. [CrossRef]
14. Hagood, J.S. Beyond the genome: Epigenetic mechanisms in lung remodeling. *Physiology* **2014**, *29*, 177–185. [CrossRef]
15. Morales, E.; Sakurai, R.; Husain, S.; Paek, D.; Gong, M.; Ibe, B.; Li, Y.; Husain, M.; Torday, J.S.; Rehan, V.K. Nebulized PPARγ agonists: A novel approach to augment neonatal lung maturation and injury repair in rats. *Pediatr. Res.* **2014**, *75*, 631–640. [CrossRef]
16. Rehan, V.K.; Wang, Y.; Patel, S.; Santos, J.; Torday, J.S. Rosiglitazone, a peroxisome proliferator-activated receptor-γ agonist, prevents hyperoxia-induced neonatal rat lung injury in vivo. *Pediatr. Pulmonol.* **2006**, *41*, 558–569. [CrossRef]
17. Aprile, M.; Cataldi, S.; Ambrosio, M.R.; D'esposito, V.; Lim, K.; Dietrich, A.; Blüher, M.; Savage, D.B.; Formisano, P.; Ciccodicola, A.; et al. PPARγΔ5, a Naturally Occurring Dominant-Negative Splice Isoform, Impairs PPARγ Function and Adipocyte Differentiation. *Cell Rep.* **2018**, *25*, 1577–1592.e6. [CrossRef]
18. Shi, C.-Y.; Xu, J.-J.; Li, C.; Yu, J.-L.; Wu, Y.-T.; Huang, H.-F. A PPARG Splice Variant in Granulosa Cells Is Associated with Polycystic Ovary Syndrome. *J. Clin. Med.* **2022**, *11*, 7285. [CrossRef]
19. Zhao, J.; Ballard, C.; Cohen, A.J.; Ringham, B.; Zhao, B.; Wang, H.; Zuspan, K.; Rebentisch, A.; Locklear, B.A.; Dahl, M.; et al. Postnatal growth restriction impairs rat lung structure and function. *Anat. Rec.* **2025**, *308*, 1051–1065. [CrossRef]
20. Weinheimer, C.; Wang, H.; Comstock, J.M.; Singh, P.; Wang, Z.; Locklear, B.A.; Goodwin, K.L.; Maschek, J.A.; Cox, J.E.; Baack, M.L.; et al. Maternal Tobacco Smoke Exposure Causes Sex-Divergent Changes in Placental Lipid Metabolism in the Rat. *Reprod. Sci.* **2020**, *27*, 631–643. [CrossRef]
21. McGovern, T.K.; Robichaud, A.; Fereydoonzad, L.; Schuessler, T.F.; Martin, J.G. Evaluation of Respiratory System Mechanics in Mice using the Forced Oscillation Technique. *J. Vis. Exp.* **2013**, *75*, e50172. [CrossRef]
22. Joss-Moore, L.; Carroll, T.; Yang, Y.; Fitzhugh, M.; Metcalfe, D.; Oman, J.; Hale, M.; Dong, L.; Wang, Z.-M.; Yu, X.; et al. Intrauterine growth restriction transiently delays alveolar formation and disrupts retinoic acid receptor expression in the lung of female rat pups. *Pediatr. Res.* **2013**, *73*, 612–620. [CrossRef]
23. Frank, L.; Sosenko, I.R. Undernutrition as a major contributing factor in the pathogenesis of bronchopulmonary dysplasia. *Am. Rev. Respir. Dis.* **1988**, *138*, 725–729. [CrossRef]
24. Ehrenkranz, R.A.; Das, A.; Wrage, L.A.; Poindexter, B.B.; Higgins, R.D.; Stoll, B.J.; Oh, W. Early nutrition mediates the influence of severity of illness on extremely LBW infants. *Pediatr. Res.* **2011**, *69*, 522–529. [CrossRef]
25. Morrow, L.A.; Wagner, B.D.; Ingram, D.A.; Poindexter, B.B.; Schibler, K.; Cotten, C.M.; Dagle, J.; Sontag, M.K.; Mourani, P.M.; Abman, S.H. Antenatal Determinants of Bronchopulmonary Dysplasia and Late Respiratory Disease in Preterm Infants. *Am. J. Respir. Crit. Care Med.* **2017**, *196*, 364–374. [CrossRef]
26. Reiss, I.; Landmann, E.; Heckmann, M.; Misselwitz, B.; Gortner, L. Increased risk of bronchopulmonary dysplasia and increased mortality in very preterm infants being small for gestational age. *Arch. Gynecol. Obstet.* **2003**, *269*, 40–44. [CrossRef]
27. Bose, C.; Van Marter, L.J.; Laughon, M.; O'Shea, T.M.; Allred, E.N.; Karna, P.; Ehrenkranz, R.A.; Boggess, K.; Leviton, A. Fetal growth restriction and chronic lung disease among infants born before the 28th week of gestation. *Pediatrics* **2009**, *124*, e450–e458. [CrossRef]
28. Ambalavanan, N.; Van Meurs, K.P.; Perritt, R.; Carlo, W.A.; Ehrenkranz, R.A.; Stevenson, D.K.; Lemons, J.A.; Poole, W.K.; Higgins, R.D. Predictors of death or bronchopulmonary dysplasia in preterm infants with respiratory failure. *J. Perinatol.* **2008**, *28*, 420–426. [CrossRef]

29. Ehrenkranz, R.A.; Dusick, A.M.; Vohr, B.R.; Wright, L.L.; Wrage, L.A.; Poole, W.K. Growth in the neonatal intensive care unit influences neurodevelopmental and growth outcomes of extremely low birth weight infants. *Pediatrics* **2006**, *117*, 1253–1261. [CrossRef]
30. Khan, M.A.; Kuzma-O'Reilly, B.; Brodsky, N.L.; Bhandari, V. Site-specific characteristics of infants developing bronchopulmonary dysplasia. *J. Perinatol.* **2006**, *26*, 428–435. [CrossRef]
31. Figueras-Aloy, J.; Palet-Trujols, C.; Matas-Barceló, I.; Botet-Mussons, F.; Carbonell-Estrany, X. Extrauterine growth restriction in very preterm infant: Etiology, diagnosis, and 2-year follow-up. *Eur. J. Pediatr.* **2020**, *179*, 1469–1479. [CrossRef] [PubMed]
32. Takahashi, A.; Bartolák-Suki, E.; Majumdar, A.; Suki, B. Changes in respiratory elastance after deep inspirations reflect surface film functionality in mice with acute lung injury. *J. Appl. Physiol.* **2015**, *119*, 258–265. [CrossRef] [PubMed]
33. Smith, L.C.; Gow, A.J.; Abramova, E.; Vayas, K.; Guo, C.; Noto, J.; Lyman, J.; Rodriquez, J.; Gelfand-Titiyevskiy, B.; Malcolm, C.; et al. Role of PPARγ in dyslipidemia and altered pulmonary functioning in mice following ozone exposure. *Toxicol. Sci.* **2023**, *194*, 109–119. [CrossRef] [PubMed]
34. Albertine, K.H. Progress in understanding the pathogenesis of BPD using the baboon and sheep models. *Semin. Perinatol.* **2013**, *37*, 60–68. [CrossRef]
35. Cantu, A.; Gutierrez, M.C.; Dong, X.; Leek, C.; Sajti, E.; Lingappan, K. Remarkable sex-specific differences at single-cell resolution in neonatal hyperoxic lung injury. *Am. J. Physiol. Lung Cell Mol. Physiol.* **2023**, *324*, L5–L31. [CrossRef]
36. Grimm, S.L.; Dong, X.; Zhang, Y.; Carisey, A.F.; Arnold, A.P.; Moorthy, B.; Coarfa, C.; Lingappan, K. Effect of sex chromosomes versus hormones in neonatal lung injury. *JCI Insight.* **2021**, *6*, e146863. [CrossRef]
37. Fougère, H.; Bilodeau, J.-F.; Lavoie, P.M.; Mohamed, I.; Rudkowska, I.; Pronovost, E.; Simonyan, D.; Berthiaume, L.; Guillot, M.; Piedboeuf, B.; et al. Docosahexaenoic acid-rich algae oil supplementation on breast milk fatty acid profile of mothers who delivered prematurely: A randomized clinical trial. *Sci. Rep.* **2021**, *11*, 21492. [CrossRef]
38. Hellström, A.; Nilsson, A.K.; Wackernagel, D.; Pivodic, A.; Vanpee, M.; Sjöbom, U.; Hellgren, G.; Hallberg, B.; Domellöf, M.; Klevebro, S.; et al. Effect of Enteral Lipid Supplement on Severe Retinopathy of Prematurity. *JAMA Pediatr.* **2021**, *175*, 359–367. [CrossRef]
39. Wendel, K.; Aas, M.F.; Gunnarsdottir, G.; Rossholt, M.E.; Bratlie, M.; Nordvik, T.; Landsend, E.C.S.; Fugelseth, D.; Domellöf, M.; Pripp, A.H.; et al. Effect of arachidonic and docosahexaenoic acid supplementation on respiratory outcomes and neonatal morbidities in preterm infants. *Clin. Nutr.* **2023**, *42*, 22–28. [CrossRef]
40. McCulley, D.; Wienhold, M.; Sun, X. The pulmonary mesenchyme directs lung development. *Curr. Opin. Genet. Dev.* **2015**, *32*, 98–105. [CrossRef]
41. Kyle, J.E.; Clair, G.; Bandyopadhyay, G.; Misra, R.S.; Zink, E.M.; Bloodsworth, K.J.; Shukla, A.K.; Du, Y.; Lillis, J.; Myers, J.R.; et al. Cell type-resolved human lung lipidome reveals cellular cooperation in lung function. *Sci. Rep.* **2018**, *8*, 13455. [CrossRef]
42. Varisco, B.M.; Ambalavanan, N.; Whitsett, J.A.; Hagood, J.S. Hagood Thy-1 signals through PPARgamma to promote lipofibroblast differentiation in the developing lung. *Am. J. Respir. Cell Mol. Biol.* **2012**, *46*, 765–772. [CrossRef]
43. Scarpato, M.; Federico, A.; Ciccodicola, A.; Costa, V. Novel transcription factor variants through RNA-sequencing: The importance of being "alternative". *Int. J. Mol. Sci.* **2015**, *16*, 1755–1771. [CrossRef] [PubMed]
44. Kilroy, G.E.; Zhang, X.; Floyd, Z.E. PPAR-gamma AF-2 domain functions as a component of a ubiquitin-dependent degradation signal. *Obesity* **2009**, *17*, 665–673. [CrossRef] [PubMed]
45. Grygiel-Gorniak, B. Peroxisome proliferator-activated receptors and their ligands: Nutritional and clinical implications—A review. *Nutr. J.* **2014**, *13*, 17. [CrossRef]
46. Torday, J.S.; Rehan, V.K. Up-regulation of fetal rat lung parathyroid hormone-related protein gene regulatory network down-regulates the Sonic Hedgehog/Wnt/betacatenin gene regulatory network. *Pediatr. Res.* **2006**, *60*, 382–388. [CrossRef]

biomolecules

MDPI

Review

The Peroxisome Proliferator-Activated Receptors of Ray-Finned Fish: Unique Structures, Elusive Functions

Evridiki Boukouvala [1] and Grigorios Krey [2,*]

[1] Veterinary Research Institute, Hellenic Agricultural Organization-DIMITRA (ELGO-DIMITRA), 57001 Thermi, Thessaloniki, Greece; boukouvala@elgo.gr
[2] Fisheries Research Institute, Hellenic Agricultural Organization-DIMITRA (ELGO-DIMITRA), 64007 Nea Peramos, Kavala, Greece
* Correspondence: krey@elgo.gr

Abstract: The Actinopterygian and specifically the Teleostean peroxisome proliferator-activated receptors (PPARs) present an impressive variability and complexity in their structures, both at the gene and protein levels. These structural differences may also reflect functional divergence from their mammalian homologs, or even between fish species. This review, taking advantage of the data generated from the whole-genome sequencing of several fish species, highlights the differences in the primary structure of the receptors, while discussing results from the literature pertaining to the functions of fish PPARs and their activation by natural and synthetic compounds.

Keywords: teleost fish; isotypes/isoforms; gene structure; functional domains; ligands/agonists

1. Introduction

Interest in peroxisome proliferator-activated receptors (PPARs) in fish resulted from the need to address issues that were, and still are, of particular concern to the aquaculture industry. Specifically, the primary challenge encountered in the expansion of fin-fish aquaculture to an industrial-level activity concerned the sourcing of lipids for the diets of the farmed species. Aquaculture traditionally relied on fish meal and oil from industrial fisheries, a resource that is presently exploited at its maximum sustainable level. To secure its own sustainability, as well as to appease environmental concerns, the industry gradually conformed to the "less fish in than fish out" principle by developing and using aquafeeds that use less fish ingredients.

Efforts to replace fish oils with vegetable oils date more than 30 years [1]. However, vegetable oils possess distinct fatty acid profiles compared to fish oils, and, consequently, when included at high levels in the fish diets, they alter the flesh fatty acid composition; most importantly, they reduce the concentration of the n-3 long-chain polyunsaturated fatty acids eicosapentaenoic acid (EPA, 20:5n-3) and docosahexaenoic acid (DHA, 20:6n-3) that characterize the marine fish oil. DHA and EPA are associated with beneficial health effects, and are considered essential components in the human diet [2]. Fish oil replacement does not only affect the nutritional quality of the farmed fish but also compromises the health and well-being of the fish themselves [3]. The use of vegetable oil-based diets has been linked to liver and heart pathological conditions, immunosuppression, and impaired intestinal function [4–6]. These adverse effects may be exacerbated by additional stressors prevalent in contemporary aquaculture systems, including high stocking densities [7] and fluctuating environmental conditions [8]. To mitigate these effects, aquaculture-related research has invested a considerable effort to the exploration of lipid metabolism in farmed fish species, particularly at the gene/transcription level. In particular, PPARs have attracted considerable attention, due to their established roles in mammalian lipid metabolism [9], which followed the landmark cloning of a mammalian PPAR [10], the identification of three distinct PPAR isotypes (PPARα, β/δ, γ) from *Xenopus laevis* [11], and the recognition of fatty acids, eicosanoids, and various hypolipidemic agents as genuine ligands for these receptors [12,13]. Further interest for PPARs in aquaculture research stemmed from the findings that these receptors, and particularly the γ isotype, played a critical role in mammalian immune and inflammatory responses [14], since the industry suffers considerable loses due to bacterial, viral, and parasitic infections, as well as from other environmental factors that compromise the immune system of the fish.

Research on fish PPARs (herein referred to as fPPARs) was initiated by the cloning of the PPARγ isotype from both plaice (*Pleuronectes platessa*) and Atlantic salmon (*Salmo salar*) [15,16]. Subsequently, the presence of three PPAR isotypes was established in the European sea bass (*Dicentrarhus labrax*), as well as in gilthead seabream (*Sparus aurata*) and plaice [17,18]. To date, the cloning and characterization of fPPARs in over 30 fresh-water and marine fish species of particular interest to aquaculture is reported in the literature.

The interest in the study of fPPARs has not been limited to aquaculture-related issues. Fish species such as zebrafish have served as model organisms for studying human diseases associated with excessive adipogenesis [19], a process in which PPARs play a central role. Furthermore, the activation of PPARs by numerous industrial chemicals, including pharmaceuticals [20], has also been exploited in toxicological and environmental research for assessing the impact of these contaminants in aquatic ecosystems [21].

To date, the function of fPPARs on metabolism, affected by nutritional, ecotoxicological, inflammatory, or stress-related factors, has been studied in more than 50 fish species. This research involved the direct cloning of the cDNAs or genes for these receptors and, in addition to the insight into their functional properties, provided evidence of important structural differences between fPPARs and their homologues from terrestrial vertebrates, both at the gene and protein levels [18]. Furthermore, it revealed the existence of paralogues for each of the three receptor isotypes (see [22–26] for first reports of fPPAR paralogues). The existence of paralogous PPAR genes within the genomes of ray-finned fish (Actinopterygii, Teleostei) aligns with the hypothesis proposing an extra round of genome duplication (third round whole-genome duplication, 3R WGD) during the early evolution of this lineage [27–29] and suggests that the role of PPARs in fish physiology and biochemistry is more intricate compared to that in higher vertebrates.

The ever-increasing number of fish species with fully sequenced genomes provides the opportunity to further discern the structure of the genes encoding these receptors and speculate on the function of duplicated PPAR isoforms in the metabolic regulation of the fish species harboring them. Therefore, this review covers mostly the unique features and variety that characterize the structure of the fPPAR genes, and, consequently, of their encoded products. The current state of fPPAR research, concerning the functional properties of the receptors, and especially their activation by natural or synthetic compounds, is also discussed.

The new information resulting from the targeted cloning of fPPARs and/or from the whole-genome sequencing of different fish species underscores the need for a common nomenclature regarding the isoforms of the receptors. Indeed, it is often difficult for the reader of the relevant literature to understand, without resorting to a sequence homology analysis, which of the PPAR isotype–isoforms is being studied. To avoid further confusion, this review will follow the nomenclature used by GenBank [30] for the fPPAR genes and their protein products. Thus, the isoforms of each of the three PPAR isotypes will be distinguished by the letter a or b, following the gene/protein name (e.g., PPARαa, PPARαb). Consequently, the PPARβ/δ isotype will be referred to as PPARδ.

2. PPAR Isotypes and Isoforms in Fish Species: Diversity Reflects Complexity

Three distinct PPAR isotypes orthologous to the mammalian and amphibian counterparts, i.e., PPARα, δ, and γ, were first identified in the marine fish species European sea bass (*Dicentrarchus labrax*), plaice (*Pleuronectes platessa*), and gilthead sea bream (*Sparus aurata*) [17,18]. However, even at that time it was evident that this subfamily of nuclear receptors may present an increased level of complexity and diversity in teleost fish, owing to the teleost-specific extra round of whole-genome duplication (3WGD, or Ts3R) that occurred in the early evolution of this lineage. This was supported by the earlier identification of two different genes homologous to the mammalian PPARδ in the genome of zebra fish (*Danio rerio*) [22] and of two PPARα genes in the puffer fish (*Fugu rubripes*) genome [23]. Indeed, Leaver et al. [24] identified four different genes for the PPARδ subtype in Atlantic salmon (*Salmo salar*), of which at least two were shown to be functional, consistent with a fourth round of whole-genome duplication (4R WGD) inferred to the Salmoniformes lineage [31]. Subsequently, through gene/cDNA cloning, or through genomic analyses, two distinct isoforms of PPARα were identified in the turbot (*Scopthalmus maximus*) [32,33], the loach (*Misgurnus anguillicaudatus*) [34], the Japanese sea bass (*Lateolabrax japonicus*) [35], the brown trout (Salmo *trutta*) [36], the tambaqui (*Colossoma macropomum*) [37], and the cod (*Gadus morhua*) [38]. Similarly, two PPARδ isotypes were identified in the tambaqui [37] and the loach [39]. The duplication of the PPARγ gene in a teleost fish was first established in the blind cave fish (*Astyanax mexicanus*) [26], while more recently through comparative genomic and phylogenetic analyses, [40] two PPARγ genes were identified in the genomes of European sardine, *Sardina pilchardus* (Clupeiformes), herring, *Clupea harengus* (Clupeiformes), red-bellied piranha, *Pygocentrus nattereri* (Characiformes), northern pike, *Esox lucius* (Esociformes), and channel catfish *Ictalurus punctatus* (Siluriformes), as well as in the salmonid species Atlantic salmon, Coho salmon (*Oncorynchus kisutch*), and rainbow trout (*Oncorynchus mykiss*).

For the purposes of this review, a non-exhaustive search of teleost fish species with sequenced genomes revealed that two PPARα paralogues are present in 34 out of the 40 different Orders examined (178 species, Table S1). According to this search, the PPARαa isoform has not as yet been identified in the Acipenseriformes, Gonorynchiformes, Osmeriformes, and Osteoglossiformes Orders, while in the Blenniiformes and Holocentriformes, it is the αb isoform that is presently missing. In contrast, PPARδ duplicates were identified in only 9 out of the 33 Orders examined (150 species) and specifically in the Anguilliformes, Characiformes, Clupeiformes, Cypriniformes, Esociformes, Gymnotiformes, Osmeriformes, Salmoniformes, and Siluriformes (Table S1). Similarly, the search for PPARγ paralogues in 41 teleost Orders (173 species) identified, in addition to the species mentioned by [40], a duplicate gene in the Osmeriform species *Osmerus eperlanus* (GenBank Access. nos. XM_062459215 and XM_062459587, Table S1). It is of interest to note that the species that retain gene duplicates for both PPARδ and γ do not belong to Neoteleostean Orders. Specifically, and according to [29], the Characiformes, Clupeiformes, and Siluriformes are phylogenetically placed in the Otocephala branch of Clupeocephala, while the Salmoniformes, Esociformes, and Osmeriformes are placed in, as yet, unnamed Euteleostean clades that diverged prior to the Neoteleosteans. As both Cypriniformes and Gymnotiformes are placed in the Otocepha branch, it is reasonable to expect that PPARγ duplicate genes will be identified in the species of these Orders. It remains, however, to be confirmed whether the presence of PPARδ and PPARγ duplicates will remain limited to only these Teleostean Orders, or will be identified in additional ones, as the assembly and annotation of fish genomes progresses.

Deciphering the potential functional divergence of the duplicated loci of the three fPPAR isotypes is a fundamental issue to our understanding of the biology of these receptors in fish species.

3. The Functional Domains of fPPARs

The proteins encoded by the fPPAR genes, just as their homologues from higher vertebrates, are organized in four functional domains: the A/B domain, located at the NH_2-terminus of the protein; the C or DNA-binding domain; the D domain or hinge region; and the E domain or ligand binding domain.

3.1. The A/B Domain

In mammals, the A/B domain contains activation function 1 (AF1) that has been shown to modulate ligand-independent PPAR activity via phosphorylation [41,42] and is encoded by a single exon, except in the case of alternative splicing, which, as in the case of the mammalian PPARγ isotype [43,44], results in a short and long A/B domain and, consequently, in a short and long receptor form (PPARγ1 and PPARγ2, respectively). In Teleost fish species, this domain is also encoded by a single exon, as splice variants have not as yet been formally identified. The function of this domain remains unexplored in the fish receptors, despite the fact that it presents interesting structural features that will be discussed below.

The examination of the protein sequence of the A/B domain in the two PPARα isoforms from the species listed in Table S1 revealed that the length of this domain ranges from 57 to 75 amino acid (aa) residues, as compared to 65 aa in the human receptor, with 57 aa being the length encountered in 70% of the species examined. In these species, the initiation methionine is followed by Ala-Gly-Asp-Leu (MAGDL motif), while in the longer-length sequences, the MVDM motif is found, as is also the case in the mammalian receptor.

In the PPARαb isoform, the A/B domain length ranges from 59 to 108 aa, being 65 aa long in 70% of the sequences examined. In all the PPARαb sequences, the MVDX motif is found, where X is an aliphatic residue. In the species with duplicate PPARα genes, the two isoforms share 40–50% sequence similarity in this domain. Notable is the low degree of similarity between isoforms in the species of the Tetraodontiformes, Siluriformes, and Gadiformes Orders (≈25%), as well as the high degree of homology (≈80%) in the two Atlantic salmon isoforms.

The PPARδa A/B domain, in the few species where this isoform has been identified, has an average length of 118 aa. Sequence identity between species is relatively low, averaging 27%.

In the PPARδb isoform and in the sequences examined, the domain length has an average length of 113 aa, but is considerably longer in the Acipeceriformes (one species, 175 aa), Clupeiformes (five species, 150–200 aa), and Polypteriformes (one species, 186 aa). Sequence identity between species, in 25 of the 35 Orders examined, is relatively high, ranging between 40 and 70%. The Clupeiformes, Cypriniformes, Esociformes, Gadiformes, Gymnotiformes, Osmeriformes, Osteoglossiformes, Salmoniformes, and Siluriformes sequences diverge considerably, exhibiting low levels of sequence identity with the other Orders or between them (<17%). As expected, the most distantly related between all sequences are those from the Acipenseriformes and Polypteriformes species. The within-species sequence identity between the two PPARδ isoforms is approximately 15%.

The fPPARγ A/B domain also varies considerably in length and, according to the sequences examined (Table S1), it ranges from 76 aa in the sterlet (*Acipenser ruthenus*, Acipenceriformes) to 332 aa in the chum salmon (*Oncorynchus keta*, Salmoniformes). The MVDT motif is present in the vast majority of the sequences examined, although it is, in most of them, preceded by a sequence containing the motif MQTP. As most of these sequences are predicted from genome analyses and have not been functionally characterized, it remains to be demonstrated which of the two motifs marks the translation initiation site, considering that the sequences encoding the two motifs are located in different and distal (>20 kb) exons (e.g., see GenBank entries XM_026350118.2 and XM_023954652.1, among others), with the upstream exon encoding only for 19 aa.

In the species with duplicate genes, moderate sequence identity is observed in this domain (33–66%) for the, tentatively labelled, PPARγa isoform between different Orders. In contrast, for the PPARγb, identity is low (<23%). It is of interest to note that in the chum salmon, there are apparently two PPARγb isoforms (GenBank: XM_052481996 and XM_052483128). Both isoforms have an A/B domain extremely rich in histidine residues, which is also characterized by the presence of multiple repeat sequences. Thus, in the PPARγb1 sequence (XM_052481996), the sequence NHSFNHPDI/R is repeated 28 times. In the PPARγb2, the sequence SHNSPHHH is tandemly repeated 11 times, while there are an additional 11 tandem repeats of the sequence SPDRSHSFNH. The rainbow trout PPARγb A/B domain is also His-rich and contains 22 tandem repeats of the SPDRSHSFNH sequence, while it has only one copy of the SHNSPHHH motif. Repeated elements have also been identified in the PPARγa isoform of the channel catfish (GenBank: XM_017450361.3), where 16 copies of the sequence QENSYRA are found. Similarly, in the northern pike PPARγa (GenBank: XM_034297489.1), 11 tandem repeats of the QPSSSLPQHH motif are present. Repeated motifs have also been reported in the Atlantic salmon A/B domain [16,45] as well as in that of the yellow catfish [46]. It has been proposed that these repeated motifs may function in coactivator recruitment [45].

3.2. The DNA-Binding Domain

The DNA-binding or C domain (DBD herein) is the most conserved region of the nuclear hormone receptors and is responsible for the recognition and binding to specific response elements in the promoters of genes regulated by these transcription factors [47,48]. As is the case in higher vertebrates, the fPAAR domain is encoded by two different exons, one for each of the two zinc fingers of the domain. All fPPAR sequences maintain the characteristic D box structure, containing three instead of five amino acids between the first two Cys residues of the second Zn finger [11].

In the PPARαa isoform, the two exons encode a sequence that ranges in length from 94 aa in the Salmoniformes to 106 aa in the Gadiformes, with 102 aa being the length encountered in the majority of sequences examined (>60%). Sequence identity between species from different Orders ranges from 67 to 94%, with the large variation being due to the sequence upstream of the first Zn finger. When just the core of the domain is considered, i.e., excluding the upstream to the first Zn finger sequences, the identity is greater than 87%, reaching 100% in some cases.

For the PPARαb isoform, the most frequently encountered domain length is of 105 aa (>50% of the sequences examined), followed by domain lengths from 102 to 104 aa. Sequence identity of the core domain between species from different Orders ranges from 90 to 100% for this isoform. In the species with duplicate genes for PPARα, sequence identity between isoforms ranges from a low of 70% in the Gadiformes (86% in the core) to a high of 91% (99% in the core) in Salmoniformes. Notable is also the high sequence identity (84%, 97% in the core) between the two isoforms in the Anguilliformes.

The DBD of the PPARδa sequences examined varies in length between 93 and 102 aa. The sequence identity between species of different Orders for the full domain is between 67 and 87%, while at the core of the domain, it is between 88 and 97%. In contrast, in the PPARδb sequences, the domain length presents a more restricted variability with 99 aa in 48% of the sequences and 97 aa in 42% of them, with the remaining having either 96 or 98 aa. Sequence identity in the full domain ranges from 71 to 99% (85 to 100% in the core). In the species retaining the two PPARδ loci, sequence identity between isoforms ranges in the full domain, from a low of 72% between the Cyprinoformes (88% in the core) to a high of 84% in the Characiformes and Osmeriformes (100% in the core).

The PPARγa DBD in the species examined varies in length from 101 to 105 aa, with the 101 aa and 102 aa lengths representing 61% and 28% of the sequences, respectively. Sequence identity between species for the full domain is 80% on average, while at the core, it is 98%.

The Clupeiformes, Siluriformes, and Characiformes PPARγb domain has a length of 102 aa; of the Salmoniformes, 108 aa; of the Esociformes, 111 aa; and of Osmeriformes, 117 aa. Sequence identity between species is 63% on average in the full domain and 82% at its core. Within species, sequence identity between isoforms is 72% for the full domain in the Clupeiformes and Esociformes, 70% in the Salmoniformes, 65% in the Osmeriformes, 61% in the Characiformes, and 59% in the Siluriformes. At the core of the domain, the identity is 85%, 97%, 92%, 85%, 78%, and 72%, respectively.

Residue deletions or insertions, relative to the corresponding human PPAR sequences, in all isoforms concern the region upstream of the first Zn finger and do not affect the Zn fingers' structure or the spacer between them.

PPARs bind as heterodimers with a retinoid X receptor (RXR) on response elements (PPREs) of the consensus sequence AGGTCA N AGGTCA (DR1-type element) [49], with residues upstream of the core element determining the polarity of the PPAR:RXR heterodimer [50]. In vitro, by means of the electrophoretic mobility shift assay (EMSA), the binding of European sea bass, gilthead sea bream, and plaice PPARαa, δa, and γa isotypes, as heterodimers with the mouse RXRβ, has been demonstrated on well-characterized mammalian PPREs (ACOx, Cyp4A6z) and on potential elements present in the plaice glutathione-S transferase A1 gene (*gsta1*) promoter [17,18]. In transient transfection experiments, PPAR holoproteins of the same species were able to drive transcription through a synthetic promoter containing the mouse Cyp4A6z PPRE [18]. The same promoter was also used to demonstrate the transactivation potential of the Atlantic salmon PPARδ isoforms, while a promoter containing the mouse ACOx PPRE was used for PPAR-dependent transcription by the torafugu pufferfish receptors [25]. Although putative PPREs have been identified in the promoters of the zebrafish fatty acid-binding protein (FABP) genes [51–53], as well as in the Δ5Δ6 fatty acid desaturase 2 (*fads2*) promoter of the rabbitfish (*Siganus canaliculatus*) [54], these elements have not been tested for their functionality in in vitro (EMSA/transactivation) assays. Point mutations on the putative PPRE identified in the promoter of the yellow catfish (*Pelteobagrus fulvidraco*) PPARα gene downregulated, but not abolished, both the basal and fenofibrate-induced expression of the reporter gene in transient transfection experiments [55]. According to the above and due to the high conservation of the DBD and the overall structure of the proteins among species and lineages, it is expected that fPPARs can recognize and bind to canonical DR1 elements. However, the study of the promoters of putative PPAR target genes in fish species and the potential receptor–DNA interactions on native promoters merits more attention.

3.3. The D Domain (Hinge Region)

The D domain, or hinge region, located between the DNA-binding and ligand binding domains, has been proposed to condition DNA-binding affinity of the receptors by interactions with DNA sequences upstream of the response element [50,56]. The domain in fish, as in higher vertebrates, is encoded by a single exon that also includes helices 1 and 2 (H1 and H2), which historically have been considered as part of the ligand binding domain of the receptors (see also below).

In PPARαa, the encoded sequence of the exon has a length of 67–69 aa, and an average between-species identity of approximately 70%. In the PPARαb, it has the invariable length of 67aa in all the sequences examined and an average identity of 84% between species. Between isoforms and within species, sequence identity is lower to that of the between-species comparison, ranging from 51% in Gadiformes to 77% in Lampriformes.

For PPARδa, the encoded sequence has a length of 68 aa in the majority of species examined, with the remaining having 69 aa. A longer domain (74 aa) is found in the Atlantic salmon β1A sequence (GenBank: NM_001123635). Between species, sequence identity averages 76%.

For PPARδb, the sequence length in the vast majority of the species (>80%) is 68 aa, with the remaining ranging from 64 to 69 aa. Between species, sequence identity averages 83%. Within species, sequence identity between isoforms is 82% in the Anguilliformes (1 species, European eel, *Anguilla anguilla*), 79% in the Clupeiformes, 74% in the Gymnotiformes (1 species, electric eel, *Electrophorus electricus*), 71% in the Salmoniformes (1 species, Atlantic salmon), and 69% in both Characiformes and Esociformes.

For PPARγ and in species of Orders with a single gene for the receptor, i.e., the tentatively labelled PPARγa isoform, the length of the encoded sequence is invariably 67 aa and exhibits a high degree of identity (>90%) between species. For the PPARγb isoform, the domain has 65 aa in the northern pike; 66 aa in the salmonoid species, with the exception of Atlantic salmon (67 aa); and 67 aa in the Characiformes, Clupeiformes, and Siluriformes species. Between species, sequence identity is 55% on average, while within species, the sequence identity is slightly higher, at 57% on average.

Residue deletions or insertions, relative to the corresponding human PPAR sequences encoded by this exon, occur upstream of H1 of the LBD. While the sequence of H2 is highly conserved between the fish and the human receptors, the sequence corresponding to H1 diverges considerably but still retains helical properties, as suggested by protein structure prediction models [46].

3.4. *The Ligand Binding Domain*

The PPAR ligand binding domain (LBD herein) is the most interesting domain of the receptors due to its unique structural characteristics. This domain, which assumes a similar structure in all three isotypes and resembles that of the other nuclear receptors, contains 13 helices and a small four-stranded β-sheet. Therefore, compared to the other nuclear receptors, the PPAR LBD contains an extra helix, located between the first β-strand and H3, which, according to the nomenclature proposed by Uppenberg et al. [57], is termed H2B or H2′ [58]. The LBD, in addition to ligand binding, provides the dimerization interface with RXR [59], as well as the ligand-dependent activation function 2 (AF-2), located at the C-terminus of the domain in H12, which is responsible for co-activator recruitment and binding. Characteristic of the PPAR LBDs is the size of the ligand cavity that is considerably larger as compared to other nuclear receptors. The ligand binding cavity is Y-shaped and is divided into two arms: Arm I that reaches towards H12 (AF-2) and Arm II that is situated between the β-sheet and H3 [58]. At the entrance of the binding site, a loop is formed between helices H2′ and H3, the so-called ω-loop, which due to its flexibility, may allow molecules of different sizes to enter the pocket. Both the large size of the binding pocket and the presence of the ω-loop may explain the multi-ligand activation property of PPARs. Although the crystal structure of a fish PPAR has not as yet been determined, structural models suggest that the fPPAR LBDs retain the overall structure of the mammalian receptors with unique, however, structural and potentially functional, features.

As previously mentioned, in the fPPARs, and as is the case for the amphibian and mammalian receptors [60,61], helices 1 and 2 of the LBD are encoded by the exon that also contains the hinge region (D domain) sequences. The remaining structural features of the domain, starting from the first β-strand of the β-sheet, are encoded by two exons in higher vertebrates. In contrast, the initially characterized genes for the PPAR isotypes from European sea bass, gilthead sea bream, plaice, and Atlantic salmon [17,18,24] revealed that in these fish species, the region corresponding to the first exon of the mammalian LBD is encoded by two exons in the α and δ isotypes (αa and δb, according to the nomenclature used herein) and by three exons in the γ isotype, resulting in "three and four exon LBDs", respectively. Unexpectedly, and according to genome sequencing data, this structural organization is not always observed in the receptors of the different Teleost Orders.

3.4.1. The fPPARα LBD

As depicted in Figure 1, the two isoforms of the PPARα isotype are encoded by genes that have distinct structure as far as the number of the LBD exons is concerned. Accordingly, the isoforms can be divided into two distinct groups.

The first group (group I) refers to species in which the LBD is encoded by three exons in the PPARαa and by two exons in the PPARαb. This arrangement is encountered in the majority of Orders/species in which duplicate PPARα genes have been identified.

In these species, the encoded LBD sequence ranges from 233 to 286 aa in the PPARαa isoform (Figure 1A) and is, thus, considerably longer than its mammalian counterpart (231 aa). This is due to the first exon, which contains an extended ω-loop, characteristic of the PPARα and γ isotypes [18]. The most extended ω-loop was observed in the Gobeiformes species *Boleophthalmus pectinirostris* and *Periophthalmus magnuspinnatus* (57 and

48 aa, respectively) and the shortest in the Anguilliformes species examined (3 aa). The ω-loop in the "group I" Orders/species contains variations (1–10 aa substitutions and/or additions or deletions of flanking residues) of the sequence PEPESGLQAGEXVPAVGCG. In most cases, the loop is preceded by an acidic residue and follows an LQ motif. Between species in different Orders, the overall LBD sequence identity is ≈90% for this isoform. It is noted that in the third exon, and in the spacer sequence between H11 and H12, the isoform suffers a one-residue deletion, as compared to the human PPARα (Ser 452).

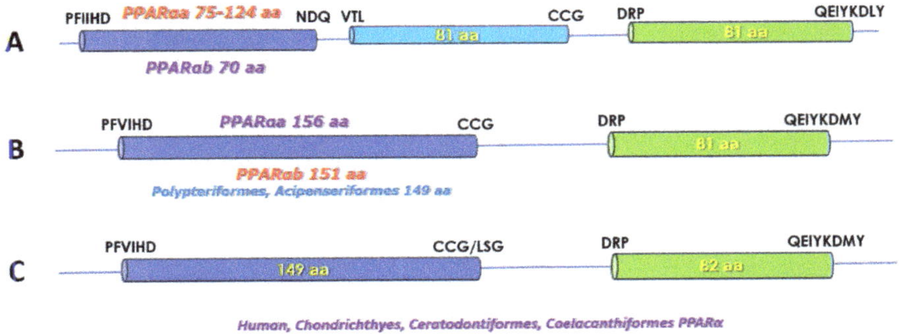

Figure 1. A schematic representation of the exons encoding the LBD of the fPPARα isoforms. (**A**) The 3-exon LBD type for the PPARαa (red) and PPARαb (mauve) isoforms. (**B**) The 2-exon LBD type for the PPARαa (mauve) and PPARαb (red) isoforms. (**C**) The LBD exons in the mammalian and non-Actinopterygian fish PPARα. Residues encoded by each exon are indicated above or on the exons. The aa sequences at the exon boundaries are also shown.

In "group I" of Teleost species, the PPARαb isoform LBD is encoded by two exons, the first of which corresponds to exons 1 and 2 of PPARαa. The domain of this isoform has an invariable length of 232 aa, and, consequently, as compared to the mammalian LBD, the ω-loop is extended by two aa residues. However, in the sturgeons and bichirs where only the PPARαb isoform has thus far been identified, the first exon encodes a 149 aa sequence, as is the case in the mammalian, lungfish, coelacanth, and chondrichthyan receptor. As is the case for the αa isoform, the one-residue deletion, at the equivalent of position Ser 452 in the human receptor, is also observed in these sequences. Sequence identity of the LBD of the PPARαb isoform between the different species is >90%. Between the two isoforms, identity in this domain does not exceed 80%, with the ω-loop not affecting this score significantly.

In the second group ("group II") of species that includes those of the Characiformes, Clupeiformes, Cypriniformes, Gymnotiformes, and Siluriformes Orders, the sequences listed in GenBank as the PPARαa isoform have two exons encoding the LBD, while three exons encode the domain in those predicted as the PPARαb isoform. According to the available sequence data, the two LBD exons in PPARαa encode invariably for 156 and 81 aa sequences, with the intron/exon boundaries positioned exactly as in the two-exon PPARαb LBD (Figure 1B). The ω-loop in this isoform type is extended by 7 aa residues as compared to the human receptor. The loop sequence is not conserved between species of different Orders but is highly conserved in species of the same Order.

In the "group II" species, the LBD of the αb isoform has an invariable length of 232 aa with the exon/intron boundaries positioned exactly as in the 3-exon PPARαa type (Figure 1A). The sequence between species for this isoform is also highly conserved (>90%) and the within-species sequence identity of the two isoforms exceeds 75%. It is of interest to note that the 2-exon PPARαa LBD sequences are more closely related to the 2-exon PPARαb LBD ones, than to those of the 3-exon PPARαa, implying that their classification as the PPARαb isoform possibly needs reconsideration. In both of these isoforms, the one-residue deletion at the terminal LBD exon is observed, as is the case in the "group I" receptors.

Note, also, that the ancient non-ray-fin fish lineages of Chondrichthyans (chimaeras, rays, skates, and sharks), Coelacanthiformes (coelacanths), and Ceratodontiformes (lungfish), which diverged from the Actinopterygii before the 3R WGD, share the same gene structure and LBD length with the mammalian receptor (Figure 1C). In the also ancient Actinopterygian, but not Teleostean, Orders of Polypteriformes (bichirs) and Acipenseriformes (sturgeons), the sequence encoded by the first LBD exon is of the same length as that of the mammalian receptor (149 aa, Figure 1B), i.e., it lacks an extended ω-loop sequence. However, in both Orders, the one-residue deletion in the second exon is observed. Therefore, it appears that the ω-loop extension event in the fPPARα coincides with the evolution of the Teleostean lineage about 350 million years ago. Also to note is the fact that all "group II" Orders, i.e., Characiformes, Clupeiformes, Cypriniformes, Gymnotiformes, and Siluriformes, phylogenetically resolve in the Otocephala clade that diverged from the Euteleosteans, to which the majority of the other Orders examined belong, about 270 million years ago [29,40].

Concerning the aa residues identified as critical for ligand binding to the receptor (reviewed in [58]), the majority of them are conserved in the fPPARα isoforms (Table 1), with most substitutions being conservative, except for two residues in the entrance sequence of the binding site. There, at the equivalent position of V324, a Cys is present in the PPARαa isoform and the Y334 is substituted by the positively charged Arg, as is also the case in some PPARαb sequences. Furthermore, a degree of variability is observed between the sequences of the fish receptors, especially in PPARαb, with potentially important consequences in ligand binding (e.g., substitution of Y314 by His, or Arg instead of Glu at position 251 in the Arm I and Arm II sites, respectively). However, these need further confirmation as they may represent sequencing errors.

Evidence for direct binding of ligands to the fPPARα LBD is limited to the European sea bass αa isoform and includes natural fatty acids, as well as the specific mammalian PPARα ligands pirinixic acid (Wy-14643) and 5,8,11,14-eicosatetraynoic acid (ETYA) [17]. The recognition of these compounds by the fish receptor is in accordance with the positional conservation of the Arm I residues of the binding pocket (Table 1). Through transient transfection experiments with the native plaice and seabream PPARαa in sea bass larval cells, Leaver et al. [18] demonstrated that the receptor is activated by a variety of natural fatty acids, as well as by Wy-14643 and ETYA, but not by the known activator of mammalian PPARα, and environmental obesogen, perfluoroctanic acid (PFOA) [62]. The receptor was also unresponsive to the mammalian PPARγ-specific ligands rosiglitazone and GW1929. Similarly, using both PPARα isoforms from the torafugu pufferfish for transfections in a goldfish cell line, Kondo et al. [25] also demonstrated the activation of the receptors by fatty acids, Wy-14643, and ETYA and, furthermore, showed a small but significant activation of PPARαb by rosiglitazone. Similar results were also reported for the medaka (*Oryzias latipes*) PPARαa [63].

To maximize the transfection assay specificity, Colliar et al. [64] used the plaice PPARαa LBD fused to the Gal4 DNA-binding domain in flathead minnow cells. In addition to confirming previous results, they showed a strong response of the receptor to bromopalmitate, gemfibrozil, and the mammalian PPARδ-specific ligand GW501506, as well as to the mono-1methyl-hexyl-phthalate. As in the previous study, PFOA failed to induce transcriptional activity of PPARαa. In the same study, the antifoulant tributyltin (TBT), a partial mammalian PPARγ agonist, was shown to be a potent antagonist of the plaice PPARαa.

With the same Gal4 fusion system, but using instead the LBDs of the two Atlantic cod PPARα isoforms for transfections in the mammalian COS-7 cell line, Søderstrøm et al. [38] also confirmed receptor activation by Wy-14643 and demonstrated that this compound is a more potent PPARαb agonist than of PPARαa. Furthermore, the cod PPARαb was shown to be activated by three carboxylated congeners of perfluoroalkyl acids (PFAAs), including PFOA.

Table 1. Residues defining the ligand binding cavity of the PPAR isotypes (adapted from [58]). hPPAR: human PPAR. Residue numbering is according to GenBank sequences XM_011530240.1, NM_006238.5, and NM_138712.5 for the hPPARα, δ, and γ isotypes, respectively. C.C. refers to the "Charge-Clamp" residues involved in interactions with co-activators. Residues in parenthesis are more rarely encountered in the fish sequences examined (at least two species of each Actinopterygian Order of Table S1, when applicable). Shaded residues indicate differences between the human and fish receptors. Arm I residues in bold are important for hydrogen bond formation with the polar head of ligands. In the fPPARγ isotype, X stands for an exon in the 4- and 2-exon type LBD.

	hPPARα	fPPARαa	fPPARαb	hPPARδ	fPPARδa	fPPARδb	hPPARγ	fPPARγa_4x	fPPARγa_2x	fPPARγb
Arm I	F273	F	F	F245	F	F	F282	F	F	F
	C276	C	C	C249	C	C	C285	C	C(F/W)	V/C(I/Y)
	Q277	Q	Q	Q250	Q	Q	Q286	Q	Q	Q
	S280	S	S	**T253**	T	T	**S289**	S	S	S(L/Q)
	Y314	Y	Y(H)	**H287**	H	H	**H323**	I	I(M)	I(M/H/F)
	I317	L	L(M)	I290	I	I	I326	L	L	L/M(N)
	F318	F	F	F291	F	F	Y327	I(T)	I(F/S)	I(T/M)
	I354	M	M	I327	I	I	F363	M(I)	M(I/L)	M(I)
	H440	H	H	**H413**	N	N	**H449**	H	H	H
	V444	V(I)	V(I)	M417	V	V	L453	I(M)	Q(E)	I
	L460	L	L	L433	L	L	L469	L	L(M)	L
	Y464	Y	Y	**Y437**	Y	Y	**Y473**	M(I/L)	M(I)	M(I)
Arm II	I241	I	I	I213	I	I(V)	I249	I	I(V)	I(V)
	L247	L	L(F)	L219	L	L	L255	L	L	L(M)
	E251	E	E(Q/R)	E223	E	E	E259	E(Q)	E(Q)	Q(E)
	L254	L	L(F)	W228	W	W	F264	C	C(F/S)	C(F/S/R)
	I272	L	I(L)	V245	V	V	I281	F(L)	F(L)	F(M/V/I)
	C275	C	C	R248	R	R	G284	S	S(R)	R(Q/S)
	M330	L	L	L303	L	L	V339	T	T	T(V/M/A)
	V332	V	V	V341	V	V	I341	I(M)	I(V)	F/M(L/I/C)
	I339	I	I	V348	V	V	M348	M	M	M(I)
	F343	F	F	F305	F	F	F352	F	F	F
	L344	L	L	L317	L	L	L353	L(I)	L	L(I/F)
	M355	M	M	I328	M	M	M364	M(L)	M	M(L/I)
Entrance	M220	M	M	M192	M	M	L228	L	L	L(I/T)
	T279	T(I/A/F)	T	T252	T(S)	T	R288	R	R	R(G/C)
	E282	E	E	E255	E	E	E291	E	E	E
	T283	T	T	T256	T	T	A292	A	A	A(S/T/C/G)
	E286	E	E(Q)	E259	E(Q)	E(Q)	E295	E	E	E(R/D/Q)
	L321	L	L	L294	L	L	L330	M(W)	M	L(W/M/V/T)
	V324	C	S/C	I297	L	I(L)	L333	L	L	L(F)
	A333	A	A	A306	A	A	S342	S(A)	S(A)	A/S(G)
	Y334	R(S)	Y(H/Q/R/S)	N307	N	N	E343	Y	Y	Y(E/S/C/D)
C.	K292	K	K	K265	K	K	K301	K	K	K
C.	E462	E	E	E435	E	E	E471	E	E	E

In agreement with the results obtained with the plaice receptor, PFOA, or indeed any of the other PFAA congeners, were not able to activate the PPARαa isoform. The authors addressed the discrepancy between the two isoforms by molecular docking analyses and concluded that this effect is due to the more favorable binding position of the agonists in PPARαb, assisted also by the reduced flexibility of the shorter ω-loop of the isoform, as opposed to that of PPARαa. Moreover, they demonstrated that binary combinations of Wy-14643 with PFAAs that exhibit non-agonistic behavior (e.g., the sulfonated congener PFOS) elicited a dose-dependent activation of both PPARα isoforms. Binary combinations of PFOA and PFOS also resulted in the synergistic activation of PPARαb but had no effect on PPARαa activity. Thus, a double-ligand model was proposed, in which the non-agonist molecule occupies a second binding site in the hydrophobic pocket formed by the β-sheet and helices 2 and 3. Accordingly, this leads to the stabilization of the ω-loop and, consequently, to a more stable LBD structure favoring interactions with the coactivator(s). Regarding this double-ligand model, in a later study by the same group, it was observed that while Wy-14643 is a more potent activator of PPARαb, eliciting an isoform response at lower ligand concentrations, the efficiency of PPARαa is three-fold higher at high ligand concentrations [65]. Thus, it would be interesting to investigate whether higher ligand concentrations "force" the entry of a second agonist molecule into the binding cavity of PPARαa, assisted also by the more flexible ω-loop of this isoform. The binding of two Wy-1463 molecules to the human PPARα LBD has been previously demonstrated [66] and results in the stabilization of the domain, and particularly of the AF-2 (H12). Simultaneously, binding of multiple ligands by the PPAR LBDs is also supported by earlier crystallographic studies of the mammalian receptors (see [67] and references therein).

3.4.2. The fPPARδ LBD

In all the fPPARδ genomic sequences examined, the LBD of the receptor is encoded by three exons (Figure 2). In the sequences listed in GenBank as the PPARδa isoform, the encoded domain contains 232 aa residues, i.e., as in the mammalian, lungfish, and coelacanth receptor (Figure 2B). The same number of residues is also encountered in the PPARδb isoform of Salmoniformes, Esociformes, Osmeriformes, and Anguilliformes. However, in the species of the Characiformes, Clupeiformes, Cypriniformes, Siluriformes, and Gymnodontiformes Orders, the domain length is one residue shorter due to a deletion in the first exon, and specifically in the region of the ω-loop (equivalent position of V232 in the human receptor, GenBank NM_006238). Interestingly, the same deletion is observed in the chondrichthyan receptor.

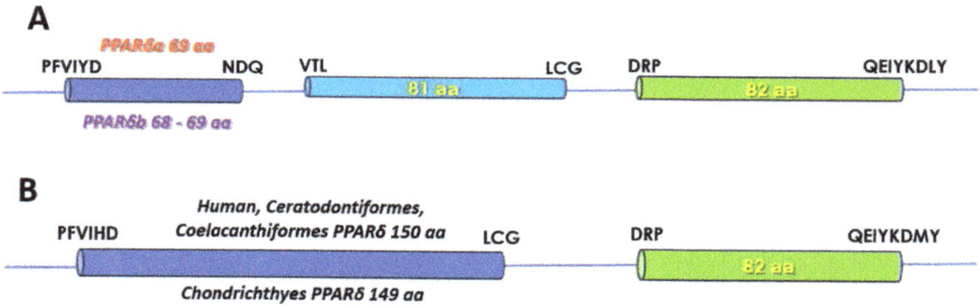

Figure 2. A schematic representation of the exons encoding the LBD of the fPPARδ isoforms. (**A**) The 3-exon Actinopterygian LBD type. (**B**) The 2-exon LBD type common to the mammalian and non-Actinopterygian fish PPARδ. Residues encoded by each exon are indicated above or on the exons. The aa sequences at the exon boundaries are also shown.

For both isoforms, between-species sequence identity in this domain is high (≥90%), as is also the case between the two isoforms within species. As is the case in fPPARα, the majority of the residues important for ligand binding are conserved in both PPARδ isoforms, with the most important difference being in the Arm I cavity of the binding site where H413 in the human receptor is substituted by Asn in both fish isoforms (Table 1).

For this fPPAR isotype, the direct interaction of ligands/activators with the receptor has only been demonstrated with the European sea bass PPARδb isoform [17], with the receptor shown to efficiently bind fatty acids and ETYA, but not the mammalian PPARα-specific ligand Wy-14643. Through transient transfection experiments with the plaice and sea bream native PPARδb, Leaver et al. [18] confirmed the above results and, further, showed that the receptor does not respond to the mammalian PPARγ-specific ligands rosiglitazone and GW1929. Interestingly, in these experiments, and in contrast to the sea bream receptor, the plaice PPARδb was poorly activated by natural fatty acids. A lack of activation by natural fatty acids was also observed with the pufferfish and the medaka receptor [25,63]. In further transfection experiments with the plaice PPARδb, but using the LBD of the receptor fused to the GAL4 DBD instead of the native protein, Colliar et al. [64] showed a significant activation of the receptor with the palmitoleic, linoleic, stearidonic, and EPA fatty acids, in addition to bromopalmitate, gemfibrozil, mono-1methyl-hexyl-phthalate, and the mammalian PPARδ-specific ligand GW501516. As was the case with the plaice PPARαa, tributyltin was found to inhibit the activity of the receptor but with reduced potency as compared to PPARαa.

In transient transfection experiments with the two Atlantic salmon isoforms, originally described as PPARβ1A and β2A, Leaver et al. [24] showed that PPARδ1A, although poorly activated by natural fatty acids, exhibited a strong response to bromopalmitate and to GW501516. In contrast, the δ2A isoform was not responsive to any of the compounds tested and, in fact, reduced the basal transcription level of the reporter gene. This repressive effect was also demonstrated in co-transfection experiments of both isoforms and in the presence of the PPARδ1A activator GW501516, leading to the hypothesis that the function of PPARδ2A, which is highly expressed in the gills of salmon, may be to prevent the activation of other PPARs by exogenous compounds or contaminants. In addition, the inactivation of the δ2A isoform by PPARδ-specific ligands could not be attributed to residues important for ligand binding, as the LBD in both isoforms is practically identical. Therefore, other structural features of the receptor, and particularly the A/B domain in which the two isoforms diverge significantly, may be responsible for this effect.

Using the cod PPARδb LBD fused to the Gal4 DBD in transfection experiments, Søderstrøm et al. [38] also demonstrated the activation of the receptor by GW501516. As in the case of the cod PPARα isoforms, and according to the double-ligand model proposed by the authors, combined exposure of the receptor's LBD to GW501516 and PFAA congeners, and especially to nonadecafluorodecanoic acid (PFDA), significantly and synergistically increased its transcriptional activity.

3.4.3. The fPPARγ LBD

Of the three fish isotypes/isoforms' LBDs, that of the fPPARγ is the most perplexing both in terms of structure and function. As depicted in Figure 3, the domain, in the majority of species examined and depending on the Order they belong, is encoded by either two or four exons. However, in the species where duplicate genes of the receptor have been identified, in the Salmoniform and Esociform species, the LBD of the isoform termed "γa" is encoded by four exons and the "γb" by two exons, while in the Osmeriform, Characiform, Clupeiform, and Siluriform ones, the domain of both isoforms is encoded by two exons.

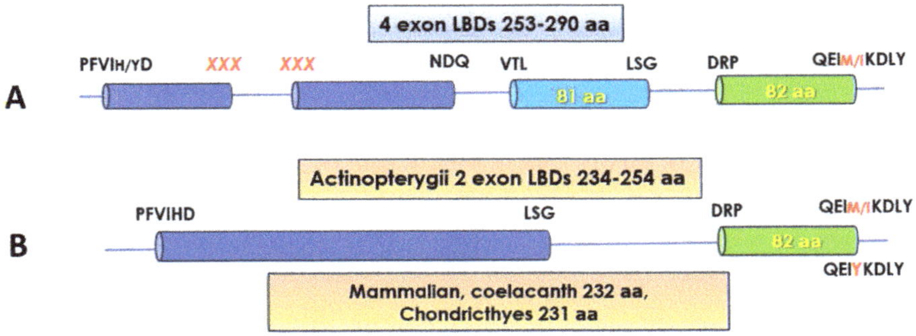

Figure 3. A schematic representation of the exons encoding the LBD of the fPPARγ isoforms. (**A**) The 4-exon LBD of Actinopterygian PPARγ. (**B**) The 2-exon LBD type common to the mammalian, Actinopterygian, and non-Actinopterygian fish PPARγ. The number of encoded residues is indicated, where applicable, on the exon representation. The aa sequences at the exon boundaries are also shown. The sequence *XXX* in the "4-exon" type (panel **A**) indicates variable residues.

The "4-exon type" LBD varies considerably in length due to the ω-loop sequence. The loop spans exons 1 and 2 of the LBD and, consequently, the flanking aa residues at these exon boundaries are not conserved (marked in red as *XXX*, Figure 3A), although a Gln stretch is often observed at the boundary of exon 1. The longest ω-loop insertion (59 aa) was observed in the Atlantic salmon receptor (GenBank: XM_014168482.2) and the shortest (24 aa) in the two Sparidae species examined (*Sparus aurata* and *Acanthopagrus latus*). The primary structure of the loop appears to be conserved between species of different Orders, as the sequence LTAGHGG(I/L/V)TG(A/V)HXGS(E/D)CGVLGM(T/A), with deletions or insertions, is observed in all but the Salmoniform, Esociform, Gadiform, and Lampriform species. This sequence bears no similarity to the ω-loop of fPPARαa. Between species, sequence identity for this domain type is >80%.

The two-exon-encoded "γb" isoform LBD of the Atlantic and chum salmon has 241 aa residues, with the ω-loop extended by 9 aa residues, and of sequence RSVLPPEEP, in both species. The same ω-loop sequence is also observed in the other Salmonoid species in which the "γb" isoform has been identified, i.e., the rainbow trout and the Sunapee trout (*Savelinus alpinus*). In contrast, in the Esociform species, northern pike, the ω-loop of the "γb" isoform is extended by only two residues. Sequence identity of this domain between the salmonoid species exceeds 90% and is at 80% between the Salmonoids and northern pike.

In the Characiform *Astyanax mexicanus* (blind cave fish), in which the LBD of both PPARγ isoforms is encoded by two exons, the "γa" has the ω-loop extended by 13 aa residues and that of "γb" by 4 aa. In the other two Characiform species (*Pygocentrus nattereri* and *Colossoma macropomum*), in which only the PPARγa isoform has thus far been identified, the ω-loop is also extended by 13 aa residues and is related to the *Astyanax* sequence. Characteristic of the loop in these species is the prevalence of charged and polar residues (R, E, and particularly Q). The two isoforms in *Astyanax mexicanus* share ≈60% sequence identity at the LBD.

The 2-exon-encoded PPARγa LBD of the Osmeriform species contains a 22 aa residue insertion in the ω-loop and that of the γb isoform 18 aa. As in the case of the Characiform PPARγa, the ω-loop in Osmeriforms is particularly rich in Gln, but also in Pro residues. The within-species sequence identity between the two isoforms' LBDs is 65%.

In the Clupeidae, the ω-loop sequence length varies between species in either isoform, from 11 to 14 aa in the "γa" and from 6 to 8 residues in the "γb". Between species, sequence identity of the entire domain, for either the "γa" or the "γb" isoform, is at 80%. Within species, identity of the two isoforms is at 63% for the sardine and 67% for the herring.

In the Siluriform species, residue insertion at the ω-loop of the PPARγa isoform accounts for the extension of the LBD sequence by 12 aa, relative to the mammalian receptor. In contrast, the loop is extended by only two aa residues in the "γb" isoform. The inserted loop sequence is highly conserved in the species of this Order. Sequence identity within species for the two isoforms is at 56%. Between species, the identity is 97% for PPARγa and 85% for PPARγb.

For the remaining species in Orders in which the PPARγ LBD is encoded by two exons and only one isoform has thus far been identified, the LBD sequence varies in length. Thus, in the Acipenseriformes and Polypteriformes, it contains 235 aa, in the Cypriniformes, 238 aa; in the Elopiformes, Osteoglossiformes, Semionontiformes, and Gonorynchiformes, 241 aa; and in the Gymnotiformes, 243 aa, with either a 3, 6, 9, or 11 aa residue insertion in the ω-loop, respectively. The loop sequence does not present a common motif between these Orders but is highly conserved in all Cypriniformes species examined. The presence of an extended ω-loop in the Acipenseriformes and Polypteriformes Orders suggests that the residue insertion event in this PPAR isotype coincides with the early Actinopterygian evolution.

Phylogenetically, it appears that the LBD of the PPARγa isoform in Osmeriformes, Characiformes, Clupeiformes, and Siluriformes is more closely related to the PPARγ LBD sequence of the Cypriniformes, Elopiformes, Gymnotiformes, Gonorynchiformes, and

Osteoglossiformes, (Figure S1). In contrast, a more distant relation is observed between the PPARγb isoform of Osmeriformes, Characiformes, and Siluriformes, as well as with the LBD sequences of the Acipenseriformes and Semionotiformes PPARγ. Despite this discrepancy, for the data presented in Table 1, the Acipenseriformes, and Semionotiformes sequences are considered as homologous to the PPARγb isoform type, while those of the Cypriniformes, Gymnotiformes, Gonorynchiformes, Elopiformes, and Osteoglosiformes, are considered as homologues to PPARγa.

In contrast to the fPPARα and δ isotypes that have been shown to be activated by their respective natural and synthetic mammalian ligands (see above), natural fatty acids as well as the mammalian high-affinity PPARγ-specific ligands (e.g., rosiglitazone, GW1929) fail to activate the fish receptor. This failure has been attributed to residues important for receptor–ligand interactions not being conserved in the fPPARγ [18,46,64,68,69]. Indeed, as shown in Table 1, the fPPARγ isotype contains more substituted residues in the ligand binding site than either the fPPARα or δ isotypes. Most importantly, these substitutions include two out of the four residues in Arm I that form hydrogen bonds with the acidic group of the ligands in the mammalian receptor, i.e., H323 and Y473. In all fish species of the 4-exon LBD PPARγa type, as well as in most of the 2-exon "γa" type examined, the equivalent position of H323 is occupied by Ile (Met in the European sardine). In the PPARγb isoform of Salmoniformes, Esociformes, Characiformes, Elopiformes, and Osteoglossiformes, Ile is also encountered at this position, while a Met residue is present in the Clupeidae and Osmeridae species. In contrast, the receptor from Acipenseriformes, Semionotiformes, and Polypteriformes, which we have arbitrarily assigned to the PPARγb isoform, retains the His residue. Most importantly, in all Actinopterygii, the Y473 is replaced by Met and in rare cases by either Leu or Ile. Being located in helix 12, Y473, through hydrogen bonding with the ligand, holds the AF-2 in an active confirmation for co-activator recruitment (see [58] and ref. therein), an effect that cannot be produced by the non-polar Met, Ile, or Leu residues. Interestingly, Y473 is conserved in the Chondrichthyan species, as well as in the coelacanth but not in the lungfish *Protopterus annectens* (XM_044078056.1) (Figure 3, Table S2). In fact, most of the ligand binding residues are conserved in the Chondrichtyan and coelacanth receptor (Table S2). Indeed, Capitão et al. [69] demonstrated that the mammalian PPARγ-specific ligand rosiglitazone can activate transcription through the LBD of little skate (*Leucoraja erinacea*). Furthermore, it was shown by both transient transfection and ligand binding assays that the little skate PPARγ can bind to the partial human PPARγ agonist tributyltin (TBT). In previous studies with plaice PPARγ, TBT, like all other natural or synthetic compounds tested, was not effective in transactivation experiments and it was indeed shown to act as a repressor of plaice PPARαa and δa [64]. A lack of activation by this compound was also observed with the zebrafish (*Danio rerio*) PPARγ [69].

In a study aimed to assess interspecies differences in response to endocrine disrupting chemicals, Garoche et al. [21] also used the zebrafish PPARγ, along with the human and *Xenopus* receptor. Interestingly, and in contrast to previous studies, the zebrafish receptor was shown to be responsive to several of the compounds tested, albeit with various efficiencies. In particular, two organophosphorus compounds, triphenyl phosphate (TPP) and tri-*o*-tolyl phosphate (ToTP), exhibited strong agonistic behavior, which was also specific to the fish receptor. Furthermore, the human LXR ligand, GW3965, also specifically and strongly activated the zebrafish PPARγ. Molecular modeling studies located the binding of the above compounds in the hydrophobic sub-pocket of the LBD formed by helices H3 and H5 and the β-sheet, in both the human and the zebrafish receptor. This binding does not allow contacts of the compound(s) with the AF-2 (H12), i.e., contacts that determine the agonistic behavior of the mammalian PPARγ ligands and, particularly, the hydrogen bond with Y473 of the receptor. However, as the equivalent Y473 position is substituted by Met in the zebrafish PPARγ, the AF-2 involvement in the activation of the zebrafish receptor appears not to be a prerequisite.

Support for the AF-2-independent activation of the fPPARγ comes from a study that demonstrated the activation of both European sardine PPARγ isoforms by TBT [40]. TBT also binds to the hydrophobic sub-pocket in the human receptor and stabilizes the LBD by bond formation between the tin atom and C285 [70]. In a previous study by the same group [69], it was reported that TBT could not activate the zebrafish PPARγ, an effect that was attributed to the substitution of C285 by Tyr. However, C285 is also substituted in the sardine receptor by an aromatic residue (Phe). The discrepancy between the two fish receptors was explained by the presence of a Met residue in the S3 β-strand of the β-sheet, only in the sardine receptor, which has the potential of bond formation with the tin moiety of TBT, albeit with lower efficacy compared to Cys. However, it should be noted that the C285 substitution in the PPARγ receptor appears to be a very rare event in the teleost fish lineages. From all fPPARγa sequences examined, this substitution was observed only in the two Danionidae species, *D. rerio* and *D. aesculapii* (Cypriniformes), and the European sardine. Consequently, if this residue is critical for receptor activation by TBT, it would be expected that all fish receptors that maintain Cys at that position would be responsive to the compound. However, as noted above, this was not the case with the plaice PPARγ. Therefore, the activation of the sardine's receptor by TBT may involve other unique structural features in the LBD of this species.

4. The Expression of the PPAR Isoforms in Fish Tissues

Fish represent the most diverse and largest group of all vertebrates that includes marine, anadromous, and fresh-water species, inhabiting environments as diverse as abyssal depths (*Pseudoliparis belyaevi* < 8000 m) and high-altitude lakes and streams (Tibetan loach, *Triplophysa stolickai*, >5000 m). Fish, being also poikilotherms, have adapted their energy homeostasis mechanisms to the environmental conditions of their habitat range and, consequently, to the dietary energy sources available. Lipids serve as the principal energy source in fish, with species utilizing different tissues to store it (liver, muscle, peritoneal cavity). Consequently, it is expected that transcription factors, as PPARs, which are involved in the regulation of lipid metabolism, will not present the same tissue expression pattern in all fish species. Indeed, this is supported by the presently available data, as in only a handful of species, a similar, but not identical, pattern of expression has been observed. Important to note is that the majority of the published reports concern the expression pattern of only one of the receptors' isoforms. As the differential expression of the PPAR isoforms can provide clues on their biological functions, the discussion here is limited to the few fish species in which the tissue expression profile of both isoforms of the isotype(s) has been examined in the same sampled specimens.

The expression of both PPARα isoforms has been studied in the turbot (*Scopthalmus maximus*) [32], the Japanese sea bass (*Lateolabrax japonicus*) [35], and the loach (*Misgurnus anguillicaudatus*) [34]. In general, the two isoforms exhibited a species-specific tissue distribution pattern. Thus, in the turbot, PPARαa was most abundantly expressed in the heart with lower expression in the kidney, gill, liver, brain, and muscle. PPARαb was also highly expressed in the heart as well as in the kidney but, in terms of transcript abundance, its expression in the heart was two-fold lower as compared to that of PPARαa. In the Japanese sea bass, PPARαa was ubiquitously expressed in the tissues tested and most abundant in the stomach and adipose tissue but with low expression in the liver. In contrast, PPARαb had a more restricted tissue distribution and was most highly expressed in the liver and moderately in the adipose tissue. Similarly, in the loach, both isoforms were detected in all tissues tested, exhibiting a somewhat different profile as PPARαa was predominantly expressed in the heart and PPARαb in the liver. In terms of absolute expression levels, the PPARαb copy numbers were, in all tissues, two-fold higher than those of PPARαa.

The expression of the PPARδ isoforms was also studied in the loach [39]. Both isoforms were found to have a wide tissue distribution, with PPARδa being most abundantly expressed in the liver and the gonads of both sexes. PPARδb was also strongly expressed in the gonads but compared to PPARδa, its expression was 10- to 100-fold lower in all the other tissues. Transcripts of this isoform were not detected in the kidney or spleen.

The expression of the two PPARδ genes of Atlantic salmon, originally described as PPARβ1A and PPARβ2A [24], differed essentially only in the liver and gill, with the 1A isoform being predominantly expressed in the liver and the 2A in the gills, while comparable levels of the two isoforms' transcripts were detected in the other tissues tested [24].

The only species in which the tissue expression of both PPARγ isoforms has thus far been studied is that of the European sardine [40]. PPARγa was found to be ubiquitously expressed in the tissues examined with its transcripts being most abundant in the brain, midgut, gill, and intestine. PPARγb was also highly expressed in the midgut, albeit at lower levels than PPARγa, and moderately expressed in the liver and kidney and at low levels in the brain. The isoform was not detected in the gills, gonads, and muscle.

Although not concerning a PPARγ isoform, it is worth noting the diverse tissue- and developmental-stage-dependent pattern of expression of the two transcripts, which result from alternative promoter usage in the PPARγ gene of the yellow catfish (*Trachysurus fulvidraco*) [46].

The expression pattern of both the PPARα and δ isoforms was also studied in the early developmental stages of the loach [34,39]. Transcripts of both PPARα and PPARδ isoforms were detected in spermatozoa and the unfertilized egg, and in most of the developmental stages examined. A different expression pattern between the PPARα isoforms was observed after the blastula stage, while the PPARδ isoforms exhibited similar expression throughout development. The expression of both PPAR isotypes in the gonads and in the early developmental stages suggested that the fish receptors, like their mammalian counterparts, are involved in germ cell energy metabolism, oocyte maturation, and organ development [39].

The differential tissue expression of the PPARα and PPARδ isoforms has also been demonstrated in several species and in response to diets containing different fatty acid profiles [34,35,37,39], as well as in response to heat stress [63,71,72], underscoring the involvement of the PPAR isotypes/isoforms in the regulation of metabolism and energy balance in fish species.

According to the above limited data, the expression of the PPAR isoforms, although presenting a differential pattern under normal or challenged conditions, still shows a significant tissue overlap. Therefore, the expression of either isoform does not seem to act as a "turn-off switch" for the other. This may imply that the isoforms are involved in the regulation of different biochemical pathways, possibly responding to different ligands/activators and/or interacting differently with the other components of the transcriptional machinery on the promoters of target genes. Alternatively, isoform-specific structural features may confer differential stability properties to the receptors, thus securing the uninterrupted regulation of target genes under conditions that render one of the isoforms less stable. Addressing these questions is crucial to understanding the functions of the duplicate PPAR genes and the advantages they may present in the physiology of fish, regarding their adaptation to different environments and to their nutritional requirements.

5. Conclusions

The study of fPPARs spans more than two decades, a period during which the results produced have significantly increased our understanding of the role of these receptors in fish physiology and biochemistry. Nevertheless, there is still a number of questions that need to be addressed in order to precisely determine their functions in the biochemical pathways in which they are involved. Presently, there is only indirect evidence for genes that are targets of fPPAR action, deduced either from the known involvement of the mammalian receptors in their regulation, or from the concomitant, with that of the receptor(s), up- or down-regulation of their expression, in response to chemical or physical stimuli that are expected to activate PPAR-driven transcription. Potential PPREs have been identified in the promoters of a few fish genes (*fabp*, *fads2*, *gsta1*), and these promoters could be used to decipher the interactions of fPPARs with the element itself, as well as with the other components of the transcriptional complex. Furthermore, studies on the molecular mechanisms/transcription factors controlling the expression of the fPPAR genes are presently lacking. Regarding the above, it is interesting that the interactions of the fish receptors with co-activators have not as yet been examined, despite the fact that the PPARγ coactivator-1α (PGC-1α) has been identified in several fish species.

As for the structural features of the receptors, two regions present particular interest for further studies, i.e., the A/B domain and the ω-loop of the LBD. Concerning the latter, its impressive variability in terms of sequence length and residue composition suggests potentially important functional differences between species, at both the ligand recognition and binding affinity levels. In addition, the Actinopterygian/Teleost-specific ω-loop sequence expansion appears to be associated with the divergence of the different lineages, while in some of the fPPARγ genes, it is directly linked to the insertion of an extra intron in the gene structure.

The most mysterious of the fish receptors is PPARγ, for which no strong natural agonist has been identified. Although this receptor may not require the involvement of AF-2 for transactivation, it still remains to be seen whether a lack of response to the mammalian PPARγ-specific ligands results solely from the substitution of Y473 in the AF-2, or whether other factors are also involved.

Supplementary Materials: The following supporting information can be downloaded at: https://www.mdpi.com/article/10.3390/biom14060634/s1, Table S1: GenBank Accession numbers of Actinopterygian PPAR isotypes/isoforms; Table S2: Residues defining the ligand binding cavity of huma, lungfish coelacanth and chondrichthyan PPARγ; Figure S1: Phylogenetic tree (Maximum Likelihood) for the LBDs of the PPARγ isoforms from different fish Orders.

Author Contributions: Conceptualization and writing—original draft preparation, G.K. and E.B. All authors have read and agreed to the published version of the manuscript.

Funding: This review received no external funding.

Acknowledgments: This work was supported by the Directorate of Agricultural Research, Hellenic Agricultural Organization-DIMITRA. The authors would like to thank Elisavet Ioannidou for critical reading of the manuscript.

Conflicts of Interest: The authors declare no conflicts of interest.

References

1. Bell, J.G.; Dick, J.R.; McVicar, A.H.; Sargent, J.R.; Thompson, K.D. Dietary sunflower, linseed and fish oils affect phospholipid fatty acid composition, development of cardiac lesions, phospholipase activity and eicosanoid production in Atlantic salmon (*Salmo salar*). *Prostag. Leukotr. Ess.* **1993**, *49*, 665–673. [CrossRef] [PubMed]
2. Calder, P.C. Very long chain omega-3 (n-3) fatty acids and human health. *Eur. J. Lipid Sci. Technol.* **2014**, *116*, 1280–1300. [CrossRef]
3. Tocher, D.R.; Betancor, M.B.; Sprague, M.; Olsen, R.E.; Napier, J.A. Omega-3 long-chain polyunsaturated fatty acids, EPA and DHA: Bridging the gap between supply and demand. *Nutrients* **2019**, *11*, 89. [CrossRef] [PubMed]
4. Mourente, G.; Good, J.E.; Bell, J.G. Partial substitution of fish oil with rapeseed, linseed and olive oils in diets for European sea bass (*Dicentrarchus labrax* L.): Effects on flesh fatty acid composition, plasma prostaglandins E-2 and F-2 alpha, immune function and effectiveness of a fish oil finishing diet. *Aquac. Nutr.* **2005**, *11*, 25–40. [CrossRef]
5. Leaver, M.J.; Villeneuve, L.A.; Obach, A.; Jensen, L.; Bron, J.E.; Tocher, D.R.; Taggart, J.B. Functional genomics reveals increases in cholesterol biosynthetic genes and highly unsaturated fatty acid biosynthesis after dietary substitution of fish oil with vegetable oils in Atlantic salmon (*Salmo salar*). *BMC Genomics* **2008**, *9*, 299. [CrossRef] [PubMed]
6. Cruz-Garcia, L.; Sánchez-Gurmaches, J.; Bouraoui, L.; Saera-Vila, A.; Pérez-Sánchez, J.; Gutiérrez, J.; Navarro, I. Changes in adipocyte cell size, gene expression of lipid metabolism markers, and lipolytic responses induced by dietary fish oil replacement in gilthead sea bream (*Sparus aurata* L.). *Comp. Biochem. Phys. A Mol. Integ. Phys.* **2011**, *158*, 391–399. [CrossRef]
7. Holhorea, P.G.; Naya-Català, F.; Belenguer, A.; Calduch-Giner, J.A.; Pérez-Sánchez, J. Understanding how high stocking densities and concurrent limited oxygen availability drive social cohesion and adaptive features in regulatory growth, antioxidant defense and lipid metabolism in farmed gilthead sea bream (*Sparus aurata*). *Front. Physiol.* **2023**, *14*, 1272267. [CrossRef]
8. Arts, M.T.; Kohler, C.C. Health and condition in fish: The influence of lipids on membrane competency and immune response. In *Lipids in Aquatic Ecosystems*; Kainz, M., Brett, M., Arts, M., Eds.; Springer: New York, NY, USA, 2009; pp. 237–256. [CrossRef]
9. Wahli, W.; Michalik, L. PPARs at the crossroads of lipid signaling and inflammation. *Trends Endocrinol. Metab.* **2012**, *23*, 351–363. [CrossRef] [PubMed]

10. Isseman, I.; Green, S. Activation of a member of the steroid hormone receptor superfamily by peroxisome proliferators. *Nature* **1990**, *347*, 645–650. [CrossRef]
11. Dreyer, C.; Krey, G.; Keller, H.; Givel, F.; Helftenbein, G.; Wahli, W. Control of the peroxisomal β-oxidation pathway by a novel family of nuclear hormone receptors. *Cell* **1992**, *68*, 879–887. [CrossRef]
12. Forman, B.M.; Chen, J.; Evans, R.M. Hypolipidemic drugs, polyunsaturated fatty acids, and eicosanoids are ligands for peroxisome proliferator-activated receptors α and δ. *Proc. Natl. Acad. Sci. USA* **1997**, *94*, 4312–4317. [CrossRef] [PubMed]
13. Krey, G.; Braissant, O.; L'Horset, F.; Kalkhoven, E.; Perroud, M.; Parker, M.G.; Wahli, W. Fatty acids, eicosanoids, and hypolipidemic agents identified as ligands of peroxisome proliferator-activated receptors by coactivator-dependent receptor ligand assay. *Mol. Endocrinol.* **1997**, *11*, 779–791. [CrossRef]
14. Zhang, X.; Young, H.A. PPAR and immune system—What do we know? *Int. Immunopharmacol.* **2002**, *2*, 1029–1044. [CrossRef] [PubMed]
15. Leaver, M.J.; Wright, J.; George, S.G. A peroxisomal proliferator-activated receptor gene from the marine flatfish, the plaice (*Pleuronectes platessa*). *Mar. Environ. Res.* **1998**, *46*, 75–79. [CrossRef]
16. Andersen, O.; Eijsink, V.G.H.; Thomassen, M. Multiple variants of the peroxisomal proliferator-activated receptor, PPARγ are expressed in the liver of Atlantic salmon (*Salmo salar*). *Gene* **2000**, *255*, 411–418. [CrossRef]
17. Boukouvala, E.; Antonopoulou, E.; Favre-Krey, L.; Diez, A.; Bautista, J.M.; Leaver, M.J.; Tocher, D.R.; Krey, G. Molecular characterization of three peroxisome proliferator-activated receptors from the sea bass (*Dicentrarchus labrax*). *Lipids* **2004**, *39*, 1085–1092. [CrossRef] [PubMed]
18. Leaver, M.J.; Boukouvala, E.; Antonopoulou, E.; Diez, A.; Favre-Krey, L.; Ezaz, M.T.; Bautista, J.M.; Tocher, D.R.; Krey, G. Three Peroxisome Proliferator-Activated Receptor Isotypes from Each of Two Species of Marine Fish. *Endocrinology* **2005**, *146*, 3150–3162. [CrossRef]
19. Den Broeder, M.J.; Kopylova, V.A.; Kamminga, L.M.; Legler, J. Zebrafish as a Model to Study the Role of Peroxisome Proliferating-Activated Receptors in Adipogenesis and Obesity. *PPAR Res.* **2015**, *2015*, 358029. [CrossRef] [PubMed]
20. Maloney, E.K.; Waxman, D.J. trans-Activation of PPARalpha and PPARgamma by structurally diverse environmental chemicals. *Toxicol. Appl. Pharmacol.* **1999**, *161*, 209–218. [CrossRef]
21. Garoche, C.; Boulahtouf, A.; Grimaldi, M.; Chiavarina, B.; Toporova, L.; den Broeder, M.J.; Legler, J.; Bourguet, W.; Balaguer, P. Interspecies Differences in Activation of Peroxisome Proliferator-Activated Receptor γ by Pharmaceutical and Environmental Chemicals. *Environ. Sci. Technol.* **2021**, *55*, 16489–16501. [CrossRef]
22. Robinson-Rechavi, M.; Marchand, O.; Escriva, H.; Bardet, P.; Zelus, D.; Hughes, S.; Laudet, V. Euteleost Fish Genomes are Characterized by Expansion of Gene Families. *Genome Res.* **2001**, *11*, 781–788. [CrossRef] [PubMed]
23. Maglich, J.M.; Caravella, J.A.; Lambert, M.H.; Willson, T.M.; Moore, J.T.; Ramamurthy, L. The first completed genome sequence from a teleost fish (*Fugu rubripes*) adds significant diversity to the nuclear receptor superfamily. *Nucleic Acids Res.* **2003**, *31*, 4051–4058. [CrossRef] [PubMed]
24. Leaver, M.J.; Ezaz, M.T.; Fontagne, S.; Tocher, D.R.; Boukouvala, E.; Krey, G. Multiple peroxisome proliferator-activated receptor beta subtypes from Atlantic salmon (*Salmo salar*). *J. Mol. Endocrinol.* **2007**, *38*, 391–400. [CrossRef] [PubMed]
25. Kondo, H.; Misaki, R.; Gelma, R.; Watabe, S. Ligand-dependent transcriptional activities of four torafugu pufferfish *Takifugu rubripes* peroxisome proliferator-activated receptors. *Gen. Comp. Endocrinol.* **2007**, *154*, 120–127. [CrossRef] [PubMed]
26. Wafer, R.; Tandon, P.; Minchin, J.E.N. The Role of Peroxisome Proliferator-Activated Receptor Gamma (PPARG) in Adipogenesis: Applying Knowledge from the Fish Aquaculture Industry to Biomedical Research. *Front. Endocrinol.* **2017**, *8*, 102. [CrossRef] [PubMed]
27. Ohno, S. *Evolution by Gene Duplication*; Springer: Berlin/Heidelberg, Germany; New York, NY, USA, 1970. [CrossRef]
28. Taylor, J.S.; Braasch, I.; Frickey, T.; Meyer, A.; Van de Peer, Y. Genome duplication, a trait shared by 22,000 species of ray-finned fish. *Genome Res.* **2003**, *13*, 382–390. [CrossRef] [PubMed]
29. Near, T.J.; Eytan, R.I.; Dornburg, A.; Kuhn, K.L.; Moore, J.A.; Davies, M.P.; Wainwright, P.C.; Friedman, M.; Smith, W.L. Resolution of ray-finned fish phylogeny and timing of diversification. *Proc. Natl. Acad. Sci. USA* **2012**, *109*, 13698–13703. [CrossRef] [PubMed]
30. Sayers, E.W.; Cavanaugh, M.; Clark, K.; Ostell, J.; Pruitt, K.D.; Karsch-Mizrachi, I. GenBank. *Nucleic Acids Res.* **2020**, *48*, D84–D86. [CrossRef] [PubMed]
31. Lien, S.; Koop, B.; Sandve, S.; Miller, J.R.; Kent, M.P.; Nome, T.; Hvidsten, T.R.; Leong, J.S.; Minkley, D.R.; Zimin, A.; et al. The Atlantic salmon genome provides insights into rediploidization. *Nature* **2016**, *533*, 200–205. [CrossRef]
32. Urbatzka, R.; Galante-Oliveira, S.; Rocha, E.; Castro, L.F.C.; Cunha, I. Tissue expression of PPAR-α isoforms in *Scophthalmus maximus* and transcriptional response of target genes in the heart after exposure to WY-14643. *Fish Physiol. Biochem.* **2013**, *39*, 1043–1055. [CrossRef]
33. Cunha, I.; Galante-Oliveira, S.; Rocha, E.; Planas, M.; Castro, L.F.C. Dynamics of PPARs, fatty acid metabolism genes and lipid classes in eggs and early larvae of a teleost. *Comp. Biochem. Physiol. B* **2013**, *164*, 247–258. [CrossRef] [PubMed]
34. Liang, X.; Gao, J.; Li, D.; Cao, X. Cloning and expressions of peroxisome proliferator activated receptor alpha1 and alpha2 (PPARa1 and PPARa2) in loach (*Misgurnus anguillicaudatus*) and in response to different dietary fatty acids. *Biochem. Biophys. Res. Commun.* **2016**, *481*, 38–45. [CrossRef] [PubMed]

35. Dong, X.; Xu, H.; Mai, K.; Xu, W.; Zhang, Y.; Ai, Q. Cloning and characterization of SREBP-1 and PPAR-α in Japanese seabass Lateolabrax japonicus, and their gene expressions in response to different dietary fatty acid profiles. *Comp. Biochem. Physiol. B* **2015**, *180*, 48–56. [CrossRef] [PubMed]
36. Madureira, T.V.; Pinheiro, I.; de Paula Freire, R.; Rocha, E.; Castro, L.F.; Urbatzka, R. Genome specific PPARαB duplicates in salmonids and insights intoestrogenic regulation in brown trout. *Comp. Biochem. Physiol. B* **2017**, *208–209*, 94–101. [CrossRef] [PubMed]
37. Ferraz, R.B.; Modesto, A.L.S.; Cunha, I.; Ozório, R.; O'Sullivan, F.L.A.; Salaro, A.L.; Monroig, O.; Casto, L.F.C. Regulation of gene expression associated with LC-PUFA metabolism in juvenile tambaqui (*Colossoma macropomum*) fed different dietary oil sources. *Aquac. Res.* **2021**, *52*, 3923–3934. [CrossRef]
38. Søderstrøm, S.; Lille-Langøy, R.; Yadetie, F.; Rauch, M.; Milinski, A.; Dejaegere, A.; Stote, R.H.; Goksøyr, A.; Karlsen, O.A. Agonistic and potentiating effects of perfluoroalkyl substances (PFAS) on the Atlantic cod (*Gadus morhua*) peroxisome proliferator-activated receptors (Ppars). *Environ. Int.* **2022**, *163*, 107203. [CrossRef]
39. Liang, X.; Zhao, Y.; Gao, J. Identification and structural characterization of two peroxisome proliferator activated receptors and their transcriptional changes at different developmental stages and after feeding with different fatty acids. *Comp. Biochem. Physiol. B* **2016**, *193*, 9–16. [CrossRef] [PubMed]
40. Páscoa, I.; Fonseca, E.; Ferraz, R.; Machado, A.M.; Conrado, F.; Ruivo, R.; Cunha, I.; Castro, L.F.C. The Preservation of PPARg Genome Duplicates in Some Teleost Lineages: Insights into Lipid Metabolism and Xenobiotic Exploitation. *Genes* **2022**, *13*, 107. [CrossRef]
41. Adams, M.; Reginato, M.J.; Shao, D.L.; Lazar, M.A.; Chatterjee, V.K. Transcriptional activation by peroxisome proliferator-activated receptor g is inhibited by phosphorylation at a consensus mitogen-activated protein kinase site. *J. Biol. Chem.* **1997**, *272*, 5128–5132. [CrossRef]
42. Pawlak, M.; Lefebvre, P.; Staels, B. General molecular biology and architecture of nuclear receptors. *Curr. Top. Med. Chem.* **2012**, *12*, 486–504. [CrossRef]
43. Tontonoz, P.; Hu, E.; Graves, R.A.; Budavari, A.I.; Spiegelman, B.M. mPPAR gamma 2: Tissue-specific regulator of an adipocyte enhancer. *Genes Dev.* **1994**, *8*, 1224–1234. [CrossRef] [PubMed]
44. Mukherjee, R.; Jow, L.; Croston, G.E.; Paterniti, J.R., Jr. Identification, characterization, and tissue distribution of human peroxisome proliferator-activated receptor (PPAR) isoforms PPARgamma2 versus PPARgamma1 and activation with retinoid X receptor agonists and antagonists. *J. Biol. Chem.* **1997**, *272*, 8071–8076. [CrossRef] [PubMed]
45. Sunvold, H.; Ruyter, B.; Østbye, T.K.; Moen, T. Identification of a novel allele of peroxisome proliferator-activated receptorgamma (PPARG) and its association with resistance to *Aeromonas salmonicida* in Atlantic salmon (*Salmo salar*). *Fish Shellfish Immunol.* **2010**, *28*, 394–400. [CrossRef] [PubMed]
46. Zheng, J.L.; Zhuo, M.Q.; Luo, Z.; Pan, Y.X.; Song, Y.F.; Huang, C.; Zhu, Q.L.; Hu, W.; Chem, Q.L. Peroxisome proliferator-activated receptor gamma (PPARγ) in yellow catfish *Pelteobagrus fulvidraco*: Molecular characterization, mRNA expression and transcriptional regulation by insulin in vivo and in vitro. *Gen. Comp. Endocrinol.* **2015**, *212*, 51–62. [CrossRef] [PubMed]
47. Rastinejad, F.; Perlmann, T.; Evans, R.M.; Sigler, P.B. Structural determinants of nuclear receptor assembly on DNA direct repeats. *Nature* **1995**, *375*, 203–211. [CrossRef] [PubMed]
48. Chandra, V.; Huang, P.; Hamuro, Y.; Raghuram, S.; Wang, Y.; Burris, T.P.; Rastinejad, F. Structure of the intact PPAR-gamma-RXR-nuclear receptor complex on DNA. *Nature* **2008**, *456*, 350–356. [CrossRef] [PubMed]
49. Kliewer, S.A.; Umesono, K.; Noonan, D.J.; Heyman, R.A.; Evans, R.M. Convergence of 9-cis retinoic acid and peroxisome proliferator signalling pathways through heterodimer formation of their receptors. *Nature* **1992**, *358*, 771–774. [CrossRef] [PubMed]
50. Ijpenberg, A.; Jeannin, E.; Wahli, W.; Desvergne, B. Polarity and specific sequence requirements of peroxisome proliferator-activated receptor (PPAR)/retinoid X receptor heterodimer binding to DNA. A functional analysis of the malic enzyme gene PPAR response element. *J. Biol. Chem.* **1997**, *272*, 20108–20117. [CrossRef] [PubMed]
51. Venkatachalam, A.B.; Sawler, D.L.; Wright, J.M. Tissue-specific transcriptional modulation of fatty acid-binding protein genes, fabp2, fabp3 and fabp6, by fatty acids and the peroxisome proliferator, clofibrate, inzebrafish (*Danio rerio*). *Gene* **2013**, *520*, 14–21. [CrossRef]
52. Laprairie, R.B.; Denovan-Wright, E.M.; Wright, J.M. Subfunctionalization of peroxisome proliferator response elements accounts for retention of duplicated fabp1 genes in zebrafish. *BMC Evol. Biol.* **2016**, *16*, 147–161. [CrossRef]
53. Laprairie, R.B.; Denovan-Wright, E.M.; Wright, J.M. Divergent evolution of cis-acting peroxisome proliferator-activated receptor elements that differentially control the tandemly duplicated fatty acid-binding protein genes, fabp1b.1 and fabp1b.2 in zebrafish. *Genome* **2016**, *59*, 403–412. [CrossRef] [PubMed]
54. Dong, Y.; Zhao, J.; Chen, J.; Wang, S.; Liu, Y.; Zhang, Q.; You, C.; Monroig, O.; Tocher, D.R.; Li, Y. Cloning and characterization of Δ6/Δ5 fatty acyl desaturase (Fad) gene promoter in the marine teleost *Siganus canaliculatus*. *Gene* **2018**, *647*, 174–180. [CrossRef] [PubMed]
55. You, W.J.; Fan, Y.-F.; Xu, Y.-H.; Wu, K.; Tan, X.-Y. Molecular characterization and functional analysis of PPARα promoter in yellow catfish *Pelteobagrus fulvidraco*. *Gene* **2017**, *627*, 106–113. [CrossRef] [PubMed]

56. Hsu, M.H.; Palmer, C.A.N.; Song, W.; Griffin, K.J.; Johnson, E. Carboxyl-terminal Extension of the Zinc Finger Domain Contributes to the Specificity and Polarity of Peroxisome Proliferator-activated Receptor DNA Binding. *J. Biol. Chem.* **1998**, *273*, 27988–27997. [CrossRef] [PubMed]
57. Uppenberg, J.; Svensson, C.; Jaki, M.; Bertilsson, G.; Jendeberg, L.; Berkenstam, A. Crystal structure of the ligand binding domain of the human nuclear receptor PPARγ. *J. Biol. Chem.* **1998**, *273*, 31108–31112. [CrossRef] [PubMed]
58. Zoete, V.; Grosdidier, A.; Michielin, O. Peroxisome proliferator-activated receptor structures: Ligand specificity, molecular switch and interactions with regulators. *Biochim. Biophys. Acta BBA-Mol. Cell Biol. Lipids* **2007**, *1771*, 915–925. [CrossRef]
59. Gampe, R.T., Jr.; Montana, V.G.; Lambert, M.H.; Miller, A.B.; Bledsoe, R.K.; Milburn, M.V.; Kliewer, S.A.; Willson, T.M.; Xu, H.E. Asymmetry in the PPARg/RXRa Crystal Structure Reveals the Molecular Basis of Heterodimerization among Nuclear Receptors. *Mol. Cell* **2000**, *5*, 545–555. [CrossRef]
60. Krey, G.; Keller, H.; Mahfoudi, A.; Medin, J.; Ozato, K.; Dreyer, C.; Wahli, W. Xenopus Peroxisome Proliferator Activated Receptors: Genomic Organization, Response Element Recognition, Heterodimer Formation with Retinoid X Receptor and Activation by Fatty Acids. *J. Steroid Biochem. Mol. Biol.* **1993**, *47*, 65–73. [CrossRef]
61. Zhu, Y.; Qi, C.; Korenberg, J.R.; Chen, X.-N.; Noya, D.; Rao, M.S.; Reddy, J.K. Structural organization of mouse peroxisome proliferator-activated receptor γ (mPPARγ) gene: Alternative promoter use and different splicing yield two mPPARγ isoforms. *Proc. Natl. Acad. Sci. USA* **1995**, *92*, 7921–7925. [CrossRef]
62. Behr, A.-C.; Plinsch, C.; Braeuning, A.; Buhrke, T. Activation of human nuclear receptors by perfluoroalkylated substances (PFAS). *Toxicol. In Vitro* **2020**, *62*, 104700. [CrossRef]
63. Kondo, H.; Misaki, R.; Watabe, S. Transcriptional activities of medaka *Oryzias latipes* peroxisome proliferator-activated receptors and their gene expression profiles at different temperatures. *Fish. Sci.* **2010**, *76*, 167–175. [CrossRef]
64. Colliar, L.; Sturm, A.; Leaver, M.J. Tributyltin is a potent inhibitor of piscine peroxisome proliferator-activated receptor α and β. *Comp. Biochem. Physiol. C* **2011**, *153*, 168–173. [CrossRef]
65. Eide, M.; Goksøyr, A.; Yadetie, F.; Gilabert, A.; Bartosova, Z.; Frøysa, H.G.; Fallahi, S.; Zhang, X.; Blaser, N.; Jonassen, I.; et al. Integrative omics-analysis of lipidmetabolism regulation by peroxisome proliferator-activated receptor a and b agonists in male Atlantic cod. *Front. Physiol.* **2023**, *14*, 1129089. [CrossRef]
66. Bernardes, A.; Souza, P.C.; Muniz, J.R.; Ricci, C.G.; Ayers, S.D.; Parekh, N.M.; Godoy, A.S.; Trivella, D.B.; Reinach, P.; Webb, P. Molecular mechanism of peroxisome proliferator-activated receptor α activation by WY14643: A new mode of ligand recognition and receptor stabilization. *J. Mol. Biol.* **2013**, *425*, 2878–2893. [CrossRef]
67. Kaupang, A.; Hansen, T.V. The PPAR Ω Pocket: Renewed Opportunities for Drug Development. *PPAR Res.* **2020**, *2020*, 9657380. [CrossRef] [PubMed]
68. Cho, H.K.; Kong, H.J.; Nam, B.H.; Kim, W.J.; Noh, J.K.; Lee, J.H.; Kim, Y.O.; Cheong, J.H. Molecular cloning and characterization of olive flounder (*Paralichthys olivaceus*) peroxisome proliferator-activated receptor γ. *Gen. Comp. Endocrinol.* **2009**, *163*, 251–258. [CrossRef] [PubMed]
69. Capitão, A.M.F.; Lopes-Marques, M.S.; Ishii, Y.; Ruivo, R.; Fonseca, E.S.S.; Páscoa, I.; Jorge, R.P.; Barbosa, M.; Hiromori, Y.; Miyagi, T.; et al. Evolutionary Exploitation of Vertebrate Peroxisome Proliferator-Activated Receptor γ by Organotins. *Environ. Sci. Technol.* **2018**, *52*, 13951–13959. [CrossRef]
70. Harada, S.; Hiromori, Y.; Nakamura, S.; Kawahara, K.; Fukakusa, S.; Maruno, T.; Noda, M.; Uchiyama, S.; Fukui, K.; Nishikawa, J.; et al. Structural basis for PPARgamma transactivation by endocrine-disrupting organotin compounds. *Sci. Rep.* **2015**, *5*, 8520. [CrossRef] [PubMed]
71. Tseng, Y.C.; Chen, R.D.; Lucassen, M.; Schmidt, M.M.; Dringen, R.; Abele, D.; Hwang, P.P. Exploring Uncoupling Proteins and Antioxidant Mechanisms under Acute Cold Exposure in Brains of Fish. *PLoS ONE* **2011**, *6*, e18180. [CrossRef]
72. Wang, X.; Zhao, T.; Ma, A. Genetic Mechanism of Tissue-Specific Expression of PPAR Genes in Turbot (*Scophthalmus maximus*) at Different Temperatures. *Int. J. Mol. Sci.* **2022**, *23*, 12205. [CrossRef]

MDPI AG
Grosspeteranlage 5
4052 Basel
Switzerland
Tel.: +41 61 683 77 34

Biomolecules Editorial Office
E-mail: biomolecules@mdpi.com
www.mdpi.com/journal/biomolecules

www.ingramcontent.com/pod-product-compliance
Lightning Source LLC
Chambersburg PA
CBHW061504210926
51850CB00008B/1818
* 9 7 8 3 7 2 5 8 5 5 2 9 2 *